Modular Texts in Molecular and Cell Biology 1

Signal Transduction

Edited by

Carl-Henrik Heldin

Professor of Molecular and Cell Biology,

Uppsala University,

Sweden

and

Mary Purton

Freelance editor,

Cambridge,

UK

Series Editors

Ralph Bradshaw and Mary Purton

Published in association with the International Union of Biochemistry and Molecular Biology

First published in 1996 by Chapman & Hall
Reprinted in 1998 by Stanley Thornes (Publishers) Ltd

Reprinted in 2002 by:
Nelson Thornes Ltd
Delta Place
27 Bath Road
CHELTENHAM
GL53 7TH
United Kingdom

02 03 04 05 06 / 10 9 8 7 6 5 4 3 2

A catalogue record for this book is available from the British Library

ISBN 0 7487 4074 0

Page make-up by WestKey Ltd

Printed in China by L. Rex

Contents

Part Two Cytoplasmic Signal Transduction

Part Three Nuclear Responses

Contributors

Michael J. Berridge
The Brabraham Institute
Laboratory of Molecular Signalling
Department of Zoology
University of Cambridge
Downing Street
Cambridge CB2 3EJ
UK

Joseph B. Bolen
Department of Oncology
Bristol-Myers Squibb
Pharmaceutical Research Institute
Princeton
NJ 08543
USA

Martin D. Bootman
The Babraham Institute
Laboratory of Molecular Signalling
Department of Zoology
University of Cambridge
Downing Street
Cambridge CB2 3EJ
UK

James Briscoe
BRM Laboratory ICRF
London
UK

Anne L. Burkhardt
Department of Oncology
Bristol-Myers Squibb
Pharmaceutical Research Institute,
Princeton
NJ 08543
USA

Mari R. Candelore
Department of Molecular
Pharmacology and Biochemistry
Merck Research Laboratories
Rahway
NJ 07065
USA

Margaret A. Cascieri
Department of Molecular
Pharmacology and Biochemistry
Merck Research Laboratories
Rahway
NJ 07065
USA

Steve Coats
Fred Hutchinson Cancer Research Center
Department of Basic Sciences
Mail Stop A3–023
1124 Columbia Street
Seattle
WA 98104
USA

Jackie D. Corbin
Department of Molecular
Physiology and Biophysics
Vanderbilt University School of Medicine

Nashville
TN 37232–0615
USA

David Cowburn
Laboratories of the Rockefeller University
1230 York Avenue
New York
NY 10021
USA

Edward A. Dennis
Department of Chemistry and Biochemistry
University of California at San Diego
La Jolla
CA 92093
USA

Vishva M. Dixit
Department of Pathology
University of Michigan Medical School
Ann Arbor
MI 48109
USA

Raymond L. Eriksen
Department of Molecular and Cellular Biology
Harvard University
16 Divinity Avenue
Cambridge
MA 02138
USA

Tung Ming Fong
Department of Molecular
Pharmacology and Biochemistry
Merck Research Laboratories
Rahway
NJ 07065
USA

Sharron H. Francis
Department of Molecular
Physiology and Biophysics
Vanderbilt University School of Medicine
Nashville
TN 37232-0615
USA

Michael P. Graziano
Department of Molecular
Pharmacology and Biochemistry
Merck Research Laboratories
Rahway
NJ 07065
USA

Brian A. Hemmings
Friedrich Miescher-Institut
P.O. Box 2543
CH-4002 Basel
Switzerland

Weidong Huang
Department of Molecular and Cellular Biology
Harvard University
16 Divinity Avenue
Cambridge
MA 02138
USA

Horst Ibelgaufts
Institut für Biochemie
LMU
Munich
Germany

W. Michael Kavanaugh
Chiron Corporation
4560 Horton Street
Emeryville

CA 94608
USA

Tadamitsu Kishimoto
Department of Medicine III
Osaka University Medical School
2–2 Yamada-oka
Suita
Osaka 565
Japan

John Kuriyan
Howard Hughes Medical Institute
The Rockefeller University
1230 York Avenue
New York
NY 10021
USA

Kelly LaMarco
Tularik, Inc.
270 E. Grand Avenue
South San Francisco
CA 94080
USA

Sally J. Leevers
Ludwig Institute for Cancer Research
University College/Middlesex Hospital Branch
91 Riding House Street
London W1P 8BT
UK

Jo Milner
Yorkshire Cancer Research Campaign
P53 Research Group
Department of Biology
University of York
York YO1 5DD
UK

Kohei Miyazono
Department of Biochemistry
The Cancer Institute
Tokyo
Japanese Foundation for Cancer Research
1–37–1 Kami-ikebukuro
Toshima-ku
Tokyo 170
Japan

Mathias Müller
Institut für Tierzucht und Genetik
VUW Vienna
and
Department of Biotechnology in Animal
Production
IFA-Tulin
Konrad Lorenzstr. 20
A-3430 Tulln
Austria

George Panayotou
Ludwig Institute for Cancer Research
University College/Middlesex Hospital Branch
91 Riding House Street
London W1P 8BT
UK

Peter J. Parker
Protein Phosphorylation Laboratory
Imperial Cancer Research Fund
44 Lincoln's Inn Fields
London WC2A 3PX
UK

Sue Goo Rhee
Laboratory of Cell Signaling
National Heart, Lung, and Blood Institute
National Institutes of Health
Bethesda

MD 20892
USA

Jim Roberts
Fred Hutchinson Cancer Research Center
Department of Basic Sciences
Mail Stop A3-023
1124 Columbia Street
Seattle
WA 98104
USA

Alexander Sorkin
Department of Pharmacology
University of Colorado Health Sciences Center
Denver
CO 80262
USA

Paul C. Sternweis
University of Texas
Southwestern Medical Center
5323 Harry Hines Boulevard
Dallas
TX 75235-9041
USA

Catherine D. Strader
Department of Molecular
Pharmacology and Biochemistry
Merck Research Laboratories
Rahway
NJ 07065
USA

Tetsuya Taga
Institute for Molecular and Cellular Biology
Osaka University
1–3 Yamada-oka
Suita
Osaka 565
Japan

Muneesh Tewari
Department of Pathology
University of Michigan Medical School
Ann Arbor
MI 48109
USA

N. K. Tonks
Cold Spring Harbor Laboratory
1 Bungtown Road
Cold Spring Harbor
NY 11724-2208
USA

Michael R. Tota
Department of Molecular
Pharmacology and Biochemistry
Merck Research Laboratories
Rahway
NJ 07065
USA

Maria d. M. Vivanco
Department of Biochemistry and Biophysics
University of California
San Francisco
CA 94143
USA

Lewis T. Williams
Chiron Corporation
4560 Horton Street
Emeryville
CA 94608
USA

James R. Woodgett
Division of Cell and Molecular Biology
Ontario Cancer Institute
610 University Avenue
Toronto
Ontario
Canada M5G 1X5

Stanislaw Zolnierowicz
Medical University of Gdansk
Faculty of Biotechnology
Department of Biochemistry
Debinki 1
80-211 Gdansk
Poland

Preface

The cells of multicellular organisms need to communicate with each other in order to co-ordinate their growth and differentiation. The mechanisms for such communication include secretion of soluble signaling molecules, as well as direct contacts between cells. Generally, extracellular signaling molecules interact with a receptor embedded within the plasma membrane (there are exceptions, however) which transduces the signal across the membrane. Then distinct intracellular signal transduction pathways are initiated that directly affect different kinds of cytoplasmic machineries or lead to the cell nucleus where the activities of specific genes are regulated.

Research on the molecular mechanisms of signal transduction has been very intense during the recent years. The purpose of this book is to give an up-to-date summary of the enormous amount of information that is now available about signal transduction. The book contains 23 chapters which have been written by world authorities in their respective fields. Each chapter summarizes what is known (and not known) about a particular aspect of signal transduction and emphasizes central themes and issues.

Part One covers the structural and functional properties of the major classes of cell surface receptors. Several types of receptors described have a common feature in that their activation leads to an increased phosphorylation of cytoplasmic proteins on tyrosine residues. This kinase activity can either reside in the receptor (as for protein tyrosine kinase receptors) or in an associated protein (for example, the cytokine classes I and II and the hemopoietic antigen receptors). Receptors for members of the transforming growth factor-beta superfamily also signal via stimulation of protein phosphorylation, but in this case with specificity for serine or theonine residues. Chapters are also included on receptors for the tumor necrosis factor superfamily, which have been linked to pathways leading to apoptosis among others, and on receptors with seven transmembrane domains that couple to G proteins. Finally, a chapter describes regulation of receptor internalization and turnover. One notable omission is coverage of the large and important topic of ion channels. This subject will be covered in depth in a later volume in the series.

Part Two focuses on components important for cytoplasmic signal transduction. The first chapter describes the structure of SH2, SH3 and PH domains; these building blocks in signaling molecules have the important function of mediating physical interactions between the molecules in a signaling chain. Separate chapters describe the well characterized Ras and MAP kinase signaling pathways. Others cover important families of signaling molecules including phospholipases, phospholnositide 3-kinase, protein kinase C and G proteins. The roles of calcium, cyclic AMP and cyclic GMP in signaling are also discussed. Two chapters describe protein tyrosine and serine/threonine phosphatases. Since many signaling pathways involve protein phosphorylation, the activities of cytoplasmic phosphatases are important for regulating the magnitude and duration of signals.

Part Three deals with nuclear responses. Many signal transduction pathways end in the nucleus and some have now been delineated all the way from the cell surface to the nucleus. Chapters in this section focus on the regulation of transcription factors by phosphorylation, the mechanisms for control of the cell cycle and structural and functional properties of the tumor suppressor p53. The steroid hormone receptors, a special class of receptors that interact with their ligand inside the cell and then translocate to the nucleus, are also described.

We hope the concise and abundantly illustrated format of the chapters will make this book useful for both students and scientists who want to know more about this exciting field of research.

Carl-Henrik Heldin
Professor of Molecular
Cell Biology
Uppsala University
Sweden

Mary Purton
Cambridge,
UK

August 1995

Part One

Cell Surface Receptors

1 Signaling through receptor tyrosine kinases

W. Michael Kavanaugh
and Lewis T. Williams

The receptor tyrosine kinases (RTKs) are a family of more than 50 different transmembrane polypeptides with a protein tyrosine kinase domain in their intracellular portion. Upon binding to their corresponding growth factors, these receptors initiate a complex series of intracellular reactions, which ultimately result in such diverse cellular responses as proliferation, differentiation, cell motility, changes in cell shape, production of extracellular matrix and transcription of specific genes. This chapter will provide a broad overview of the principles of RTK signaling, with examples taken from recent studies. More detailed information on specific molecules and pathways can be found in other chapters in this volume, and in several recent excellent reviews.[1–7]

Although the members of the RTK family can be classified into at least 14 different subgroups based on the details of their structural organization (Figure 1.1), all share common structural and functional features. With the exception of the insulin receptor, all RTKs are single polypeptide chains which transverse the membrane once. Thus, the RTKs can be divided into three structural regions: an extracellular segment, a short transmembrane domain and an intracellular portion.

The extracellular domain is responsible for binding of receptor to its corresponding growth factor. The variety of structural motifs that are found in this region (Figure 1.1) reflect the different structural strategies adopted by different family members for the specific and high-affinity recognition of their corresponding ligands. These motifs may also participate in receptor dimerization (see below).

Although the transmembrane segments of these receptors are thought

Signal Transduction. Edited by Carl-Henrik Heldin and Mary Purton. Published in 1996 by Chapman & Hall. ISBN 0 412 70810 8

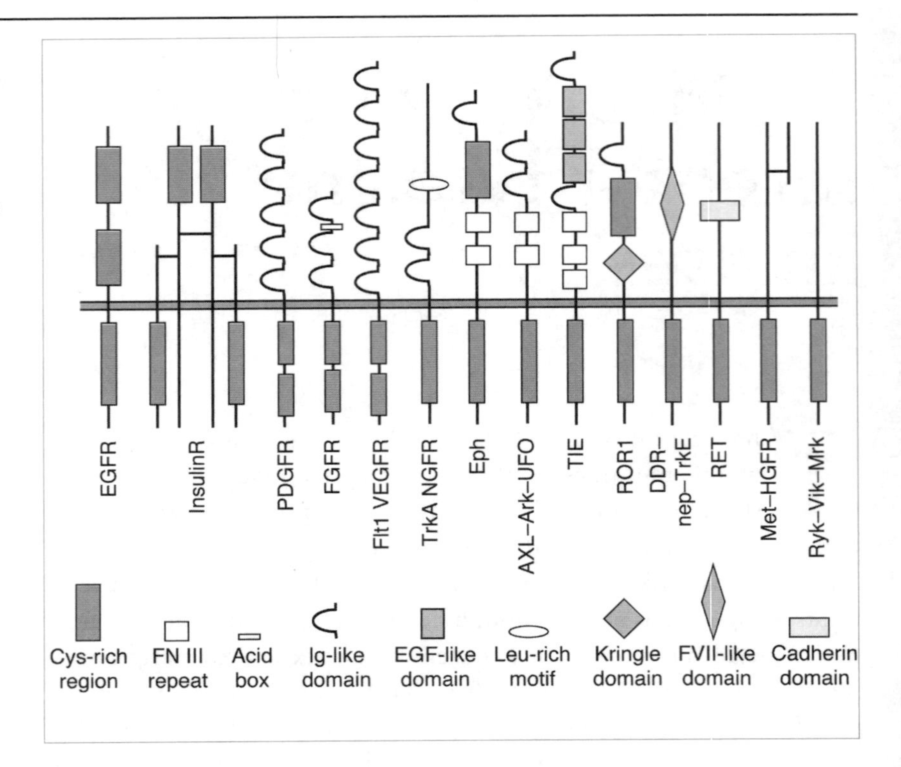

Figure 1.1 The vertebrate receptor tyrosine kinase family. A proposed classification scheme based on structural features. Motifs in the extracellular domains are indicated in the legend; black boxes represent the tyrosine kinase domains. Receptors representative of each class are indicated; multiple names that have been given to apparently identical proteins are included. Abbreviations: Cys-rich, cysteine-rich; FN III, fibronectin type III repeat; Ig-like, immunoglobulin-like domain; Leu-rich, leucine-rich domain; FVII-like, factor VII-like domain.

to function primarily as passive lipid anchors, there is evidence that the transmembrane domain can influence receptor function. For example, a single amino acid substitution of Val664 to Glu in the transmembrane domain of the c-Neu/c-ErbB2 RTK produces an oncogenic receptor which dimerizes in the absence of ligand and is constitutively activated. Transmembrane mutations of the fibroblast growth factor (FGF) receptor are associated with clinical disease.[8]

The hallmark feature which distinguishes the RTK family from other receptor classes is the presence of a tyrosine kinase domain in the intracellular portion. Although the overall amino acid sequence of the tyrosine kinase domain is conserved among members of the RTK family, in three subfamilies the kinase domain is interrupted by an inserted sequence (Figure 1.1). These kinase insert regions participate in the recruitment of cytoplasmic molecules into receptor-based signaling complexes.

Initiation of signaling: dimerization and autophosphorylation

The first event in the action of a growth factor is initiated by its binding to the receptor (Figure 1.2). Many of the growth factors that bind to the receptor tyrosine kinase family are dimeric molecules. For example, platelet-derived growth factor (PDGF) is a dimer composed of two A or B chains joined by disulfide bridges. These ligands are therefore able to bind to two receptors at the same time, and thereby promote the formation of stable receptor dimers. Although detailed studies of dimer formation have not been carried out on all members of the RTK family, it is likely that some form of oligomerization is necessary for signaling to proceed for all of these receptors. For example, artificially induced clustering of receptors in the Eph family activates signaling.[9] Some evidence suggests that members of the Axl/Ark family can cluster and be activated by homophilic interactions between the extracellular domains of neighboring receptor molecules.[10] The insulin receptor can be viewed as a variation on this theme, where α and β chains already exist as dimers in the absence of ligand. Presumably binding of insulin induces a conformational change in the relationship of the subunits which initiates activation.

Dimerization, by bringing two receptor molecules into close proximity, promotes autophosphorylation, in which each member of the pair trans-phosphorylates its partner (Figure 1.2).[7] In some studies, autophosphorylation appears to potentiate the tyrosine kinase activity of the receptor. Recently, X-ray crystallographic data for the tyrosine kinase domain of the insulin receptor suggested a structural basis for this observation.[11] The catalytic site of this RTK in its unphosphorylated

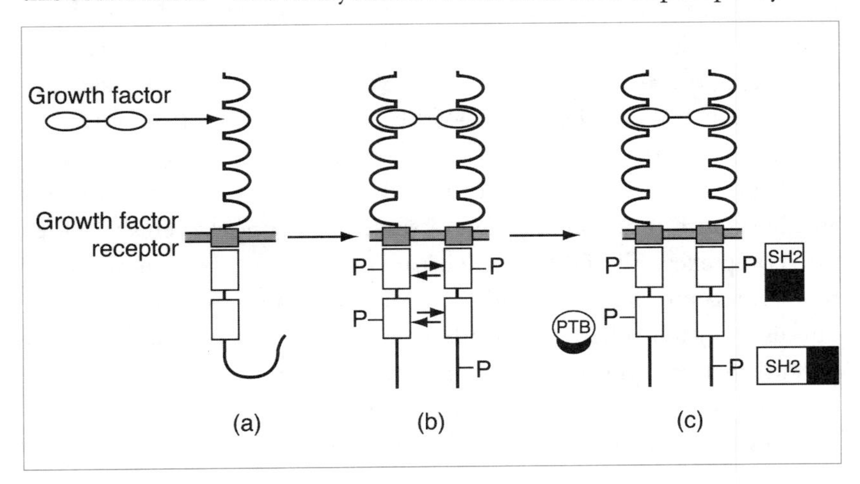

Figure 1.2 Initiation of signaling by receptor tyrosine kinases. (a) Ligand binding. A growth factor binds to the extracellular domain of its receptor; since most growth factors are dimeric molecules, stable receptor dimers are formed. (b) Receptor activation. Dimerization promotes autophosphorylation, with each member of the pair trans-phosphorylating its partner. For some RTK receptors, autophosphorylation induces a conformational change. (c) Formation of signaling complexes. Intracellular proteins containing SH2 or PTB domains then bind to phosphotyrosines and adjacent residues, creating signaling complexes.

state is obstructed by an 'activation' loop containing tyrosine residues 1158, 1162 and 1163. Tyr1162 is hydrogen-bonded to the catalytic site. These are known autophosphorylation sites and biochemical data have suggested that phosphorylation of these tyrosine residues precedes activation of kinase activity, autophosphorylation of other sites and phosphorylation of substrates. Thus, it has been proposed that phosphorylation of Tyr1162 causes a conformational change in which the activation loop is displaced from the catalytic site, allowing access to other substrates. It is unknown whether this mechanism is shared by all members of this family. The PDGF receptor, like most RTKs, contains a tyrosine residue in an analogous position to Tyr1162 (Tyr 825, mouse β receptor) within its tyrosine kinase domain. Upon binding of PDGF, Tyr825 is phosphorylated, the receptor undergoes a conformational change that is dependent on kinase activity and the tyrosine kinase activity of the PDGF receptor toward exogenous substrates is activated.[2]

Formation of signaling complexes

SH2 domains

A second consequence of receptor autophosphorylation is the creation of binding sites on the receptor for signaling molecules that interact specifically with phosphotyrosine (Figure 1.2). The first class of protein domains to be recognized that specifically bind tyrosine-phosphorylated targets were the SH2 domains (Chapter 9). Receptor autophosphorylation serves as a switch for assembling SH2 domain-containing molecules into receptor-based signaling complexes (Figure 1.2). SH2 domains are present in dozens of intracellular proteins, and often more than one SH2 domain is present in the same molecule (Table 1.1). In some cases, the SH2 domain-containing protein has a known enzymatic function, such as phopholipase C-γ (PLC-γ), the Ras GTPase activating protein (Ras-GAP), or the tyrosine-specific phosphatase SH-PTP2 (Syp). However, other proteins have no known catalytic activity and are thought to function as adapters, since they bring proteins together into complexes. All of these proteins are often tyrosine kinase substrates themselves, and therefore bind not only to autophosphorylated RTKs, but also to each other and to other phosphorylated proteins. Thus, a potentially large number of protein–protein interactions are induced by activation of the receptor tyrosine kinase.

These complex interactions are highly specific. The specificity of an

Table 1.1 Examples of Signaling Proteins Containing Modular Binding Domains

Protein	Structural organization[a]	Activity[a]
I. Proteins with known activities		
PLC-γ	PH–SH2–SH2–SH3–PH–catalytic domain	PtdIns(4,5)P_2 hydrolysis
GAP	SH2–SH3–SH2–PH–catalytic domain	Ras GTPase activator
SH-PTP1 and 2	SH2–SH2–catalytic domain	Tyrosine phosphatase
Src	SH3–SH2–catalytic domain	Tyrosine kinase
Fps	SH–2–catalytic domain	Tyrosine kinase
Syk	SH2–SH2–catalytic domain	Tyrosine kinase
VAV	PH–SH3–SH2–SH3	Ras GNEF
STAT proteins	Leucine repeats–SH3–SH2	Transcription factor
II. Proteins with no apparent intrinsic activity (adapters)		
p85	SH3–SH2–SH2	(Bound to PI3 kinase)
Grb2/Sem5	SH3–SH2–SH3	
Snc	PTB–SH2	
Nck	SH3–SH3–SH3–SH2	
Crk	SH2–SH3	
CrkII	SH2–SH3–SH3	

[a] Abbreviations: PH, pleckstrin homology domain; PtdIns(4,5)P2, phosphoinositol bisphosphate; GNEF, guanine nucleotide exchange factor; PTB, phosphotyrosine binding domain.

individual SH2 domain for its target is determined by the first 1–3 residues carboxy-terminal to the phosphotyrosine.[5] For example, the 85 kDa subunit of phosphatidylinositol 3′-kinase (PI3K), the first signaling molecule to be shown to associate with an RTK,[12] contains two SH2 domains. The carboxy-terminal p85 SH2 domain recognizes the motif pYXXM (where pY represents phosphotyrosine and X represents any amino acid) in the mouse β PDGF receptor at residues 708 and 719,[13,14] and in other RTKs and signaling molecules, but does not bind to phosphotyrosines in other sequence contexts. The structural features of the SH2 domains that account for their binding specificity is reviewed in Chapter 9. Using a phosphopeptide library approach, the predicted optimal binding motifs for many SH2 domains have been recently determined.[15]

PTB domains

Recently, a novel protein domain has been identified which, like SH2 domains, binds phosphotyrosine within a specific sequence context, but differs from SH2 domains in structure and mechanism (Table 1.2). The phosphotyrosine binding (PTB) domain was first identified in the amino terminus of the adapter protein Shc and in a related molecule, Sck.[16] The PTB domain binds specifically to the tyrosine-phosphorylated form of several signaling proteins and is thought to represent an alternative to SH2 domains for mediating the assembly of signaling complexes during tyrosine kinase signaling. Although the minimal size has not yet been determined, deletion experiments have shown that 186 residues are sufficient for high-affinity binding to phosphotyrosine- containing proteins.[16] The structure of the PTB domain is different from SH2 domains. PTB domains do not contain functional Phe–Leu–Val–Arg–Glu–Ser ('FLVRES') sequences which constitute part of the phosphotyrosine binding site in SH2 domains.[16] Further, the PTB domain of Shc recognizes residues amino-terminal to the phosphotyrosine within the motif NXXpY, in contrast to SH2 domains, which bind to carboxy-terminal residues.[17,18] Finally, mutational and structural studies suggest that a specific conformation of the NXXpY motif is necessary for high-affinity binding, unlike SH2 domain target motifs, which bind in an extended conformation.[17] A similar protein domain which binds to tyrosine phosphorylated targets has been described in the protein IRS-1,[18] and computer analysis of gene databases has revealed a diverse group of proteins containing PTB-like sequences.[19,20] This suggests that Shc and Sck may be members of a larger family of PTB domain-containing proteins. The solution structure of the Shc PTB domain has recently been determined. Surprisingly, it is very similar to that of pleckstrin homology (PH) domains (see below). This suggests that PTB and PH domains may have functional similarities.[51]

SH3 domains

One or more SH3 domains are often found in proteins that also contain SH2 domains (Table 1.1). These domains also mediate protein–protein interactions (Chapter 9). The understanding of SH3 domain function is less complete than that for SH2 domain function. Several *in vivo* targets for SH3 domains have been identified based on expression cloning, mutagenesis studies and screening of synthetic peptide libraries.[5] All contain proline-rich sequences, and particularly the motif PXXP. Biochemical and genetic approaches in a number of different systems have demonstrated that SH3 domain-mediated interactions are important for

Table 1.2 Comparison of SH2 and PTB domains

Parameter	SH2	PTB
Size	~100 aa	~186 aa
Distribution	>100 proteins	Shc, Sck IRS-1, others
FLVRES sequence	Yes	No
Domain structure	Central β-sheet; flanking helixes	Similar to PH domains
Target binding site:		
Length	Five residues	~12 residues
p-Tyr specific	Yes	Yes
Residues recognized	Carboxy-terminal	Amino-terminal
Structure	Extended	β turn

proper signaling by RTKs. For example, the adapter protein Grb2/Sem-5 contains two SH3 domains which bind to the guanine nucleotide exchange factor Sos. The Grb2–Sos complex can influence Ras activity and therefore activation of the Ras/Raf/Mek/MAP kinase signaling pathway (Chapters 10 and 11). Mutations in these SH3 domains which prevent Grb2/Sem-5 binding to Sos blocks vulval development in *Caenorhabditis elegans*. Whether SH3 domain-mediated interactions are regulated during RTK signaling is unknown.

Other mechanisms for complex formation

Another domain, known as the pleckstrin-homology (PH) domain, has been identified in many signaling proteins, often together with SH2 and SH3 domains[21] (Chapter 9). Although the overall sequence similarity between PH domains is low, they may be structurally very similar. The function of PH domains is not known, but they are believed to mediate protein–protein or protein–lipid interactions. Another example of regulated protein–protein interactions during RTK signaling involves the Ras superfamily of small GTP-binding proteins. Downstream targets can bind preferentially to the GTP-bound, activated form of these proteins. Examples include the interaction of Ras with the Raf-1 serine/threonine kinase[22] and with the guanine nucleotide exchange factor RalGDS or the related molecule RGL.[23,24]

Mechanisms of signal propagation

Although the mechanisms by which the formation of signaling complexes regulates downstream pathways are not entirely defined, there is experimental support for three general mechanisms. An individual molecule may use one or more of these mechanisms to propagate the intracellular signal.

Re-distribution of signaling proteins to new subcellular compartments

Many signaling molecules are either integral plasma membrane proteins (e.g. RTKs) or are associated with the membrane through specific localization sequences (e.g. Ras, Src). Thus, the formation of signaling complexes may serve to recruit cytoplasmic enzymes to the plasma membrane and into proximity with their membrane-associated substrates. Examples include PLC-γ, GAP and PI3K. Similarly, it has been proposed that one function of complex formation of receptor and non-receptor tyrosine kinases with Shc, Grb2 and Sos is to bring Sos into contact with membrane-bound Ras.[25] A third example is the interaction between activated Ras and the serine kinase Raf-1. One function of activated Ras may be to bring Raf-1 to the membrane, where Raf-1 kinase activity is activated by as yet unknown mechanisms.[26]

Facilitation of tyrosine phosphorylation and modulation of enzymatic activities

Many SH2- and PTB-domain containing proteins bind to tyrosine kinases with high affinities. Thus, complex formation may facilitate tyrosine phosphorylation of substrates. *In vitro*, tyrosine kinases phosphorylate targets more efficiently when functional SH2 domains are present. The function of tyrosine phosphorylation is either to create a new set of binding sites for SH2 and PTB domains, or directly to modulate enzyme function. For example, the enzymatic activity of PLC-γ is activated by tyrosine phosphorylation.[27,28]

Allosteric activation of targets

It has been reported that binding of PI3K and SH-PTP2 to synthetic phosphopeptides containing recognition sequences for their SH2 domains directly activates their enzymatic activities.[29] The magnitude of this activation is modest (maximum of several-fold), and its physiological significance is unknown. Circular dichroism (CD) and NMR data suggest that minor conformational change may occur within the SH2 domain upon binding of peptides.[30] Alternatively, binding of peptides may release SH2 domains from intermolecular interactions with internal phosphotyrosines which normally inhibit enzymatic function.

Major pathways in RTK signaling

The PLC-γ pathway

One of the first signaling pathways to be identified in RTK signaling involves PLC-γ (see Chapter 12). PLC-γ hydrolyzes PtdIns (4,5)P_2 into diacylglycerol, an activator of protein kinase C, and inositol-1,4,5-triphosphate, which mobilizes calcium from intracellular stores. PLC-γ associates with most, but not all, RTKs through its SH2 domains at specific sequence motifs (Figure 1.3). This association is thought to facilitate its phosphorylation and subsequent activation. Profilin may participate in regulation of this pathway by binding to PtdIns (4,5)P_2

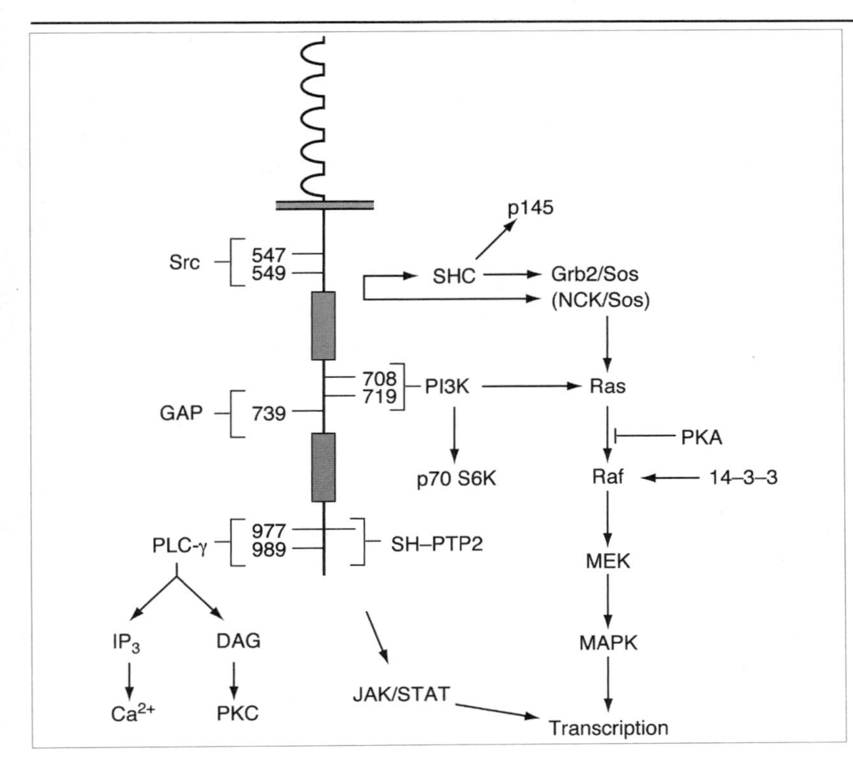

Figure 1.3 Major pathways in PDGF receptor signaling. The phosphotyrosines on the mouse PDGF β receptor which bind phosphatidylinositol 3′-kinase (PI3K), phospholipase C-γ (PLC-γ), GTPase activating protein (GAP), Src and SH-PTP2 are indicated. No binding sites are shown for SHC, NCK or GRB2. Abbreviations: IP_3, inositol triphosphate; DAG, diacylglycerol; p70 S6K, p70 S6 kinase; PKA, protein kinase A (cAMP-dependent protein kinase); others as in text.

and preventing interaction with unphosphorylated PLC-γ.[27] The relationship of the PLC-γ pathway to mitogenesis is unclear. Activation of PLC-γ is not necessary for mitogenesis in response to FGF in L6 myoblasts.[31] In some systems, mutation of the PLC-γ binding site on the PDGF receptor partially impairs mitogenesis,[32] while in others little effect is observed. Similarly, the PLC-γ pathway has been implicated as involved in cell migration in response to PDGF in some cells, but not in others.

PI3K signaling

PI3K catalyzes the addition of phosphate to the 3′-position of phosphatidylinositol and other polyphosophoinositols (Chapter 13). PI3K was the first signaling molecule to be shown to associate physically with RTKs[12] and served as a paradigm for dissecting the function of SH2 domains.[4,14] PI3K is composed of regulatory (p85) and catalytic (p110) subunits, each of which is a member of a larger family of homologous proteins. p85 is an adapter protein which binds to the motif pYXXM in many, but not all, RTKs through its two SH2 domains. p85 also contains an SH3 domain, the *in vivo* target(s) of which are unknown, and proline-rich sequences which may bind the SH3 domains of other molecules.

PI3K activity may be regulated by translocation to the membrane, where its substrates are located, or, less likely, by allosteric activation. Tyrosine and serine/threonine phosphorylation of p85 has also been shown *in vivo* and may regulate PI3K in some systems.[33,34]

Binding of p85 and PI3K activity to receptors and oncogenes correlates strongly with transformation or mitogenesis.[4] Mutation of the p85 binding sites in the PDGF receptor impairs mitogenesis in response to PDGF in some cell types.[32,35] Studies with mutant receptors suggest that PI3K is also apparently required for PDGF-induced membrane ruffling and chemotaxis.[36] Recent evidence suggests that PI3K may interact with and signal through Ras (Figure 1.3). Ras and PI3K form a complex.[37] Mutant PDGF receptors which do not associate with PI3K do not activate Ras in some cell types.[38] Further, introduction of a constitutively activated PI3K into cells activates Ras, and induces c-Fos transcription, Raf-1 and MAP kinase activation and cellular responses such as *Xenopus* oocyte maturation and membrane ruffling in a Ras-dependent manner.[39,40] One report suggested that PI3K may also lie downstream of Ras.[37] PI3K also participates in Ras-independent pathways, such as the activation of Akt and p70 S6 kinases.[41]

How PI3K signals is unknown. Presumably the 3′-phosphorylated products of this enzyme act as second messengers or affect membrane function. In addition to lipid kinase activity, p110 also has protein serine kinase activity. The relationship of protein to lipid kinase activity in PI3K signaling is unknown, although serine phosphorylation of p85 by p110 may regulate PI3K activity.[34] A family of molecules related to p110 has been discovered in yeast, *Drosophila* and mammalian cells. One of these proteins, VPS34p in *Saccharomyces cerevisiae*, regulates vesicular transport during protein sorting,[42] suggesting that mammalian PI3K may also participate in regulation of membrane function. Other family members contain protein domains not found in P13K p110, suggesting that individual members of this family may have different functions.

Ras pathways

Ras is the prototypical member of the superfamily of small GTPases (Chapter 10). A variety of evidence has implicated Ras as a pivotal signaling molecule in RTK-mediated proliferation.[6] Ras cycles between an inactive, GDP-bound state and an activated, GTP-bound state. Its activity, therefore, is regulated by proteins which influence the relative proportions of these bound guanine nucleotides. Guanine nucleotide exchange factors (GNEFs) activate Ras by increasing the exchange of GTP for GDP. GTPase-activating proteins (GAPs) could potentially downregulate Ras activity by promoting the conversion of bound GTP to GDP.

Biochemical and genetic evidence suggests that Ras can be activated by a GNEF known as Son-Of-Sevenless (Sos), first identified in *Drosophila.* Sos contains proline-rich sequences which bind to the SH3 domains of the adapter protein Grb2. Grb2, through its SH2 domain, associates with activated RTKs or with tyrosine-phosphorylated Shc. Shc, in turn, associates through its SH2 and PTB domains with RTKs, with non-receptor tyrosine kinases such as Src, and with other, as yet unidentified proteins.[16] The adapter protein Nck also binds to Sos.[43] This cascade of protein–protein interactions is thought to transmit the signal from autophosphorylated receptor to Ras in part by re-distributing Sos to the plasma membrane, where Ras is located,[25] although other mechanisms are possible (Figure 1.3). A variety of other potential Ras GNEFs have been identified.[6]

Genetic and biochemical evidence supports a role for GAP in the negative regulation of Ras activity, although a positive role in signal transduction is also postulated.[6] GAP associates with a number of RTKs through its SH2 domains, and is tyrosine phosphorylated in response to growth factor stimulation. The functional consequences of receptor binding and phosphorylation are unclear, as neither is necessary for regulation of Ras activity. GAP also associates with other tyrosine-phosphorylated proteins, including p62 and p190; their signaling function is also unknown.

Several targets of Ras have been identified. Dominant negative Ras inhibits activation of the serine/threonine kinase Raf-1, while activated forms of Ras activate Raf-1.[6,22] The amino-terminal regulatory region of Raf-1 binds to the effector region of activated Ras, and this interaction is necessary but not sufficient for Raf activation.[6,22] As mentioned previously, the purpose of the Ras–Raf interaction may be to translocate Raf to the membrane where it is activated by unknown mechanisms. Recently, a separate region of the amino terminus of Raf-1 has been found to bind to the 14-3-3 family of proteins, and 14-3-3 proteins can induce Raf-1 activation and maturation in *Xenopus* oocytes.[44] Activation of Ras engages a phosphorylation cascade, from Raf-1 to Mek (MAP kinase kinase) to the MAP kinase family (Chapter 11). Some members of the MAP kinase family can activate transcription factors such as Elk-1 and Jun by phosphorylation. A Ras-independent kinase cascade activated by 'stressful' stimuli as well as RTKs has recently been described (Chapter 11).

Other targets of Ras have been found by yeast two-hybrid screening. These include the GNEF RalGDS and a related molecule, RGL.[23,24] The functions of these molecules are unknown. Interaction of Raf with Ras, but not Ras with RalGDS, is inhibited by cAMP and protein kinase A activation,[45] suggesting that Ras may interact selectively with different signaling targets depending on the physiological state of the cell.

Other signaling pathways

Many other molecules have been implicated in signaling by RTKs, but their signaling pathways are less understood (Figure 1.3). For example, Src, a non-receptor tyrosine kinase which contains the prototypical SH2 and SH3 domains, is critical for signaling by the cytokine receptors and probably also plays a role in signaling by RTKs. Src family members associate with RTK receptors and are activated in response to growth factors.[46] Expression of a dominant negative form of Src inhibits PDGF-induced DNA synthesis.[47]

Similarly, the tyrosine-specific phosphatases SH-PTP1 and SH-PTP2, which contain SH2 domains, are implicated in RTK signal transduction. These proteins bind to and are phosphorylated by activated RTKs.[48] Genetic evidence suggests that the *Drosophila* homologue Corkscrew functions in signaling by the RTK Torso.[49] The substrates of these phosphatases remain to be identified.

Recently, a novel, Ras-independent pathway has been described involving the STAT (signal transducer and activator of transcription) transcription factor family (Chapter 3). STAT proteins contain SH2 and SH3 domains and can bind activated RTKs. These proteins are tyrosine phosphorylated, dimerize and translocate to the nucleus where they induce transcription of a specific subset of genes. STAT proteins have been implicated in EGF, PDGF and CSF-1 signaling. The Jak/Tyk family kinases are involved in activation of STAT proteins in response to interferon and may also be involved in RTK signaling.

Summary

Although our understanding of RTK signaling has greatly improved in the past five years, many significant questions remain. First, it can be stated that no one signaling pathway has been traced completely from the receptor to its end point, and that therefore there are likely to be many more molecules and protein–protein interactions remaining to be identified and characterized. Second, little is known about protein sorting and trafficking during signaling. The large number of protein–protein interactions which have been identified could represent a single large protein complex or numerous smaller subsets with specific and distinct functions in specific subcellular compartments. A third enigma is how a single growth factor can elicit very different cellular responses depending on the cell type and environment. For example, FGF can induce proliferation, neurite outgrowth or mesoderm induction in different systems. Analogously, EGF induces proliferation in PC12 cells, but NGF and FGF induce neurite outgrowth in the same cells, despite the fact that these growth factors have many overlapping signaling pathways. Some evidence suggests that the timing of signal transduction may influence outcome: PC12 cells proliferate if the MAP kinase pathway is transiently stimulated, but differentiate when this pathway is continuously stimulated.[50] Finally, the mechanisms by which RTK signaling is turned off are obscure. Internalization with degradation, dephosphorylation of phosphorylated substrates by phosphatases and down-regulation by serine/threonine phosphorylation have been implicated in this process.

Acknowledgements

The authors wish to thank Wendy Fantl and Inga Malveaux for their assistance.

References

The authors apologize for omitting many important references owing to limitations of space.

1 van der Geer, P., Hunter, T. and Lindberg, R.A. (1994) Receptor protein-tyrosine kinases and their signal transduction pathways. *Annu. Rev. Cell. Biol.*, **10**, 251–337.

2 Fantl, W.J., Johnson, D.E. and Williams, L.T. (1993) Signaling by receptor tyrosine kinases. *Annu. Rev. Biochem.*, **62**, 453–81.

3 Schlessinger, J. and Ullrich, A. (1992) Growth factor signaling by receptor tyrosine kinases. *Neuron*, **9**, 383–91.

4 Cantley, L.C., Auger, K.R., Carpenter, C., Duckworth, B., Graziani, A., Kapeller, R. and Soltoff, S. (1991) Oncogenes and signal transduction. *Cell*, **64**, 281–302.

5 Pawson, T. (1995) Protein modules and signaling networks. *Nature*, **373**, 573–80.

6 Pronk, G.J. and Bos, J.L. (1994) The role of p21ras in receptor tyrosine kinase signaling. *Biochim. Biophys. Acta*, **1198**, 131–47.

7 Heldin, C.-H. (1995) Dimerization of cell surface receptors in signal transduction. *Cell*, **80**, 213–23.

8 Rousseau, F., Bonaventure, J., Legeai-Mallet, L. *et al.* (1994) Mutations in the

gene encoding fibroblast growth factor receptor-3 in achondroplasia. *Nature*, **371**, 252–4.

9 Davis, S., Gale, N.W., Aldrich, T.H. *et al.* (1994) Ligands for EPH-related receptor kinases that require membrane attachment or clustering for activity. *Science*, **266**, 816–9.

10 Bellosta, P., Costa, M., Lin, D.A. and Basilico, C. (1995) The receptor tyrosine kinase ARK mediates cell aggregation by homophillic binding. *Mol. Cell. Biol.*, **15**, 614–25.

11 Hubbard, S.R., Wei, L., Ellis, L. and Hendrickson, W.A. (1994) Crystal structure of the tyrosine kinase domain of the human insulin receptor. *Nature*, **372**, 746–54.

12 Coughlin, S.R., Escobedo, J.A. and Williams, L.T. (1989) Role of phosphatidylinositol kinase in PDGF receptor signal transduction. *Science*, **243**, 1191–4.

13 Klippel, A., Escobedo, J.A., Fantl, W.J. and Williams, L.T. (1992) The C-terminal SH2 domain of p85 accounts for the high affinity and specificity of the binding of phosphatidylinositol 3-kinase to phosphorylated platelet-derived growth factor beta receptor. *Mol. Cell. Biol.*, **12**, 1451–9.

14 Escobedo, J.A., Kaplan, D.R., Kavanaugh, W.M. *et al.* (1991) A phosphatidylinositol-3 kinase binds to platelet-derived growth factor receptors through a specific receptor sequence containing phosphotyrosine. *Mol. Cell. Biol.*, **11**, 1125–32.

15 Zhou, S., Shoelson, S.E., Chaudhuri, M. *et al.* (1993) SH2 domains recognize specific phosphopeptide sequences. *Cell*, **72**, 767–78.

16 Kavanaugh, W.M. and Williams, L.T. (1994) An alternative to SH2 domains for binding tyrosine-phosphorylated proteins. *Science*, **266**, 1862–5.

17 Kavanaugh, W.M., Turck, C.W. and Williams, L.T. (1995) Binding of PTB domains to signaling proteins through a motif containing phosphotyrosine. *Science*, **268**, 1177–9.

18 Gustafson, T.A., He, W., Craparo, A. *et al.* (1995) Phosphotyrosine-dependent interaction of SHC and IRS-1 with the NPEY motif of the insulin receptor via a novel (non-SH2) domain. *Mol. Cell. Biol.* **15**, 2500–8.

19 Kavanaugh, W.M. and Williams, L.T., unpublished observations.

20 Bork, P. and Margolis, B. (1995) A phosphotyrosine interaction domain. *Cell*, **80**, 693–4.

21 Musacchio, A., Gibson, T., Rice, P. *et al.* (1993) The PH domain: a common piece in the structural pathwork of signaling proteins. *Trends Biochem. Sci.*, **18**, 343–8.

22 Avruch, J., Zhang, X. and Kyriakis, J.M. (1994) Raf meets Ras: completing the framework of a signal transduction pathway. *Trends Biochem. Sci.*, **19**, 279–83.

23 Hofer, F., Fields, S., Schneider, C. and Martin, G.S. (1994) Activated Ras interacts with the Ral guanine nucleotide dissociation stimulator. *Proc. Natl. Acad. Sci. USA*, **91**, 11089–93.

24 Kikuchi, A., Demo, S.D., Ye, Z.-H. *et al.* (1994) ralGDS family members interact with the effector loop of ras p21. *Mol. Cell. Biol.*, **14**, 7483–91.

25 Aronheim, A., Engelberg, D., Li, N. *et al.* (1994) Membrane targeting of the nucleotide exchange factor Sos is sufficient for activating the Ras signaling pathway. *Cell*, **78**, 949–61.

26 Leevers, S.J., Paterson, H.F. and Marshall, C.J. (1994) Requirement for Ras in Raf activation is overcome by targeting Raf to the plasma membrane. *Nature*, **369**, 411–4.

27 Goldschmidt-Clermont, P.J., Kim, J.W., Machesky, L.M. *et al.* (1991) Regulation of phospholipase C-γ1 by profilin and tyrosine phosphorylation. *Science*, **251**, 1231–3.

28 Nishibe, S., Wahl, M.I., Hernandez-Sotomayor, S.M.T. *et al.* (1990) Increase of the catalytic activity of phospholipase C-γ1 by tyrosine phosphorylation. *Science*, **250**, 1253–6.

29 Backer, J.M., Myers, M.G., Jr, Shoelson, S.E. *et al.* (1992) Phosphatidylinositol 3′-kinase is activated by association with IRS-1 during insulin stimulation. *EMBO J.* **11**, 3469–79.

30 Shoelson, S.E., Sivaraja, M., Williams, K.P. *et al.* (1993) Specific phosphopeptide binding regulates a conformational change in the PI 3-kinase SH2 domain associated with enzyme activation. *EMBO J.*, **12**, 795–802.

31 Peters, K.G., Marie, J., Wilson, E. *et al.* (1992) Point mutation of an FGF receptor abolishes phosphatidylinositol turnover and Ca^{2+} flux but not mitogenesis. *Nature*, **358**, 678–81.

32 Valius, M. and Kazlauskas, A. (1993) Phospholipase C-gamma 1 and phosphatidylinositol 3 kinase are the downstream mediators of the PDGF receptor's mitogenic signal. *Cell*, **73**, 321–34.

33 Kavanaugh, W.M., Klippel, A., Escobedo, J.A. and Williams, L.T. (1992) Modification of the 85-kilodalton subunit of phosphatidylinositol-3 kinase in platelet-derived growth factor-stimulated cells. *Mol. Cell. Biol.*, **12**, 3415–24.

34 Carpenter, C.L., Auger, K.R., Duckworth, B.C. *et al.* (1993b) A tightly associated serine/threonine protein kinase regulates phosphoinositide 3-kinase activity. *Mol. Cell. Biol.*, **13**, 1657–65.

35 Fantl, W.J., Escobedo, J.A., Martin, G.A. *et al.* (1992) Distinct phosphotyrosines on a growth factor receptor bind to specific molecules that mediate different signaling pathways. *Cell*, **69**, 413–23.

36 Wennstrom, S., Siegbahn, A., Yokote, K. *et al.* (1994) Membrane ruffling and chemotaxis transduced by the PDGF-β receptor require the binding site for phosphatidylinositol 3′ kinase. *Oncogene*, **9**, 651–60.

37 Rodriguez-Viciana, P., Warne, P.H., Dhand, R. *et al.* (1994) Phosphatidylinositol-3-OH kinase as a direct target of Ras. *Nature*, **370**, 527–532.

38 Satoh, T., Fantl, W.J., Escobedo, J.A. *et al.* (1993) Platelet-derived growth factor receptor mediates activation of ras through different signaling pathways in different cell types. *Mol. Cell. Biol.*, **13**, 3706–13.

39 Hu, Q., Klippel, A., Muslin, A.J. *et al.* (1995) Ras-dependent induction of cellular responses by constitutively active phosphatidylinositol-3 kinase. *Science*, **268**, 110–12.

40 Fantl, W.J., Cramer, L.P., Klippel, A. *et al.* (1995) PI3-kinase and Rac1 are on the same pathway that regulates lamellipodia formation in fibroblasts.

41 Chung, J., Grammer, T.C., Lemon, K.P. *et al.* (1994) PDGF-dependent regulation of pp70-S6 kinase is coupled to receptor-dependent binding and activation of phosphatidylinositol 3-kinase. *Nature*, **370**, 71–5.

42 Schu, P.V., Takegawa, K., Fry, M.J. *et al.* (1993) Phosphatidylinositol 3-kinase encoded by yeast VPS34 gene essential for protein sorting. *Science*, **260**, 88–91.

43 Hu, Q., Milfay, D. and Williams, L.T. (1995) Binding of NCK to SOS and activation of Ras-dependent gene expression. *Mol. Cell. Biol.*, **15**, 1169–74.

44 Morrison, D. (1994) 14-3-3: modulators of signaling proteins? *Science*, **266**, 56–7.

45 Kikuchi, A. and Williams, L.T. (1996) Regulation of interaction of *Ras* p21 with RalGDS and Raf-1 by cyclic AMP-dependent protein kinase. *J. Biol. Chem.* **271**, 588–94.

46 Kypta, R.M., Goldberg, Y., Ulug, E.T. and Courtneidge, S.A. (1990) Association between the PDGF receptor and members of the *src* family of tyrosine kinases. *Cell*, **62**, 481–92.

47 Twamley-Stein, G.M., Pepperkok, R., Ansorge, W. and Courtneidge, S.A. (1993) The Src family tyrosine kinases are required for platelet-derived growth factor-mediated signal transduction in NIH 3T3 cells. *Proc. Natl. Acad. Sci. USA*, **90**, 7696–700.

48 Feng, G.S., Hui, C.C. and Pawson, T. (1993) SH2-containing phosphotyrosine phosphatase as a target of protein-tyrosine kinases. *Science*, **259**, 1607–11.

49 Perkins, L.A., Larsen, I. and Perrimon, N. (1992) Corkscrew encodes a putative protein tyrosine phosphatase that functions to transduce the terminal signal from the receptor tyrosine kinase torso. *Cell*, **70**, 225–36.

50 Marshall, C.J. (1995) Specificity of receptor tyrosine kinase signaling: transient versus sustained extracellular signal-regulated kinase activation. *Cell*, **80**, 179–85.

51 Zhou, M.-M., Ravichandran, K.S., Olejniczak, E.T. *et al.* (1995) Structure and ligand recognition of the phosphotyrosine binding domain of Shc. *Nature*, **378**, 584–92.

2 Signal transduction through class I cytokine receptors

Tetsuya Taga and
Tadamitsu Kishimoto

In the development and homeostasis of multicellular organs such as the immune and hematopoietic systems, the growth and differentiation of cells are under the precise control of various kinds of protein mediators known as cytokines. Molecular cloning of the receptors for these cytokines has revealed a large family of structurally related integral membrane proteins referred to as the class I cytokine receptor family[1,2] (the distantly related class II cytokine receptor family is discussed in Chapter 3). Members of this family contain a conserved domain of around 200 amino acid residues in their extracellular domains as well as two or three short conserved motifs in the cytoplasmic region.[3,4] The signaling mechanism of the class I cytokine receptor family involves the homo- or heterodimerization of family members.[5–8] One partner in these dimers is generally one of three common signal-transducing subunits, which may help explain the functional redundancy of cytokines.[6,9] Cytokine stimulation induces tyrosine phosphorylation of cellular proteins including the receptors themselves.[5–8] Class I cytokine receptors lack their own tyrosine kinase domain but associate with cytoplasmic protein tyrosine kinases, mostly of the Jak family.[10] Their activation after ligand-induced dimerization of receptor components is reminiscent of the growth factor-mediated dimerization and consequent activation of the receptor tyrosine kinases[11] (Chapter 1). This chapter will also cover the mechanism of cytoplasmic signaling following dimerization of class I cytokine receptors, by highlighting the recently identified, and still progressively expanding, family of latent cytoplasmic transcription factors, STATs.[12,13]

Signal Transduction. Edited by Carl-Henrik Heldin and Mary Purton. Published in 1996 by Chapman & Hall. ISBN 0 412 70810 8

Structural features of class I cytokine receptors

The molecular cloning of most of the receptors for cytokines regulating the immune and hematopoietic systems has allowed their structural analysis. The majority are members of the class I cytokine receptor family, which includes interleukin-2 (IL-2)R (β-chain), IL-3R, IL-4R, IL-5R, IL-6R, IL-7R, IL-9R, IL-11R, IL-12R, erythropoietin (EPO)R, granulocyte colony stimulating factor (G-CSF)R, granulocyte-macrophage colony stimulating factor (GM-CSF)R, leukemia inhibitory factor (LIF)R, ciliary neurotrophic factor (CNTF)R, thrombopoietin (TPO)R, prolactin (PRL)R and growth hormone (GH)R.[1,2,5–8] The signal-transducing receptor components gp130, βc (common β, initially designated as human KH95 or mouse AIC2B) and γc (common γ, originally identified as the IL-2R γ-chain) also belong to this family. The conserved domain of ≅200 amino acid residues in the extracellular region of class I cytokine receptors is composed of two fibronectin type III modules, each of which consists of seven β-strands positioned antiparallel so as to form a 'barrel-like' shape[1,14] (Figure 2.1a and b). A trough formed between two 'barrel-like' modules is believed to function as a ligand binding pocket. Four positionally conserved cysteine residues in the amino-terminal half and a WSXWS (where X is any amino acid residue) motif in the carboxy-terminal end are considered to be important to keep this ternary structure, since mutations in them abolish the binding capability of the receptors.[15] Most class I cytokine receptors also contain two short stretches of conserved amino acid residues (box 1 and box 2) in their cytoplasmic regions, close to the membrane.[3,4] gp130, LIF-R and G-CSFR also contain another conserved motif (box 3) in the middle of the cytoplasmic region[4,16] (Fig. 2.1c).

Multichain class I receptor systems and shared signal transducing components

The apparent enigma that expression of cloned receptor cDNAs conferred only low-affinity cytokine-binding sites whereas cells normally responsive to these cytokines displayed both high- and low-affinity binding sites[5] suggested the existence of an additional receptor component required for the formation of high-affinity binding sites and for signal transduction. The first molecular demonstration of such a multichain class I cytokine receptor system was provided by the identification

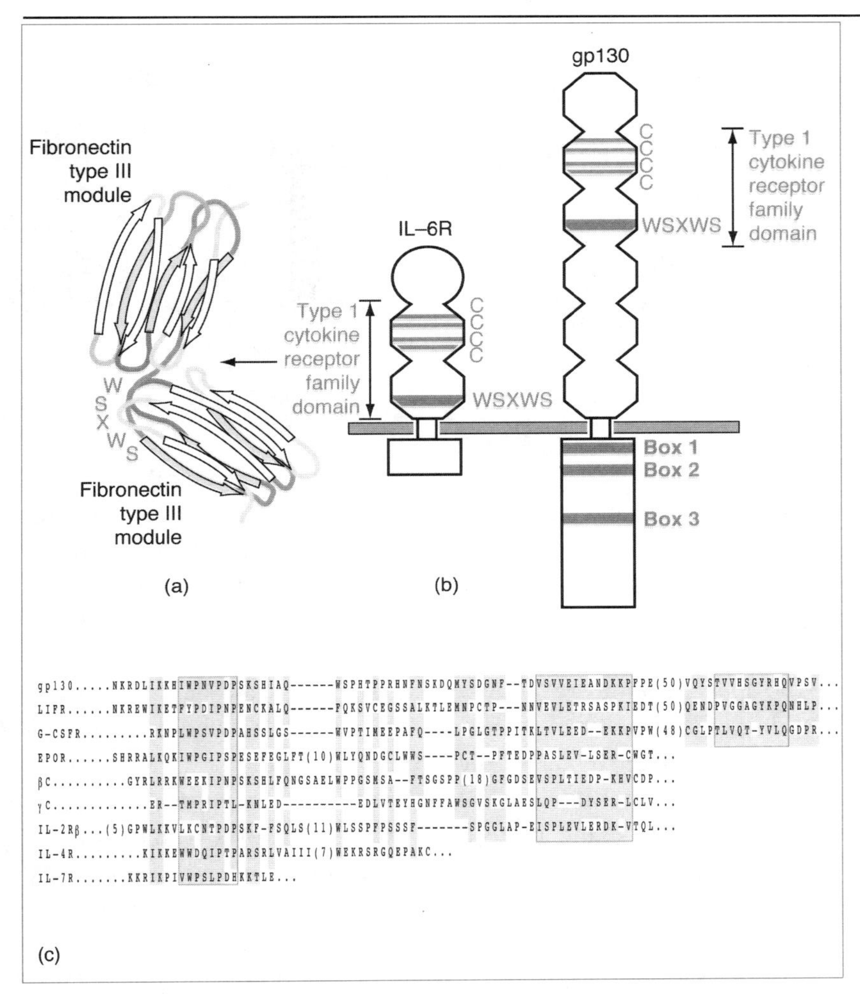

Figure 2.1 Structural features of class I cytokine receptors. (a) Predicted ternary structure of the extracellular conserved domain. This domain is composed of two barrel-like shapes, each of which comprises seven β-strands. (b) Schematic structure of IL-6R and its associated signal transducer, gp130. A hexagon represents the fibronectin type III module and ovals the immunoglobulin-like domains. (c) Amino acid sequence comparison of the conserved cytoplasmic motifs, box 1, box 2 and box 3, in class I cytokine receptors. Homologous residues at conserved positions (at a frequency of over 50% among aligned sequences) are tinted (nonpolar, A,I,L,M,F,P and W; polar but uncharged, N,C,Q,G,S,T, and Y; negatively charged, D and E; and positively charged, R, H and K). Gaps have been introduced to maximize homology (– and numbers in parentheses).

and cloning of the IL-6R-associated membrane protein gp130, which is critical for the formation of high-affinity IL-6-binding sites and for the signal transduction.[17,18] All known class I cytokine receptors are believed to form homo- or heterodimers (or, in a few cases, trimers) upon binding of their cognate cytokines. The family can be divided into three subsets on the basis of their ability to form complexes with gp130, βc and γc, the common receptor components critical for signal transduction.[5–8]

The cytokines IL-6, IL-11, LIF, OM, CNTF and cardiotrophin-1 (CT-1) are structurally and functionally similar.[9,19–21] They function in various cell systems, exerting pleiotropic biological effects with significant overlap. The functional redundancy of these cytokines is now well explained by the nature of the receptor complexes for these ligands:

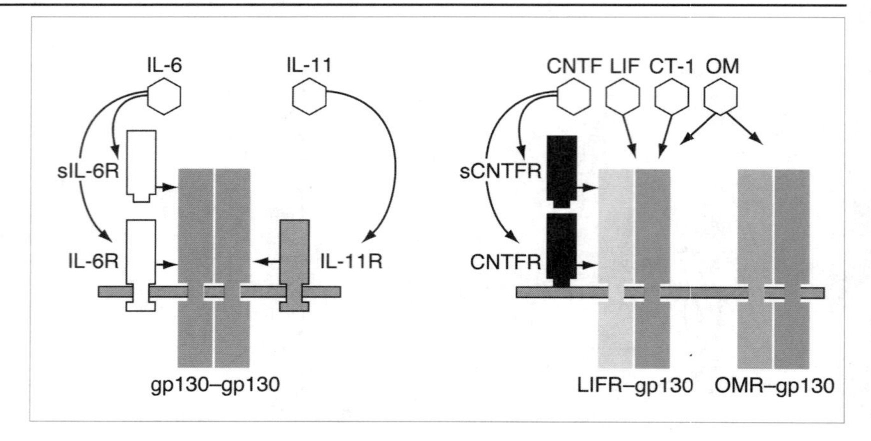

Figure 2.2 Receptor systems sharing gp130 as a common signal transducer. IL-6, when bound to either a membrane-anchored or a soluble form of its receptor (IL-6R or sIL-6R), induces homodimerization of gp130. The IL-11–IL-11R complex is suggested to induce the same type of homodimer. Heterodimerization of LIFR and gp130 is induced by the CNTF–CNTFR complex, the CNTF–sCNTFR complex, LIF, CT-1 or OM. OM also induces another type of heterodimer composed of OMR and gp130.

they all share gp130 as a signal-transducing receptor component[6,22–25] (Figure 2.2). For instance, IL-6 binds to IL-6R and the ligand–receptor complex then associates with gp130, allowing it to homodimerize.[26] IL-6R has a relatively short cytoplasmic domain that is not required for signaling[17] and so the juxtaposition of the cytoplasmic regions of the two gp130 molecules induced as a result of homodimerization of gp130 appears to initiate a downstream signaling cascade. LIF, another member of this subset of cytokines, binds to LIFR, whose structure is similar to that of gp130. LIFR then forms a heterodimer with gp130.[23,27] Formation of the LIFR–gp130 heterodimer is also triggered by a complex of CNTF and its receptor, CNTFR.[27] CT-1 and OM are also suggested to induce the LIFR–gp130 heterodimer,[25] although the latter cytokine is also proposed to form a different type of heterodimer composed of OM-specific receptor (OMR) and gp130.[28] gp130 is further utilized as a critical component in the IL-11 receptor complex.[24] From the close structural similarity of IL-6R and IL-11R, the gp130 homodimer might be a candidate complex for IL-11 signaling.[29] This model also explains why the phenotypes of mice lacking either IL-6, LIF or CNTF are much less severe than that expected from the pleiotropic functions of each cytokine.[30–32] In contrast, the deletion of gp130 or LIFR is lethal[33] (K. Yoshida *et al.*, unpublished data). Although gp130 is found in most tissues, ligand-specific receptor components display a more limited expression, suggesting that cellular responsiveness is largely determined by the regulated expression of ligand-specific receptors. Regulation of the expression of the different cytokines may also determine the place and occasion where gp130 functions and explain why these cytokines with overlapping biological functions also have distinct effects. IL-6R and CNTFR, whose cytoplasmic regions are dispensable for signal transduction,[17,34] are considered to function simply in recruiting respective ligands to the cell surface where the critical receptor components exist, leading to the formation of gp130 homodimer or LIFR–gp130 heterodimer, respectively. The ligand-binding capabilities of IL-6R and CNTFR are also retained in extracellular soluble form (sIL-6R and sCNTFR) and these soluble receptors also induce homo- or heterodimers of gp130.[17,26,27,34] More importantly, the IL-6–sIL-6R complex confers IL-6 responsiveness of the cells on which gp130, but not IL-6R, is expressed. This complex mimics not only IL-6 but also all other gp130-stimulatory cytokines in cells expressing gp130[22,35] by triggering gp130-mediated signaling pathways through the formation of gp130 homodimers. The same is true for the CNTF–sCNTFR complex; it acts even on cells which do not express CNTFR (and are thus normally nonresponsive to CNTF) if the cells are expressing both gp130 and LIFR.[34] Physiological agonistic functions of these soluble receptors have been reported.[34,36]

Although GM-CSF, IL-3 and IL-5 differ slightly in their effects on hematopoietic progenitor cells, they induce many similar effects on common target cells.[7] These three cytokines compete with each other for their cellular binding sites. It is now clear that receptor complexes for these cytokines consist of a ligand-specific subunit (GM-CSFR, IL-3R and IL-5R) and a common chain (referred to as βc) which has no ability to bind any of the three cytokines[7] (Figure 2.3). Although the cytoplasmic regions of the ligand-specific chains are relatively short, they are essential for signaling since their truncation leads to loss of cellular responsiveness.[37] Thus, it is likely that juxtaposition of the cytoplasmic regions of the receptor components by ligand-induced

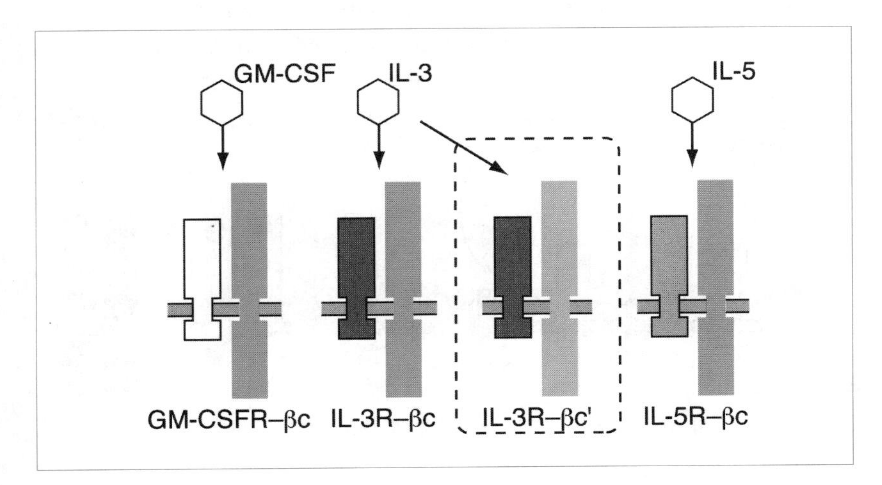

Figure 2.3 Receptor systems sharing βc as a common signal transducer. In human cells, GM-CSF, IL-3 and IL-5 induce heterodimers composed of βc and their respective receptor. In addition, in mouse cells, IL-3 also induces heterodimer of IL-3R and a molecule closely related to βc (depicted as βc′).

heterodimerization appears to be important for signal transduction in this subset of receptors.

The last subset of class I cytokine receptors uses as a common signal transducer the γc chain[8](Figure 2.4), which was initially identified as the third component of the functional IL-2 receptor complex (IL-2Rγ).[38] The high-affinity IL-2-binding site consists of three polypeptide chains, α, β and γ (β and γ belong to the class I cytokine receptor family). A combination of IL-2Rβ and IL-2Rγ (γc) subunits exhibits intermediate affinity for IL-2 and is sufficient to transduce IL-2 signals.[39] The α subunit is believed to help in recruiting IL-2 to this functional receptor complex. The γc-chain is now known also to be utilized in the receptor complexes for IL-4, IL-7, IL-9 and IL-15.[40–42] The IL-15 receptor complex also contains the IL-2Rβ subunit.[42] Inactivation of the γc-chain gene has been indicated to account for X-linked severe combined immunodeficiency (XSCID) in humans.[43] The XSCID phenotype seems to reflect the loss of all cytokine functions mediated through γc–receptor complexes.

Receptor dimerization and activation of associated Jak kinases

Members of the class I cytokine receptor family do not possess obvious enzymatic motifs in their structure. Although none of these class I cytokine receptors have an intrinsic tyrosine kinase domain, they and various cytoplasmic proteins are phosphorylated on tyrosine residues after cytokine stimulation.[44–46] A likely model is that cytoplasmic protein tyrosine kinases are associated with class I cytokine receptors (including signal-transducing components) and become activated by juxtaposition upon ligand-induced dimerization of such receptor components. In

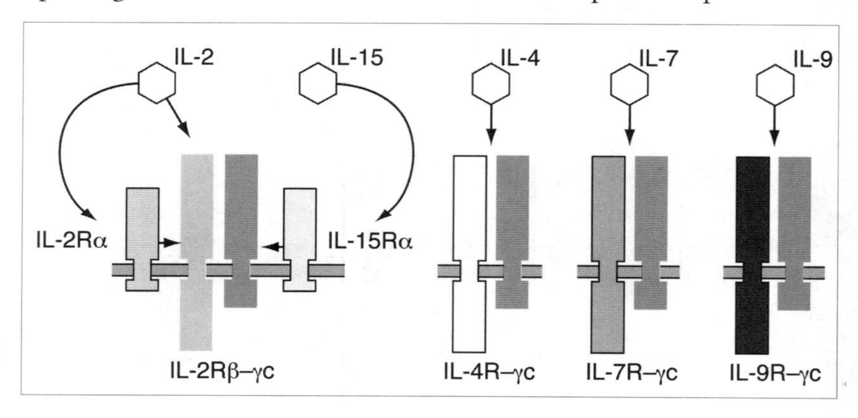

Figure 2.4 Receptor systems sharing γc as a common signal transducer. The heteromeric complex composed of IL-2Rβ and γc is induced by IL-2, the IL-2–IL-2Rα complex and the IL-15–IL-15Rα complex. Other receptors specific to IL-4, IL-7 and IL-9 also utilize γc.

support of this idea, even EPOR, G-CSFR, IL-12R and GHR, which are not considered to require an additional receptor component for signaling, form homodimers after binding of their respective ligands.[4,47–49] Several lines of evidence have demonstrated that cytoplasmic tyrosine kinases in the Jak family are associated with members of the class I cytokine receptor family.[50–55] The ligand-stimulated homo- or heterodimerization of receptor components triggers the activation of Jak-family tyrosine kinases already bound to the cytoplasmic domains of the receptors. The increase in the local concentration, or the juxtaposition, of Jak kinases may allow them to phosphorylate each other in addition to the receptor components. This model is conceptually analogous to the well-established model of transactivation of receptor tyrosine kinases following growth factor-induced receptor dimerization (Chapter 1). To date, four different members of the Jak family, Jak1, Jak2, Jak3 and Tyk2, are known to be involved in class I cytokine receptor signaling.[10,56] The nature of their involvement may differ in individual receptor complexes. In the case of the receptor–gp130 complexes, the gp130 protein is capable of associating with three different Jak family members, Jak1, Jak2 and Tyk2[55,57] (M. Narazaki *et al.*, unpublished data). These three kinases are all activated after treatment of cells by cytokines acting via the formation of gp130-containing complexes. By contrast, the single receptor chains EPOR and GHR have been shown to be associated with Jak2.[50,51,58] Thus, in cells stimulated by either EPO or GH, homodimerization of the respective receptor leads to the transphosphorylation of the Jak2 protein but not other members of the Jak family. In the γc–receptor complexes, two different Jaks are required: in the functional IL-2 receptor complex, Jak1 binds to the IL-2Rβ chain and Jak3 to the γc component.[56,59] The presence of either Jak1 or Jak3 alone does not allow the cells to respond to IL-2. The selective recruitment of different Jak kinases to the two receptor components is proposed to occur in the receptor complexes for IL-2, IL-4, IL-7 and IL-15; Jak1 is considered to associate with the ligand-specific receptors and Jak3 with γc.[60] The members of the Jak family tyrosine kinases that are activated following stimulation of different class I cytokine receptors are summarized in Table 2.1. Mutational analyses of the receptor components have shown that the conserved box 1 motif is important for the association of Jak kinases to the receptor[58,61] (M. Narazaki *et al.*, unpublished data). Amino acid substitutions in this motif abolish tyrosine phosphorylation of Jak kinases following cytokine stimulation.[3,16,53,62] Although much attention has

Table 2.1 Jak- and STAT-family members involved in class I cytokine receptor signaling

Receptor–signal-transducer complex	Jak-family tyrosine kinases[a]	STAT-family transcription factors[a]	References
gp130–gp130 (LIFR–gp130)	Jak1, Jak2, TYK2	STAT3, STAT1[b]	53,55,57,65,66
GM-CSFR–βc (IL-3R–βc) (IL-5R–βc)	Jak1, Jak2	STAT5	52,54,70
IL-2Rβ–γc[c] (IL-7R–γc)	Jak1, Jak3	STAT5	56,59,60,71
IL-4R–γc	Jak1, Jak3	STAT6	68
G-CSFR–G-CSFR	Jak1, Jak2	STAT3	66,81
EPOR–EPOR	Jak2	STAT5	51,58,70,71,80
GHR–GHR	Jak2	STAT1	50,58,82
IL-12R–?	Jak2, TYK2	STAT4	83,84
PRLR–?		STAT5	70,71,85

[a] Cytoplasmic protein tyrosine kinases in the Jak family and cytoplasmic latent transcription factors in the STAT family that are activated by stimulation of class I cytokine receptors and have been molecularly cloned are listed.
[b] STAT1 is activated to a lesser extent and in a fewer cells than STAT3.
[c] Members in the γc-sharing group do not always exhibit the identical pattern of STAT activation, probably because of the presence of distinct STAT-docking sites in the ligand-binding chains (see also text and ref. 60 for the details); the STAT species activated by IL-4 appears to be different from that by IL-2, IL-7 and IL-15.

been given to the Jak kinases, other types of cytoplasmic tyrosine kinases have also been observed to be activated. After gp130 stimulation, kinases such as Yes and Hck are activated in certain cell types.[46,63] IL-2 stimulation is shown to induce activation of such kinases as Lck and Syk[45,64] and GM-CSF stimulation leads to Fes activation.[44]

Cytokine signaling pathways via STATs and MAP kinase

One of the consequences of the cytoplasmic signal transduction pathways triggered by receptor oligomerization is the initiation of gene transcription via activation of transcription factors. Recent studies on this aspect of cytokine signaling have indicated that ligand stimulation of the class I cytokine receptors leads to tyrosine phosphorylation and functional activation (in terms of DNA-binding capability) of the latent cytoplasmic transcription factors known as signal transducers and activators of transcription or STATs.[55,57,60,65–71] p91 (STAT1α), p84 (STAT1β) and p113 (STAT2) are members of the STAT family that were originally identified as components of the interferon-stimulated gene factor 3 complex that have been demonstrated in the class II cytokine receptor signaling (Chapter 3). Newly identified members of

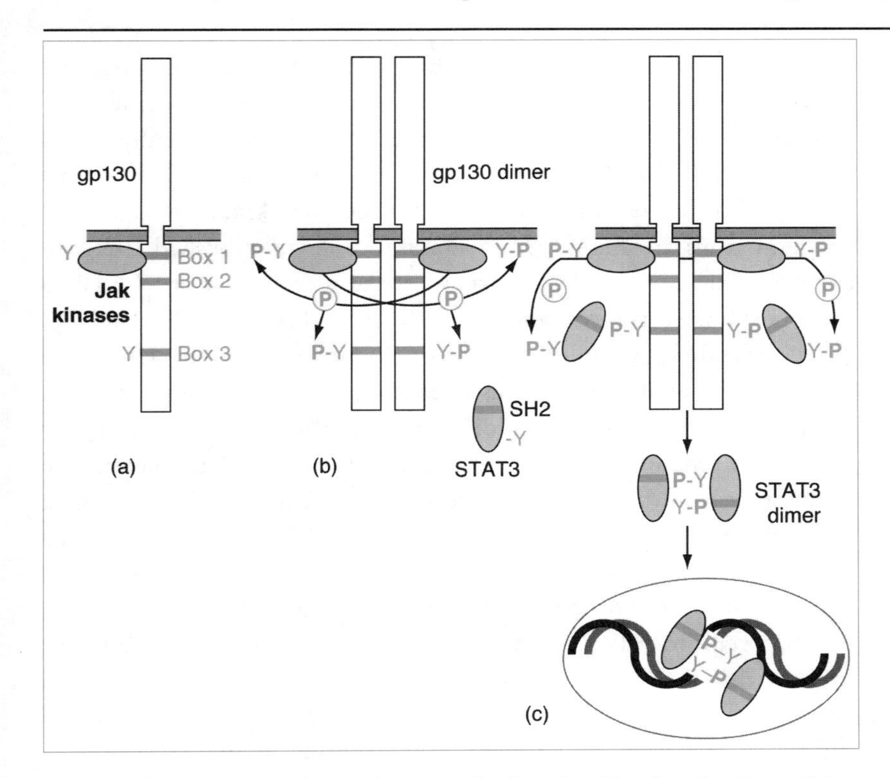

Figure 2.5 Class I cytokine receptor signaling from the receptor dimerization to the activation of latent cytoplasmic transcription factors, STATs, via Jak-family tyrosine kinases. The IL-6-induced gp130 homodimer is taken as an example. (a) Prior to stimulation, Jak kinases are associated with gp130 at its box 1 motif. (b) Dimerization of gp130 triggers the activation of associated Jak kinases, leading to tyrosine phosphorylation (P-Y) of the distal part of gp130 including the box 3 motif. (c) STAT3 is recruited to the phosphotyrosine-containing motif via its SH2 domain, and tyrosine phosphorylated by the juxtaposed Jak kinases. Tyrosine-phosphorylated STAT3 proteins then form homodimers via intermolecular SH2–phosphotyrosine interactions, and are translocated to the nucleus, resulting in the transcriptional activation of the target genes. The full activation of STAT3 requires its serine phosphorylation, presumably by MAP kinase.[86,87] The box 2 motif does not seem to be directly involved in these steps, but is critical for transmission of the signal(s) required for growth regulation, in coordination with box 1.

the STAT family have been shown to be involved in the class I cytokine receptor signaling (Table 2.1). In the case of receptor–gp130 complexes, STAT3 (also known as acute phase response factor, APRF,[72] because of its binding to the promoter region of the acute phase protein genes in hepatocytes) and, to a lesser extent and in fewer types of cells, STAT1α become activated after cytokine stimulation.[65–67] Since all STATs contain an SH2 domain (Chapter 9), it is likely that tyrosine phosphorylation of STAT3 allows this molecule to form stable homodimers, acquiring the ability to bind to the target DNA sequence (Figure 2.5). In this model, the Jak proteins that are already associated with gp130 become activated by homodimerization of gp130. This leads to phosphorylation on tyrosine residues in the distal part of gp130 (presumably at the box 3 motif), providing a docking site for the SH2 domain of STAT3. The STAT3 protein recruited to gp130 serves as a substrate for the resident Jak kinases. It is speculated that the tyrosine-phosphorylated STAT3 proteins then leave gp130 and form a homodimer via two intermolecular SH2-phosphotyrosine linkings. This might occur if, for example, the SH2 domain of STAT3 had a higher affinity for the phosphotyrosine motif in STAT3 than for that in gp130. G-CSFR is the only known receptor that has the box 3 motif apart from gp130 and LIFR.[4,16] This could explain why stimulation of cells by G-CSF induces strong STAT3 activation similar to that observed with the gp130-stimulatory cytokines.[65,66] From experiments using COS cells transiently over-expressing STAT3 together with either Jak1, Jak2 or Tyk2, it appears that STAT3 can be a direct substrate for these Jak kinases if their local concentrations are high[57] (M. Narazaki *et al.*, unpublished data). In ordinary cells, however, the purpose of the recruitment of STAT3 to the phosphotyrosine-containing gp130 is to bring STAT3 and Jak kinases together. Although the use of the members of the Jak family exhibits few variations and thus does not seem to be unique to each receptor system (Table 2.1), different receptors show much wider variations in the usage of the STAT family members. This may be due to differences in the amino acid sequence motif of the STAT docking site in different receptor species. In gp130, a YXXQ motif is proposed as a STAT3 docking motif,[57] and this motif is also present in G-CSFR. By contrast, IL-2Rβ has been suggested to have a STAT-docking tyrosine residue specific for STAT5 and IL-4R has one for STAT6.[60,68,71] This may explain why stimulation by IL-2 and IL-4 induces differential activation of distinct sets of STAT proteins, despite the shared use of the γc-chain and the γc-associated Jak3 protein in their receptor complexes.

Other downstream signaling molecules containing an SH2 domain are also recruited by the tyrosine-phosphorylated class I cytokine receptors.[73] These cytoplasmic signaling molecules can themselves be activated through tyrosine phosphorylation by the juxtaposed kinases or might serve as adapters, attracting further downstream molecules. Stimulation of class I cytokine receptors has been shown to result in the recruitment and/or tyrosine phosphorylation of such signaling molecules as the p85 subunit of PI3 kinase, Vav, Shc, insulin receptor kinase substrate 1 (IRS1)-related protein (4PS), SH2-containing protein tyrosine phosphatase 1 (SH-PTP1 or HCP/PTP-1C) and SH-PTP2 (or Syp/PTP-1D).[37,56,74,75] Some of these molecules are suggested as adapters which link tyrosine-phosphorylated receptors and activation of Ras. The Ras/MAP kinase signaling pathway is often activated following stimulation of class I cytokine receptors: the ratio of GTP-bound Ras to GDP-bound Ras has been reported to increase;[76] tyrosine phosphorylation and activation of c-Raf-1, a serine/threonine kinase that interacts with Ras-GTP, has been observed;[77] and hyperphosphorylation of MAP kinase and upregulation of its Ser/Thr kinase activity has also been noted.[78] A precise mechanism that links the receptor dimerization to GTP modification of Ras and to activation

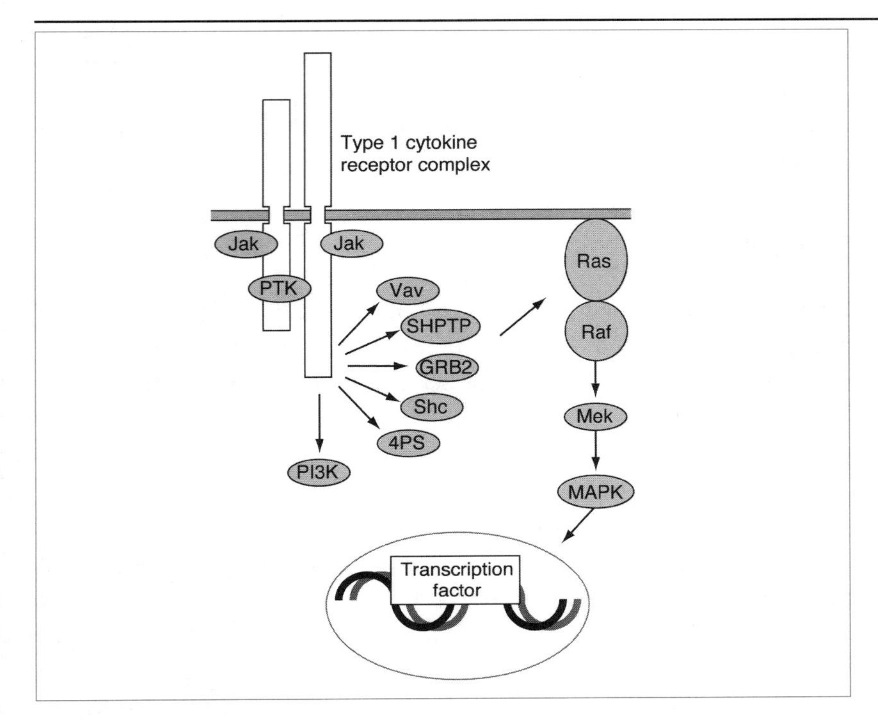

Figure 2.6 Signaling pathways from class I cytokine receptors. In this cartoon, the STAT pathway is omitted (see Figure 2.5). Transcription factors in this figure include NF-IL6 and c-Fos. A carboxy-terminal part of the cytoplasmic region of class I cytokine receptors is suggested to be important for the activation of the Ras/MAP kinase cascade. Abbreviations: MAPK, MAP kinase; MEK, MAP kinase kinase; PTK, protein tyrosine kinase; PI3K, phosphoinositide 3′-kinase; 4PS, insulin receptor kinase substrate 1-related protein; SHPTP, SH2-containing protein tyrosine phosphatase.

of MAP kinase has not yet been elucidated, although involvement of 'adapter molecules' such as Shc has been implicated[37] (Figure 2.6). Shc is postulated to be recruited to a phosphotyrosine residue in the receptor and attracts a complex consisting of Grb, another adapter molecule, and Sos, a guanine nucleotide exchange factor that promotes the activated (GTP-bound) form of Ras. For some class I cytokine receptors, it has been shown that activation (or modification) of the signaling molecules involved in these events requires the carboxy-terminal part of the class I receptor components.

Both of the two membrane-proximal cytoplasmic motifs, box 1 and box 2, are required for mediating the cytokine-induced growth response. As mentioned earlier, the box 1 motif is thought to be important for the association of class I cytokine receptors with Jak kinases[58,61] (M. Narazaki *et al.*, unpublished data). No cytoplasmic molecule which interacts exclusively with box 2 has been reported, although this motif (box 2), in concert with box 1, obviously plays a critical role in mediating the growth signal.[3,4,37,48,61,79–81] STAT proteins probably do not interact with box 2, since the STAT-recruiting tyrosine residues identified so far are usually located in the middle of the cytoplasmic domain, and thus distal from box 2.[57,60,68,71]. It is probable that signals that mediate the pleiotropic biological functions of cytokines are generally transmitted from different sites of the cytoplasmic region of class I cytokine receptors via distinct intracellular signaling pathways and that co-ordination among these pathways may be important in determining cell fate following cytokine stimulation.

Summary

Since the discovery of the class I cytokine receptor family, the signaling mechanism by which receptors in this family exert pleiotropic and redundant biological functions has been extensively studied. As we have discussed here, the functional redundancy is now explained by the presence of three subsets of class I cytokine receptors, each sharing common signal-transducing receptor components. The presence of different signaling pathways transmitted from distinct sites of the cytoplasmic region of receptors may partly account for the functional pleiotropy. The mechanism of activation of receptor-associated tyrosine kinases of the Jak family following ligand-induced homo- or heterodimerization of the receptors now answers the enigma of how the class I cytokine receptors which have no intrinsic tyrosine kinase domain mediate the cytoplasmic signals that appear to be initiated by tyrosine phosphorylation. The remaining question is how the multiple cytoplasmic signaling pathways co-ordinate to determine the fate of cytokine-stimulated cells, in particular growth promotion, growth inhibition and differentiation.

References

Space restrictions have prevented us from citing all the relevant literature. We apologize to those colleagues whose work is not listed here.

1 Bazan, J.F. (1990) Structural design and molecular evolution of a cytokine receptor superfamily. *Proc. Natl. Acad. Sci. USA*, **87**, 6934–8.
2 Taga, T. and Kishimoto, T. (1990) Immune and hematopoietic cell regulation: cytokines and their receptors. *Curr. Opin. Cell. Biol.*, **2**, 174–180.
3 Murakami, M., Narazaki, M., Hibi, M. *et al.* (1991) Critical cytoplasmic region of the IL-6 signal transducer, gp130, is conserved in the cytokine receptor family. *Proc. Natl. Acad. Sci. USA*, **88**, 11349–53.
4 Fukunaga, R., Ishizaka-Ikeda, E., Pan, C.-X. *et al.* (1991) Functional domains of the granulocyte colony-stimulating factor receptor. *EMBO J.*, **10**, 2855–65.
5 Taga, T. and Kishimoto, T. (1992) Cytokine receptors and signal transduction. *FASEB J.*, **6**, 3387–96.

6 Kishimoto, T., Taga, T. and Akira, S. (1994) Cytokine signal transduction. *Cell*, **76**, 253–62.

7 Miyajima, A., Kitamura, T., Harada, N. *et al.* (1992) Cytokine receptors and signal transduction. *Annu. Rev. Immunol.*, **10**, 295–331.

8 Taniguchi, T. and Minami, Y. (1993) The IL-2/IL-2 receptor system: a current overview. *Cell*, **73**, 5–8.

9 Kishimoto, T., Akira, S. and Taga, T. (1992) Interleukin-6 and its receptor: a paradigm for cytokines. *Science*, **258**, 593–7.

10 Ziemiecki, A., Harpur, A.G. and Wilks, A.F. (1994) MAP kinase protein tyrosine kinases: their role in cytokine signaling. *Trends Cell Biol.*, **4**, 207–12.

11 Heldin, C.-H. (1995) Dimerization of cell surface receptors in signal transduction. *Cell*, **80**, 213–23.

12 Ihle, J.N. and Kerr, I. (1995) Jaks and Stats in signaling by the cytokine receptor superfamily. *Trends Genet.*, **11**, 69–74.

13 Taga, T. and Kishimoto, T. (1995) Signaling mechanisms through cytokine receptors that share signal transducing receptor components. *Curr. Opin. Immunol.*, **7**, 17–23.

14 Patthy, L. (1990) Homology of a domain of the growth hormone/prolactin receptor family with type III modules of fibronectin. *Cell*, **61**, 13–14.

15 Yawata, H., Yasukawa, K., Natsuka, S. *et al.* (1993) Structure–function analysis of human IL-6 receptor: dissociation of amino acid residues required for IL-6-binding and for IL-6 signal transduction through gp130. *EMBO. J.*, **12**, 1705–12.

16 Baumann, H., Symes, A.J., Comeau, M.R. *et al.* (1994) Multiple regions within the cytoplasmic domains of the leukemia inhibitory factor receptor and gp130 cooperate in signal transduction in hepatic and neuronal cells. *Mol. Cell. Biol.*, **14**, 138–46.

17 Taga, T., Hibi, M., Hirata, Y. *et al.* (1989) Interleukin-6 triggers the association of its receptor with a possible signal transducer, gp130. *Cell*, **58**, 573–81.

18 Hibi, M., Murakami, M., Saito, M. *et al.* (1990) Molecular cloning and expression of an IL-6 signal transducer, gp130. *Cell*, **63**, 1149–57.

19 Yang, Y.-C. (1993) Interleukin 11: an overview. *Stem Cells*, **11**, 474–86.

20 Rose, T.M. and Bruce, A.G. (1991) Oncostatin M is a member of a cytokine family that includes leukemia-inhibitory factor, granulocyte colony-stimulating factor, and interleukin 6. *Proc. Natl. Acad. Sci. USA*, **88**, 8641–5.

21 Pennica, D., King, K.L., Shaw, K.J. *et al.* (1995) Expression cloning of cardiotrophin 1, a cytokine that induces cardiac myocyte hypertrophy. *Proc. Natl. Acad. Sci. USA*, **92**, 1142–6.

22 Taga, T., Narazaki, M., Yasukawa, K. *et al.* (1992) Functional inhibition of hematopoietic and neurotrophic cytokines by blocking the interleukin 6 signal transducer gp130. *Proc. Natl. Acad. Sci. USA*, **89**, 10998–1001.

23 Ip, N.Y., Nye, S.H., Boulton, T.G. *et al.* (1992) CNTF and LIF act on neuronal cells via shared signaling pathways that involved the IL-6 signal transducing receptor component gp130. *Cell*, **69**, 1121–32.

24 Yin., T., Taga, T., Tsang, M.L.-S. *et al.* (1993) Involvement of interleukin-6 signal

transducer gp130 in interleukin-11 mediated signal transduction. *J. Immunol.*, **151**, 2555–61.

25 Pennica, D., Shaw, K.J., Swanson, T.A. *et al.* (1995) Cardiotrophin-1: biological activities and binding to the leukemia inhibitory factor receptor/gp130 signaling complex. *J. Biol. Chem.*, **270**, 10915–22.

26 Murakami, M., Hibi, M., Nakagawa, N. *et al.* (1993) IL-6-induced homodimerization of gp130 and associated activation of a tyrosine kinase. *Science*, **260**, 1808–10.

27 Davis, S., Aldrich, T.H., Stahl, N. *et al.* (1993) LIFRβ and gp130 as heterodimerizing signal transducers of the tripartite CNTF receptor. *Science*, **260**, 1805–8.

28 Thoma, B., Bird, T.A., Friend, D.J. *et al.* (1994) Oncostatin M and leukemia inhibitory factor trigger overlapping and different signals through partially shared receptor complexes. *J. Biol. Chem.*, **269**, 6215–22.

29 Hilton, D.J., Hilton, A.A., Raicevic, A. *et al.* (1994) Cloning of a murine IL-11 receptor α chain; requirement for gp130 high affinity binding and signal transduction. *EMBO J.*, **13**, 4765–75.

30 Kopf, M., Baumann, H., Freer, G. *et al.* (1994) Impaired immune and acute-phase responses in interleukin-6-deficient mice. *Nature*, **368**, 339–42.

31 Stewart, C.L., Kaspar, P., Brunet, L.J. *et al.* (1992) Blastocyst implantation depends on maternal expression of leukaemia inhibitory factor. *Nature*, **359**, 76–9.

32 Masu, Y., Wolf, E., Holtmann, B. *et al.* (1993) Disruption of the CNTF gene results in motor neuron degeneration. *Nature*, **365**, 27–32.

33 Ware, C.B., Horowitz, M.C., Renshaw, B.R. *et al.* (1995) Targeted disruption of the low-affinity leukemia inhibitory factor receptor gene causes placental, skeletal, neural and metabolic defects and results in perinatal death. *Development*, **121**, 1283–99.

34 Davis, S., Aldrich, T.H., Ip, N.Y. *et al.* (1993) Released form of CNTF receptor α component as a soluble mediator of CNTF responses. *Science*, **259**, 1736–9.

35 Yoshida, K., Chambers, I., Nichols, J. *et al.* (1994) Maintenance of the pluripotential phenotype of embryonic stem cells through direct activation of gp130 signaling pathways. *Mech. Dev.*, **45**, 163–71.

36 Narazaki, M., Yasukawa, K., Saito, T. *et al.* (1993) Soluble forms of the interleukin-6 signal-transducing receptor component gp130 in human serum possessing a potential to inhibit signals through membrane-anchored gp130. *Blood*, **82**, 1220–6.

37 Sato, N., Sakamaki, K., Terada, N. *et al.* (1993) Signal transduction by the high-affinity GM-CSF receptor: two distinct cytoplasmic regions for the common β subunit responsible for different signaling. *EMBO J.*, **12**, 4181–9.

38 Takeshita, T., Asao, H., Ohtani, K. *et al.* (1992) Cloning of the γ chain of the human IL-2 receptor. *Science*, **257**, 379–82.

39 Nakamura, Y., Russell, S.M., Mess, S.A. *et al.* (1994) Heterodimerization of the IL-2 receptor β- and γ-chain cytoplasmic domains is required for signaling. *Nature*, **369**, 330–3.

40 Russell, S.M., Keegan, A.D., Harada, N. *et al.* (1993) Interleukin-2 receptor γ chain: a functional component of the interleukin-4 receptor. *Science*, **262**, 1880–3.

41 Kondo, M., Takeshita, T., Higuchi, M. *et al.* (1994) Functional participation of the IL-2 receptor γ chain in IL-7 receptor complexes. *Science*, **263**, 1453–4.

42 Giri, J.G., Ahdieh, M., Eisenman, J. *et al.* (1994) Utilization of the β and γ chains of the IL-2 receptor by the novel cytokine IL-15. *EMBO J.*, **13**, 2822–30.

43 Noguchi, M., Yi, H., Rosenblatt, H.M. *et al.* (1993) Interleukin-2 receptor γ chain mutation results in X-linked severe combined immunodeficiency in humans. *Cell*, **73**, 147–57.

44 Hanazono, Y., Chiba, S., Sasaki, K. *et al.* (1993) c-fps/fes protein-tyrosine kinase is implicated in a signaling pathway triggered by granulocyte-macrophage colony-stimulating factor and interleukin-3. *EMBO J.*, **12**, 1641–6.

45 Hatakeyama, M., Kono, T., Kobayashi, N. *et al.* (1991) Interaction of the IL-2 receptor with the src-family kinase $p56^{lck}$: identification of novel intermolecular association. *Science*, **252**, 1523–8.

46 Schieven, G.L., Kallestad, J.C., Brown, T.J. *et al.* (1992) Oncostatin M induces tyrosine phosphorylation in endothelial cells and activation of $p62^{yes}$ tyrosine kinase. *J. Immunol.*, **149**, 1676–82.

47 Watowich, S.S., Yoshimura, A., Longmore, G.D. *et al.* (1992) Homodimerization and constitutive activation of the erythropoietin receptor. *Proc. Natl. Acad. Sci. USA*, **89**, 2140–4.

48 Fukunaga, R., Ishizaka-Ikeda, E. and Nagata, S. (1993) Growth and differentiation signals mediated by different regions in the cytoplasmic domain of granulocyte colony-stimulation factor receptor. *Cell*, **74**, 1–20.

49 De Vos, A.M., Ultsch, M. and Kossiakoff, A.A. (1992) Human growth hormone and extracellular domain of its receptor: crystal structure of the complex. *Science*, **225**, 306–12.

50 Argetsinger, L.S., Campbell, G.S., Yang, X. *et al.* (1993) Identification of JAK2 as a growth hormone receptor-associated tyrosine kinase. *Cell*, **74**, 237–44.

51 Witthuhn, B.A., Quelle, F.W., Silvennoinen, O. *et al.* (1993) JAK2 associates with the erythropoietin receptor and is tyrosine phosphorylated and activated following stimulation with erthropoietin. *Cell*, **74**, 227–36.

52 Silvennoinen, O., Witthuhn, B., Quelle, F.W. *et al.* (1993) Structure of the JAK2 protein tyrosine kinase and its role in IL-3 signal transduction. *Proc. Natl. Acad. Sci. USA*, **90**, 8429–33.

53 Narazaki, M., Witthuhn, B.A., Yoshida, K. *et al.* (1994) Activation of JAK2 kinase mediated by the interleukin 6 signal transducer gp130. *Proc. Natl. Acad. Sci. USA*, **91**, 2285–89.

54 Quelle, F.W., Sato, N., Witthuhn, B.A. *et al.* (1994) JAK2 associates with the βc chain of the receptor for granulocyte-macrophage colony-stimulating factor, and its activation requires the membrane-proximal region. *Mol. Cell. Biol.*, **14**, 4335–41.

55 Lütticken, C., Wegenka, U.M., Yuan, J. *et al.* (1994) Association of transcription factor APRF and protein kinase Jak1 with the interleukin-6 signal transducer gp130. *Science*, **263**, 89–92.

56 Miyazaki, T., Kawahara, A., Fujii, H. *et al.* (1994) Functional activation of

Jak1 and Jak3 by selective association with IL-2 receptor subunits. *Science*, **266**, 1045–7.

57 Stahl, N., Farruggella, T.J., Boulton, T.G. *et al.* (1995) Choice of STATs and other substrates specified by modular tyrosine-based motifs in cytokine receptors. *Science*, **267**, 1349–53.

58 Tanner, J.W., Chen, W., Young, R.L. and Longmore, G.D. (1995) The conserved box 1 motif of cytokine receptors is required for association with JAK kinases. *J. Biol. Chem.*, **270**, 6523–30.

59 Russel, S.M., Johnston, J.A., Noguchi, M. *et al.* (1994) Interaction of IL-2Rβ and γc chains with Jak1 and Jak3: implications for XSCID and XCID. *Science*, **266**, 1042–5.

60 Lin, J.-X., Migone, T.-S., Tsang, M. *et al.* (1995) The role of shared receptor motifs and common Stat proteins in the generation of cytokine pleiotropy and redundancy by IL-2, IL-4, IL-7, IL-13, and IL-15. *Immunity*, **2**, 331–9.

61 VanderKuur, J.A., Wang, X., Zhang, L. *et al.* (1994) Domains of the growth hormone receptor required for association and activation of JAK2 tyrosine kinase. *J. Biol. Chem.*, **269**, 21709–17.

62 Goujon, L., Allevato, G., Simonin, G. *et al.* (1994) Cytoplasmic sequences of the growth hormone receptor necessary for signal transduction. *Proc. Natl. Acad. Sci. USA*, **91**, 957–61.

63 Ernst, M. and Gearing, D.P. (1994) Functional and biochemical association of Hck with the LIF/IL-6 receptor signal transducing subunit gp130 in embryonic stem cells. *EMBO J.*, **13**, 1574–84.

64 Minami, Y., Nakagawa, Y., Kawahara, A. *et al.* (1995) Protein tyrosine kinase Syk is associated with and activated by the IL-2 receptor: possible link with the c-myc induction pathway. *Immunity*, **2**, 89–100.

65 Akira, S., Nishio, Y., Inoue, M. *et al.* (1994) Molecular cloning of APRF, a novel IFN-stimulated gene factor 3 p91-related transcription factor involved in the gp130-mediated signaling pathway. *Cell*, **77**, 63–71.

66 Tian, S.-S., Lamb, P., Seidel, H.M. *et al.* (1994) Rapid activation of the STAT3 transcription factor by granulocyte colony-stimulating factor. *Blood*, **84**, 1760–4.

67 Zhon, Z., Wen, Z. and Darnell, J.E., Jr (1994) Stat3: a STAT family member activated by tyrosine phosphorylation in response to epidermal growth factor and interleukin-6. *Science*, **264**, 95–8.

68 Hou, J., Schindler, U., Henzel, W.J. *et al.* (1994) An interleukin-4 induced transcription factor: IL-4 Stat. *Science*, **265**, 1701–6.

69 Rothman, P., Kreider, B., Azam, M. *et al.* (1994) Cytokines and growth factors signal through tyrosine phosphorylation of a family of related transcription factors. *Immunity*, **1**, 457–68.

70 Gouilleux, F., Pallard, C., Dusanter-Fourt, I. *et al.* (1995) Prolactin, growth hormone, erythropoietin and granulocyte-macrophage colony stimulating factor induce MGF-Stat5 DNA binding activity. *EMBO J.*, **14**, 2005–13.

71 Wakao, H., Harada, N., Kitamura, T. *et al.* (1995) Interleukin 2 and erythropoietin activate STAT5/MGF via distinct pathways. *EMBO J.*, **14**, 2527–36.

72 Wegenka, U.M., Bushmann, J., Lütticken, C. *et al.* (1993) Acute-phase response factor, a nuclear factor binding to acute-phase response elements, is rapidly activated by interleuin-6 at the posttranslation level. *Mol. Cell. Biol.*, **13**, 276–88.

73 Sonyang, Z., Shoelson, S.E., Chaudhuri, M. *et al.* (1993) SH2 domains recognize specific phosphopeptide sequences. *Cell*, **72**, 762–78.

74 Wang, L.-M., Keegan, A.D., Li, W. *et al.* (1993) Common elements in interleukin 4 and insulin signaling pathways in factor-dependent hematopoietic cells. *Proc. Natl. Acad. Sci. USA*, **90**, 4032–6.

75 Damen, J.E., Mui, A.L., Puil, L. *et al.* (1993) Phosphatidylinositol 3-kinase associates, via its Src homology 2 domains, with the activated erythropoietin receptor. *Blood*, **81**, 3204–10.

76 Satoh, T., Nakafuku, M. and Kaziro, Y. (1992) Function of Ras as a molecular switch in signal transduction. *J. Biol. Chem.*, **267**, 24149–52.

77 Kanakura, Y., Druker, B., Wood, K.W. *et al.* (1991) Granulocyte-macrophage colony-stimulating factor and interleukin-3 induce rapid phosphorylation and activation of the proto-oncogene Raf-1 in a human factor-dependent myeloid cell line. *Blood*, **77**, 243–8.

78 Nakajima, T., Kinoshita, S., Sasagawa, T. *et al.* (1993) Phosphorylation at threonine-235 by a ras-dependent mitogen-activated protein kinase cascade is essential for transcription factor NF-IL-6. *Proc. Natl. Acad. Sci. USA*, **90**, 2207–11.

79 Da Silva, L., Howard, O.M., Rui, H. *et al.* (1994) Growth signaling and JAK2 association mediated by membrane-proximal cytoplasmic regions of prolactin receptors. *J. Biol. Chem.*, **269**, 18267–70.

80 He, T.C., Jiang, N., Zhuang, H. *et al.* (1995) The extended box 2 subdomain of erythropoietin receptor is nonessential for Jak2 activation yet critical for efficient mitogenesis in FDC-ER cells. *J. Biol. Chem.*, **269**, 18291–4.

81 Nicholson, S.E., Oates, A.C., Harpur, A.G. *et al.* (1994) Tyrosine kinase JAK1 is associated with the granulocyte-colony-stimulating factor receptor and both become tyrosine-phosphorylated after receptor activation. *Proc. Natl. Acad. Sci. USA*, **91**, 2985–8.

82 Meyer, D.J., Campbell, G.S., Cochran, B.H. *et al.* (1994) Growth hormone induces a DNA binding factor related to the interferon-stimulated 91-kDa transcription factor. *J. Biol. Chem.*, **269**, 4701–4.

83 Bacon, C.M., McVicar, D.W., Ortaldo, J.R. *et al.* (1995) Interleukin 12 (IL-12) induces tyrosine phosphorylation of JAK2 and TYK2: differential use of Janus family tyrosine kinases by IL-2 and IL-12. *J. Exp. Med.*, **181**, 399–404.

84 Szabo, S.J., Jacobson, N.G., Dighe, A.S. *et al.* (1995) Developmental commitment to the Th2 lineage by extinction of IL-2 signaling. *Immunity*, **2**, 665–75.

85 Campbell, G.S., Argetsinger, L.S., Ihle, J.N. *et al.* (1994) Activation of JAK2 tyrosine kinase by prolactin receptors in Nb2 cells and mouse mammary gland explants. *Proc. Natl. Acad. Sci. USA*, **91**, 5232–6.

86 Wen, Z., Zhon, Z. and Darnell, J.E. Jr. (1995) Maximal activation of transcription by Stat1 and Stat3 requires both tyrosine and serine phosphorylation. *Cell*, **82**, 241–50.

87 Zhan, X., Blenis, J., Li, H.-C. *et al.* (1995) Requirement of serine phosphorylation for formation of STAT–promoter complex. *Science*, **267**, 1990–4.

3 Signaling through cytokine class II receptors

Mathias Müller,
James Briscoe and
Horst Ibelgaufts

Cytokines are soluble proteins that form an integrated network of cell–cell communication and humoral interactions regulating growth, differentiation and survival of all cells. Cytokines are divided into subfamilies on the basis of their biological activities and properties. For example, interferons (IFNs) are involved in response to viral infection, monokines, chemokines and lymphokines in inflammation and immunity and colony-stimulating factors in hemopoiesis.[1] Cytokines bind to cell-surface receptors and transduce a signal to the nucleus to induce transcription of a specific set of genes. Different cytokines activate different (but often overlapping) sets of genes. The notable amino acid sequence homology between the cytokine receptors has led to the identification of a cytokine receptor superfamily consisting of a large class I subgroup (see Chapter 2) and the distantly related class II receptors, namely the receptors for IFNs and interleukin (IL) 10. This chapter will concentrate on the main aspects of the receptors and signal transduction pathways used by these cytokines (for additional comprehensive reviews, see refs 2–7).

IFNs constitute a family of polypeptides originally identified by their ability to induce an antiviral state, but they also have antigrowth properties, affect cell function, promote differentiation and are important mediators in immune responses. IFNs are split into two groups. The type I IFNs include, in humans, the 15 IFN-α subtypes, one IFN-β and two IFN-ω. They are acid-stable and have similar protein structure and biological activities.[7] IFN-γ, a type II IFN, is acid-labile, has a different protein structure and is encoded by a single gene in humans.[3] IL-10 also has multiple activities and is involved in the proliferation and differentiation of B cells, T cells and mast cells. It also inhibits the synthesis of a variety of cytokines in a broad spectrum of cells.[8]

Signal Transduction. Edited by Carl-Henrik Heldin and Mary Purton. Published in 1996 by Chapman & Hall. ISBN 0 412 70810 8

In common with many other cytokines (see Chapter 2), cytokines binding to class II receptors activate the Jak–STAT signal transduction pathway which consists of the Jaks (Janus kinases), a family of membrane-associated cytoplasmic protein tyrosine kinases, and the STATs (signal transducers and activators of transcription), a family of SH2-domain-containing proteins. STATs are activated by tyrosine phosphorylation in receptor complexes containing Jaks and receptor components. Unlike many cell surface receptors (see Chapters 1 and 4), most cytokine receptors have no intrinsic tyrosine kinase activity. Tyrosine phosphorylated STATs, either alone or in combination with other proteins, migrate to the nucleus where they bind to specific response elements in the promoters of inducible genes. Thus the Jak–STAT path provides a very direct route to transcriptional activation.

The current understanding of signaling through cytokine class II receptors has emerged mainly from studying the IFN system. Many laboratories have contributed to this, the main developments being the identification of (1) receptors, (2) DNA response elements in the promoters of inducible genes and their corresponding DNA binding factors and (3) signal transduction pathways connecting the receptors with the DNA elements.

Structure of class II cytokine receptors

The class II subgroup of the cytokine receptor family consists of receptors for the IFNs[9–14], IL-10[15,16] and tissue factor (coagulation factor III).[17] These receptors are multimeric and their exact composition remains to be fully characterized. However, a number of components have been identified and cloned by varying methods, including cross-linking of labeled ligand to cell surface receptors, raising receptor-specific antibodies, immunoscreening and functional cloning. All the cloned components share structural similarities in their extracellular and intracellular domains (Figure 3.1). A characteristic 200 amino acid extracellular domain (see Figure 3.1) is predicted to form two sets of seven β-strands reminiscent of the fibronectin III module.[18] This structure is believed to form the ligand-binding pocket of the receptor. Recent amino acid homology comparisons between the IFN receptors have identified two conserved intracellular motifs (see Figure 3.1) which might serve as binding sites for intracellular effector proteins, i.e. the Jaks and STATs.[19]

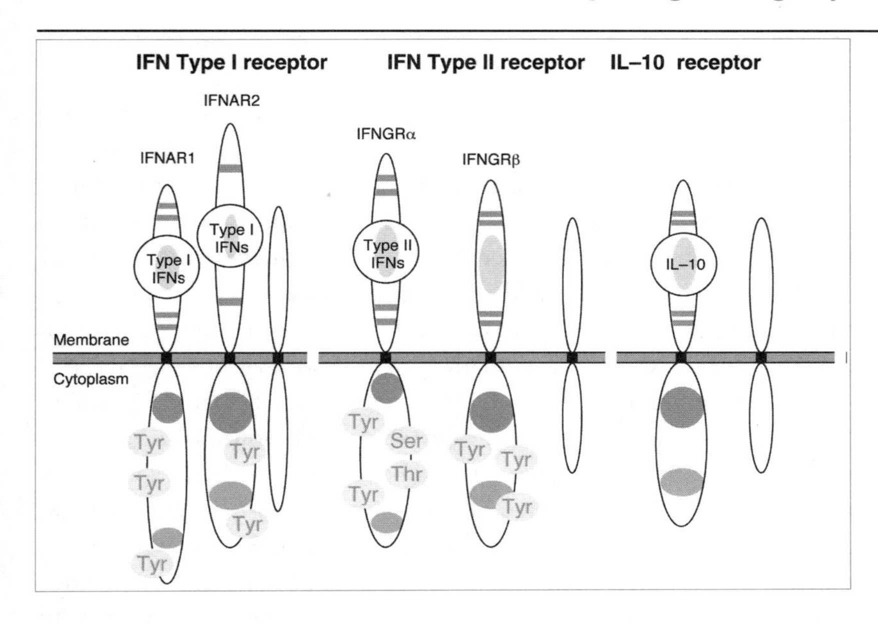

Figure 3.1 Structure of cytokine class II receptor chains. The multimeric receptors are composed of heterologous subunits with common motifs. The smaller unstructured receptor subunits represent those required in addition. In the extracellular domains, horizontal bars mark the position of conserved cysteine residues and gray ovals the region of 200 amino acid residues (see review [18]) believed to form the ligand-binding pocket. Ligands are depicted by open circles. Both cloned IFN-α/β receptor chains bind type I IFNs, albeit to various degrees (see text). For the IFN-γ and IL-10 receptors, only one binding subunit has been characterized. Phosphorylation sites on the receptor chains on tyrosine (Tyr), serine (Ser) and threonine (Thr) residues are indicated by black circles. The two conserved intracellular motifs shown might serve as docking sites for Jaks (membrane-proximal region) and STATs (membrane-distal region).[19] Most of the exact docking sites have to be established experimentally. For the IFNGR, a crucial tyrosine residue has been identified which is required for the recruitment of STAT1.[26] For details of the structure and evolution of the cytokine receptor superfamily, see refs 7 and 17–19 and Chapter 2.

Type I IFN receptor(s)

All type I IFNs share some receptor components since they bind to cells in a competitive manner (see reviews[7,20]). Two genes of the human type I IFN receptor(s) have been cloned and designated *IFNAR1* and *IFNAR2*. The *IFNAR1* gene encodes transmembrane glycoprotein of 557 amino acid residues, which, although it does not bind most IFN-α subtypes and IFN-β with high affinity, is required for signal transduction.[11] The *IFNAR2* gene encodes four splice variants of transmembrane proteins binding most IFN-α subtypes and IFN-β and differing in the cytoplasmic tails.[14,55] A long alternative product of this gene (*IFNAR2-2*, 551 amino acid residues) functionally complements an IFN-α/β receptor-defective mutant,[55] a further member of a series of IFN-unresponsive cell lines (see below). In addition to the cloned components, monoclonal antibodies have been used to identify proteins and complexes of 100, 110, 130 and 210 kDa (see review[7]).

IFNAR1 and IFNAR2 have been shown to be constitutively associated with Jak family protein tyrosine kinases Tyk2 and Jak1, respectively.[14,21,22] Tyrosine phosphorylation of the receptor subunits in response to IFN-α/β has been described,[7] providing potential SH2 docking sites (see Chapter 9). Moreover, antibodies raised against IFNAR1 co-immunoprecipitate a 95 kDa protein which is specifically tyrosine phosphorylated in response to IFN-β.[21] These findings begin to define a molecular basis for the observed differences in the biological activities of IFN-α and IFN-β.

Type II IFN receptor

Two transmembrane components of the IFN-γ receptor (IFNGR) have been cloned. A 90 kDa receptor chain is both necessary and sufficient for ligand binding but not for signal transduction.[13] A second component, denoted accessory factor 1 (AF-1), does not bind IFN-γ but is required for signaling.[10,12] According to the nomenclature of class I cytokine receptors, the ligand-binding subunit is termed α-chain and the signal transducing subunit(s) β- or γ-chains (see review[23]). There is evidence for the existence of a family of accessory factors.[24]

The association of the IFNGR α-chain with the Jak family members Jak1 and Jak2 (see below) has been demonstrated by co-immunoprecipitation experiments.[25] The receptor chain itself is rapidly and transiently phosphorylated on a number of tyrosine residues in the cytoplasmic tail[25,26] and one of these has been shown to be required for IFN-γ-signaling.[26] Slow and stable phosphorylation on serine/threonine residues is also observed (see review[3]), but does not seem to be essential for signal transduction.

IL-10 receptor

Human and murine cDNAs encoding 90–110 kDa polypeptides that bind IL-10 with high affinity have been isolated by expression cloning.[15,16] The structural relationship between the IL-10 and the IFN receptors and the fact that IL-10 inhibits IFN-γ-mediated macrophage activation suggest an interaction of both cytokines on the basis of shared receptor subunits and/or common intracellular signal transduction components (see below).[16,27,28]

IFN-responsive promoter elements and transcription factors

Our understanding of IFN signaling began with studies of IFN-inducible genes. IFN-α/β and IFN-γ induce the transcription of an overlapping set of genes.[3,20,29] Two DNA elements present in the promoters of inducible genes were defined by progressive deletion/mutation of reporter constructs and studies of IFN-dependent DNA–protein interactions. These two elements, termed ISRE (IFN-stimulable response element) and GAS (gamma activation site), are able to control gene expression in the absence of new protein synthesis.[2] They therefore represent the primary IFN-response elements. The responses

Table 3.1 DNA elements and DNA binding factors involved in the Jak-STAT pathway used by cytokine class II receptors[a]

Cytokine class II receptor	Activated Jak family member	Activated STAT protein	DNA binding factor	DNA response element	Consensus sequence of DNA response element
Type I IFNs	Jak1, Tyk2	STAT1, STAT2, STAT3	ISGF3	ISRE	AGTTTCNNTTTCNY
			AAF	GAS	TT^{A}/$_{C}$YNNAA^{A}/$_{G}$
Type II IFN	Jak1, Jak2	STAT1, STAT3	GAF	GAS	TT^{A}/$_{C}$YNNAA^{A}/$_{G}$
IL-10	Jak1, Tyk2	STAT1, novel STAT(s)	ND	GAS	TT^{A}/$_{C}$YNNAA^{A}/$_{G}$

[a] ISGF3 consists of a STAT1/STAT2 heterodimer and p48, AAF was shown to be STAT1/STAT1 or STAT3/STAT3 homodimers and STAT1/STAT3 heterodimers and GAF is a STAT1/STAT1 homodimer (see text and Figures 3.3 and 3.4). AAF, alpha-activated factor; GAF, gamma-activated factor; GAS, gamma-activated sequence; ISGF3, IFN stimulated gene factor 3; ISRE, IFN stimulable response element; N, any nucleotide; ND, not determined; Y, pyrimidine nucleotide.

of many IFN-γ inducible genes, including the MHC class I and class II genes, are secondary and are governed by different DNA elements.[3,20]

The consensus sequences of the ISRE and GAS elements are shown in Table 3.1. Although these two motifs were originally identified as sequences mediating either IFN-α/β (ISRE) or IFN-γ (GAS) responses, it has become clear that there is an overlap between both elements in that they respond to different degrees to either type of IFN (see reviews[2,30]).

Two multi-component transcription factors were characterized which bind to the two DNA elements in response to IFNs (Table 3.1). ISGF3 (IFN-stimulated gene factor 3) was identified as the major primary transcriptional activator binding to ISREs in response to IFN-α/β, and GAF/AAF (gamma-activated factor/alpha-activated factor) as the GAS-binding component induced by either type of IFNs (see review[31]). The genes for the component polypeptides of ISGF3 and GAF/AAF have been purified and cloned and this led to the identification of the first STATs (see below). ISGF3 consists of STAT1, STAT2 (formerly termed ISGF3α or p91/p84 and p113, respectively) and p48 (formerly ISGF3γ), a sequence-specific DNA binding protein related to the IRF/myb family (see review[31]). Recent data suggest that all components of ISGF3 contact DNA; STAT1 and p48 interact precisely with the ISRE while STAT2 makes general contact.[32] AAF/GAF was shown to be a homodimer of STAT1 or STAT3 and a STAT1/STAT3 heterodimer (Table 3.1; Figures 3.3 and 3.4); however, depending on the promoter, additional factors may also be involved[33,34] (see reviews[2,30,31]).

Treatment of cells with IL-10 results in the activation of GAS-binding proteins which contain STAT1, STAT3 and/or other STAT-related polypeptides.[27,28]

Mutant cell lines and mice deficient in signaling components

A genetic approach allowing dissection of the IFN signaling pathway has been developed in a human fibrosarcoma cell line. It is based on the expression of a drug-selectable marker and a cell-surface marker under the control of an IFN-inducible promoter.[29,35] Recessive mutants in eight complementation groups affecting the IFN signal transduction have been obtained using these cell lines (see reviews[29,31]). Genetic complementation with genomic DNA or cDNAs demonstrated the requirement for STATs and led to the identification of the role of Jaks in IFN signal transduction.[35–39] These cell lines also facilitate the structure/function analysis of Jaks and STATs. Mice harbouring null mutations (knockout mice) in Jak/STAT loci will provide the means to assess *in vivo* the role of these proteins in gene expression, signal transduction and protein function and their involvement in normal and disease states. The currently available knockout mice include those carrying disruptions of IFNAR1, the IFNGR α-chain and STAT1.[40,41,56,57]

STATs: a novel family of transcription factors

The novel family of STATs[31] currently consists of six mammalian members which participate in numerous cytokine signaling pathways[42,43] (and see Chapters 2 and 21). Recently, a STAT homologous protein was found in *Drosophila*.[58] The STAT proteins contain several conserved regions scattered throughout their entire length, including a highly conserved SH2 domain, an SH3 domain and a conserved tyrosine residue carboxy-terminal of the SH2 domain. Further features are depicted in Figure 3.2. SH2 domains bind phosphotyrosine residues and SH3 domains contain binding sites for proline-rich motifs. Both domains are found in many signaling molecules[44] (see Chapter 9).

STAT activation requires protein tyrosine phosphorylation at the conserved carboxy-terminal tyrosine.[31] The non-phosphorylated STATs

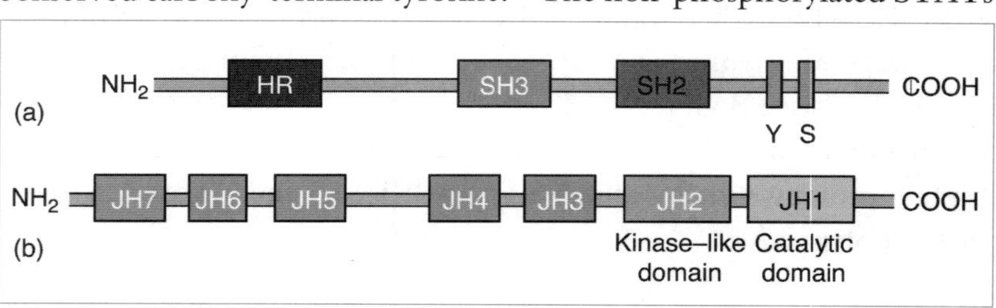

Figure 3.2 Prototype structures of the STAT polypeptides (a) and the Jak family (b). In (a), the brightening gray towards the carboxyl terminus indicates the lack of similarity of STATs in that region. HR, heptad repeat; JH, Jak homology domain; S, serine residue(s) that are phosphorylated and modulate the STAT–DNA interaction; Y, tyrosine residue that is phosphorylated upon activation. For further details, see Chapter 21 and refs 31,43,51 and 52.

exist as monomers in the cytoplasm and upon phosphorylation they form homo- or heterodimers through interactions of their SH2 domains and the phosphotyrosine residues (see Figures 3.3 and 3.4 and Table 3.1).[45]

The tyrosine phosphorylation of the STATs occurs at the cell membrane in association with receptor components. STAT1 is recruited to a single phosphotyrosine in the IFNGR α-chain via its SH2 domain.[26] STAT2 is probably constitutively associated with the IFNAR1 subunit.[21] The SH2 domain of STAT2, although required, is not sufficient for this.[46] Activation of STAT1 at the IFNAR, however, depends on the presence of STAT2 and receptor engagement.[39,47]

Recent data suggest that phosphorylation of serine residues may be required for, or modulates, the formation of STAT–DNA complexes.[48,49,50] More than 15 related GAS sequences have been identified to date[2] which are bound by all characterized STAT proteins (see below and Chapter 21), with the exception of STAT2.

Jaks: a family of non-receptor protein tyrosine kinases

The Jak family is a novel family of cytoplasmic kinases which are involved in signaling of a wide range of cytokines.[42,43] The acronym Jak stands for Janus kinase, which refers to the two-faced Roman god of gates and doorways and the Jaks' structural feature of two kinase domains. Alternatively, Jak is an abbreviation for 'just another kinase,' reflecting the identification of Jaks by PCR/homology cloning without knowing their biological function.[51] In mammals there are four family members (Jak1, Jak2, Jak3 and Tyk2) and in *Drosophila* a single Jak homolog has been characterized.[43] Jaks uniquely contain two kinase-homology domains (JH1 and JH2) (Figure 3.2). The carboxy-terminal kinase domain has all the motifs associated with protein tyrosine kinases and catalytic activity, while the adjacent domain lacks several residues known to be essential for catalytic activity. The functions of this domain and the remainder of the amino-terminal sequences (Figure 3.2) remain to be established.[52]

The phenotypes of the mutant cell lines lacking Jak1, Jak2 or Tyk2 show that a pair of Jaks is required for both type I IFN and type II IFN signaling (see Table 3.1 and Figures 3.3 and 3.4). Jak1 is required for signaling through both types of receptor, whereas Tyk2 is necessary for IFN-α/β responses and Jak2 for IFN-γ-signaling.[35,36,38] The Jaks are phosphorylated on tyrosine and their kinase activity is increased in response to IFNs (see reviews[4,30,31]). There is an interdependence

Figure 3.3 Current model for the IFN-α/β signaling pathway. The larger receptor subunits represent those already cloned but there is at least one additional subunit. Tyrosine phosphorylation events are shown by black circles with P. The STAT1/STAT2 heterodimerization which results in formation of ISGF3 seems to be the major path in the primary response to IFN-α/β (indicated by blue arrows). The identity of AAF2 with STAT3 homodimer is unclear and STAT1/STAT3 heterodimers, which are formed *in vivo*, have not been shown. The observed differences in the activation of the Jak–STAT pathway by IFN-α and IFN-β and possible branchpoints to other signaling events have not been depicted for clarity. Further details are provided in the text and in refs 2,7,30 and 31.

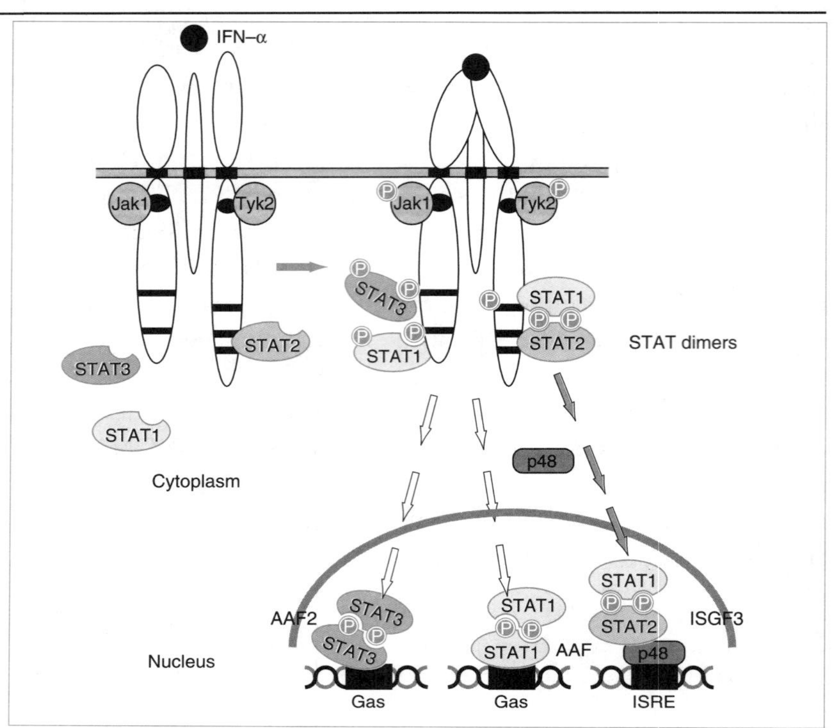

between pairs of kinases: if either of a pair is missing, the remaining kinase is not activated. It appears, therefore, that there is no simple hierarchical activation cascade for Jaks resembling that described for MAP kinases (see Chapter 11). Jaks also play an important structural role in the formation of IFN–receptor complexes. This notion is supported by the finding that mutants lacking the kinase domain of Jak1 or Tyk2 can partially restore functional type I or type II IFN receptor complexes[53] (and J. Briscoe *et al.*, in preparation).

Jaks associate with conserved motifs in the membrane proximal region of cytokine type I receptors (see Chapter 2). Constitutive association of Tyk2 with IFNAR1 and Jak1 with IFNAR2 was found for the type I IFN receptor.[14,21,22] At the type II IFN receptor, Jak1 was found to be bound to IFNGR α-chain and Jak2 was shown to co-immunoprecipitate with this chain in a ligand-dependent manner.[25,26]

Signaling through class II cytokine receptors

On the basis of the studies described above, the following model accounts for signal transduction in response to cytokines utilizing class II cytokine receptors (Figures 3.1, 3.3 and 3.4). The initial event is ligand-dependent

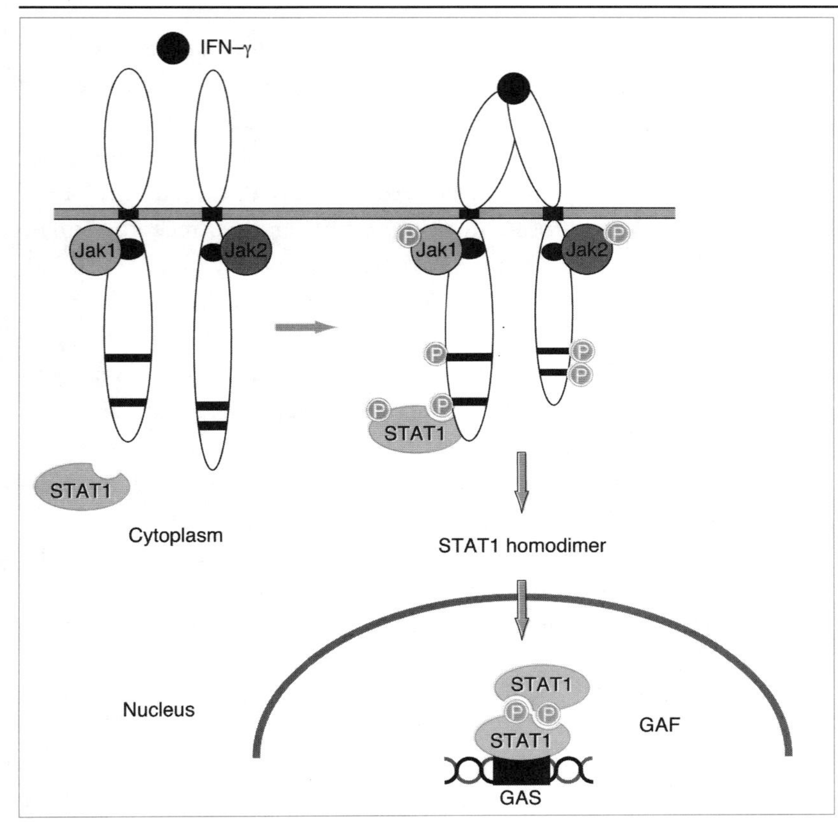

Figure 3.4 The current model of the IFN-γ signaling pathway. See legend of Figure 3.3 for an explanation of symbols. Further explanation is provided in the text and in refs 3,30 and 31.

dimerization of the multiple receptor chains. Receptor aggregation brings the Jaks into close proximity through their constitutive association with receptor chains. This results in tyrosine phosphorylation (probably by cross-phosphorylation) and increases their catalytic activity. This model is conceptually identical with that describing the activation of receptor protein tyrosine kinases following receptor engagement (see Chapter 1). At the IFN receptors, pairs of Jaks are required and the Jaks have a structural role in the formation of the receptor complex in addition to their enzymatic function. Activation of Jaks correlates with tyrosine phosphorylation of the receptor components.

The phosphotyrosine residues in the receptor complex recruit STATs through their SH2 domains which, in turn, are phosphorylated at the conserved carboxy-terminal tyrosine residue. The activation of STATs facilitates dimerization of the proteins by SH2–phosphotyrosine interactions. Homo- or heterodimers of the STATs translocate to the nucleus, where they either directly interact with promoter elements (GAS motifs) or combine with a DNA-binding protein (ISRE motifs). SH2 domain swap experiments between STAT1 and STAT2 indicate that the selection of a particular STAT at a receptor and also specific STAT dimerization depend on the SH2 domain.[46] A kinase-negative Jak protein still allows the generation of a functional receptor complex and STAT activation is sufficiently carried out by the remaining Jak (see above). Taken together, this indicates that SH2 domains in STAT proteins and tyrosine-based docking motifs in the receptor–kinase complexes[54] play a crucial role in cytokine specificity.

Although the reported findings provide a framework for understanding the mechanisms underlying the functions of the cytokine receptor superfamily, several issues remain unanswered. Little is known about the biochemistry and structure/function of Jaks. Likewise, the exact substrates for Jaks are unknown. The functional relationships between the identified components and other cytoplasmic kinases and phosphatases or other signaling molecules that have been implicated in specific cytokine signaling begin to emerge. For example, the protein-tyrosine phosphatase(s) negatively and/or positively regulating the signaling events are currently being characterized.[59,60] The increased knowledge of intracellular signaling, together with crystallographic receptor studies,[61] will facilitate the development of novel therapeutics.

Acknowledgement

We thank I.M. Kerr for his invaluable comments and kind help.

References

1 Ibelgaufts, H. (1995) *Dictionary of Cytokines*, VCH, Weinheim.

2 Schindler, C. and Darnell, J.E. (1995) Transcriptional responses to polypeptide ligands: the JAK-STAT pathway. *Annu. Rev. Biochem.*, **64**, 621–51.

3 Farrar, M.A. and Schreiber, R.D. (1993) The molecular cell biology of interferon-γ and its receptor. *Annu. Rev. Immunol.*, **11**, 571–611.

4 Ihle, J.N., Witthuhn, B.A., Quelle, F.W. *et al.* (1995) Signaling through the hematopoietic cytokine receptors. *Annu. Rev. Immunol.*, **13**, 369–98.

5 Rose-John, S. and Heinrich, P.C. (1994) Soluble receptors for cytokines and growth factors: generation and biological function. *Biochem. J.*, **300**, 281–90.

6 Taniguchi, T. (1995) Cytokine signaling through nonreceptor protein tyrosine kinases. *Science*, **268**, 251–5.

7 Uzé, G., Lutfalla, G. and Mogensen, K.E. (1995) α and β interferons and their receptor and their friends and relations. *J. Interferon Cytokine Res.*, **15**, 3–26.

8 Howard, M. and O'Garra, A. (1992) Biological properties of interleukin 10. *Immunol Today*, **13**, 198–200.

9 Lutfalla, G., Gardiner, K. and Uzé, G. (1993) A new member of the cytokine receptor gene family maps on chromosome 21 at less than 35 kb from IFNAR. *Genomics*, **16**, 366–73.

10 Soh, J., Donnelly, R.J., Kotenko, S. *et al.* (1994) Identification and sequence of

an accessory factor required for activation of the human interferon γ receptor. *Cell*, **76**, 793–802.

11 Uzé, G., Lutfalla, G. and Gresser, I. (1990) Genetic transfer of a functional human interferon α receptor into mouse cells: cloning and expression of its cDNA. *Cell*, **60**, 225–34.

12 Hemmi, S., Böhni, R., Stark, G. *et al.* (1994) A novel member of the interferon receptor family complements functionality of the murine interferon γ receptor in human cells. *Cell*, **76**, 803–10.

13 Aguet, M., Dembic, Z. and Merlin, G. (1988) Molecular cloning and expression of the human interferon-γ receptor. *Cell*, **55**, 273–80.

14 Novick, D., Cohen, B. and Rubinstein, M. (1994) The human interferon alpha/beta receptor: characterization and molecular cloning. *Cell*, **77**, 391–400.

15 Liu, Y., Wei, S.H., Ho, A.S. *et al.* (1994) Expression cloning and characterization of a human IL-10 receptor. *J. Immunol.* **152**, 1821–9.

16 Ho, A.S.Y., Liu, Y., Khan, T.A. *et al.* (1993) A receptor for interleukin 10 is related to interferon receptors. *Proc. Natl. Acad. Sci. USA*, **90**, 11267–71.

17 Bazan, J.F. (1991) WKS motifs and the cytokine receptor framework of tissue factor. *Trends Biochem. Sci.*, **16**, 329.

18 Bazan, J.F. (1990) Haemopoietic receptors and helical cytokines. *Immunol. Today*, **11**, 350–4.

19 Mullersman, J.E. and Pfeffer, L.M. (1995) A novel cytoplasmic domain in interferon receptors. *Trends Biochem. Sci.*, **20**, 55–6.

20 Pestka, S., Langer, J.A., Zoon, K.C. and Samuel, C.E. (1987) Interferons and their action. *Annu. Rev. Biochem.*, **56**, 727–7.

21 Abramovich, C., Shulman, L.M., Ratovitski, E. *et al.* (1994) Differential tyrosine phosphorylation of the IFNAR chain of the type I interferon receptor and of an associated surface protein in response to IFN-α and IFN-β. *EMBO J.*, **13**, 5871–7.

22 Colamonici, O., Yan, H., Domanski, P. *et al.* (1994) Direct binding to and tyrosine phosphorylation of the α subunit of the type I interferon receptor by $p135^{tyk2}$ tyrosine kinase. *Mol. Cell. Biol.*, **14**, 8133–42.

23 Stahl, N. and Yancopoulos, G.D. (1993) The alphas, betas, and kinases of cytokine receptor complexes. *Cell*, **74**, 587–90.

24 Cook, J.R., Emanuel, S.L., Donnelly, R.J. *et al.* (1994) Sublocalization of the human interferon-gamma receptor accessory factor gene and characterization of accessory factor activity by yeast artificial chromosome fragmentation. *J. Biol. Chem.*, **269**, 7013–8.

25 Igarashi, K.-I., Garotta, G., Ozmen, L. *et al.* (1994) Interferon-γ induces tyrosine phosphorylation of interferon-γ receptor and regulated association of protein tyrosine kinases, Jak1 and Jak2, with its receptor. *J. Biol. Chem.*, **269**, 14333–6.

26 Greenlund, A.C., Farrar, M.A., Viviano, B.L. and Schreiber, R.D. (1994) Ligand-induced IFNγ receptor tyrosine phosphorylation couples the receptor to its signal transduction system (p91). *EMBO J.*, **13**, 1591–1600.

27 Finbloom, D.S. and Winestock, K.D. (1995) IL-10 induces the tyrosine

phosphorylation of tyk2 and Jak1 and the differential assembly of STAT1 α and STAT3 complexes in human T cells and monocytes. *J. Immunol.*, **155**, 1079–90.

28 Lehman, J., Seegert, D., Strehlow, I. *et al.* (1994) IL-10-induced factors belonging to the p91 family of proteins bind to IFN-gamma-responsive promoter elements. *J. Immunol.*, **153**, 165–72.

29 Pellegrini, S. and Schindler, C. (1993) Early events in signaling by interferons. *Trends Biochem. Sci.*, **18**, 338–42.

30 Müller, M., Ibelgaufts, H. and Kerr, I.M. (1994) Interferon response pathways – a paradigm for cytokine signaling? *J. Viral. Hepatitis*, **1**, 87–103.

31 Darnell, J.E. Jr., Kerr, I.M. and Stark, G.R. (1994) Jak–STAT pathways and transcriptional activation in response to IFNs and other extracellular signaling proteins. *Science*, **264**, 1415–21.

32 Qureshi, S.A., Salditt-Georgieff, M. and Darnell, J.E. (1995) Tyrosine-phosphorylated stat1 and stat2 plus a 48-kDa protein all contact DNA in forming interferon-stimulated-gene factor 3. *Proc. Natl. Acad. Sci. USA*, **92**, 3829–33.

33 Barahmand-pour, F., Meinke, A., Eilers, A. *et al.* (1995) Colony-stimulating factors and interferon-γ activate a protein related to MGF-Stat5 to cause formation of the differentiation-induced factor in myeloid cells. *FEBS Lett.*, **360**, 29–33.

34 Seegert, D.S., Strehlow, I., Klose, B. *et al.* (1994) A novel, IFN-α-regulated, DNA-binding protein participates in the regulation of the IFP53/tryptophanyl-tRNA synthetase gene. *J. Biol. Chem.*, **269**, 8590–5.

35 Watling, D., Guschin, D., Müller, M. *et al.* (1993) Complementation by the protein tyrosine kinase JAK2 of a mutant cell line defective in the interferon-γ signal transduction pathway. *Nature*, **366**, 166–70.

36 Müller, M., Briscoe, J., Laxton, C. *et al.* (1993) The protein tyrosine kinase JAK1 complements a mutant cell line defective in the interferon-α/β and -γ signal transduction pathways. *Nature*, **366**, 129–35.

37 Müller, M., Laxton, C., Briscoe, J. *et al.* (1993) Complementation of a mutant cell line: central role of the 91 kDa polypeptide of ISGF3 in the interferon-α and -γ signal transduction pathways. *EMBO J.*, **12**, 4221–8.

38 Velazquez, L., Fellous, M., Stark, G.R. and Pellegrini, S. (1992) A protein tyrosine kinase in the interferon α/β signaling pathway. *Cell*, **70**, 313–22.

39 Leung, S., Qureshi, S.A., Kerr, I.M. *et al.* (1995) Role of STAT2 in the alpha interferon signaling pathway. *Mol. Cell. Biol.*, **15**, 1312–7.

40 Müller, U., Steinhoff, U., Reis, L.F.L. *et al.* (1994) Functional role of type I and type II interferons in antiviral defense. *Science*, **264**, 1918–21.

41 Huang, S., Hendriks, W., Althage, A. *et al.* (1993) Immune response in mice that lack the interferon-γ receptor. *Science*, **259**, 1742–5.

42 Briscoe, J., Guschin, D. and Müller, M. (1994) Just another signaling pathway. *Curr. Biol.*, **4**, 1033–5.

43 Ihle, J.N. and Kerr, I.M. (1995) Jaks and Stats in signaling by the cytokine receptor superfamily. *Trends Genet.*, **11**, 69–74.

44 Pawson, T. (1995) Protein modules and signaling networks. *Nature*, **373**, 573–80.

45 Shuai, K., Horvath, C.M., Huang, L.H.T. *et al.* (1994) Interferon activation of the transcription factor Stat91 involves dimerization through SH2-phosphotyrosyl peptide interactions. *Cell*, **76**, 821–8.
46 Heim, M.H., Kerr, I.M., Stark, G.R. and Darnell, J.E., Jr (1995) Contribution of STAT SH2 groups to specific interferon signaling by the Jak–STAT pathway. *Science*, **267**, 1347–9.
47 Improta, T., Schindler, C., Horvath, C.M. *et al.* (1994) Transcription factor ISGF-3 formation requires phosphorylated Stat91 protein, but Stat113 protein is phosphorylated independently of Stat91 protein. *Proc. Natl. Acad. Sci. USA*, **91**, 4776–80.
48 Lütticken, C., Coffer, P., Yuan, J. *et al.* (1995) Interleukin-6-induced serine phosphorylation of transcription factor APRF: evidence for a role in interleukin-6 target gene induction. *FEBS Lett.*, **360**, 137–43.
49 Zhang, X., Blenis, J., Li, H.C. *et al.* (1995) Requirement of serine phosphorylation for formation of STAT-promoter complexes. *Science*, **267**, 1990–4.
50 Wen, Z.L., Zhong, Z. and Darnell, J.E. (1995) Maximal activation of transcription by Stat1 and Stat3 requires both tyrosine and serine phosphorylation. *Cell.* **82,** 241–50
51 Wilks, A.F. and Harpur, A.F. (1994) Cytokine signal transduction and the Jak family of protein tyrosine kinases. *BioEssays*, **16**, 313–20.
52 Ziemiecki, A., Harpur, A.G. and Wilks, A.F. (1994) Jak protein tyrosine kinases: their role in cytokine signalling. *Trends Cell Biol.*, **4**, 207–11.
53 Velazquez, L., Mogensen, K.E., Barbieri, G. *et al.* (1995) Distinct domains of the protein tyrosine kinase tyk2 required for binding of interferon-α/β and for signal transduction. *J. Biol. Chem.*, **270**, 3327–34.
54 Stahl, N., Farruggella, T.J., Boulton, T.G. *et al.* (1995) Choice of STATs and other substrates specified by modular tyrosine-based motifs in cytokine receptors. *Science*, **267**, 1349–53.
55 Lutfalla, G., Holland, S.J., Cinato, E. *et al.* (1995) Mutant U5A cells are complemented by an interferon-alpha beta receptor subunit generated by alternative processing of a new member of a cytokine receptor gene cluster. *EMBO J.*, **14**, 5100–8.
56 Durbin, J.E., Hackenmiller, R., Simon, M.C. and Levy, D.E. (1996) Targeted disruption of the mouse Stat1 gene results in compromised innate immunity to viral disease. *Cell*, **84**, 433–50.
57 Meraz, M.A., White, J.M., Sheehan, K.C.F. *et al.* (1996) Targeted disruption of the STAT1 gene in mice reveals unexpected physiologic specificity in the JAK-STAT signaling pathway. *Cell*, **84**, 431–42.
58 Ihle, J.N. (1996) STATs: Signal transducers and activators of transcription. *Cell*, **84,** 331–4.
59 David, M., Chen, H.Y.E., Goelz, S. *et al.* (1995) Differential regulation of the alpha/beta interferon-simulated Jak/Stat pathway by the SH2 domain-containnig tyrosine phosphatase SHPTP1. *Mol. Cell. Biol.* **15**, 7050–8.
60 Yetter, A. Uddin, S., Krolewski, J.J. *et al.* (1995) Association of the interferon-

dependent tyrosine kinase Tyk-2 with the hematopoietic cell phosphatase. *J. Biol. Chem.*, **270**, 18179–82.

61 Walter, M.R., Windsor, W., Nagabhushan, T.L. *et al.* (1995) Crystal structure of a complex between interferon-gamma and its soluble high-affinity receptor. *Nature*, **376**, 230–5

4 Signaling through hematopoietic antigen receptors

Anne L. Burkhardt
and Joseph B. Bolen

Binding of antigen to antigen receptor on the surface of lymphocytes can lead to rapid activation of intracellular signal transduction. Depending on the type of cell, the signals received following antigen binding lead to profound biological responses such as cellular growth, cellular death, activation, anergy or differentiation. Tyrosine phosphorylation of cellular proteins is one of the earliest detectable biochemical events following antigen engagement. Within the past several years, significant insight has been gained into the biochemical mechanisms involved in antigen receptor signal transduction, including the identification of protein tyrosine kinases (PTKs) and their substrates. Several comprehensive reviews on the subject of PTKs in lymphocyte signaling and development have been published recently.[1–3]

Multichain immune recognition receptors

Hematopoietic cells express a large variety of receptors that receive signals from the extracellular environment. Among these cell surface receptors, the subset of antigen receptors are unique in that they possess the capacity to bind specifically to a very large diversity of ligands (antigens) and that ligand binding to those receptors initiates a specific immune response.

Structural homology of antigen receptors

Three different types of antigen receptors are shown in Figure 4.1. All contain more than one polypeptide chain, earning them the designation multichain immune recognition receptors (MIRRs).[4] The identification of a number of common structural features has diminished the imposing nature of these complexes. Every MIRR has a ligand-binding extracellular

Signal Transduction. Edited by Carl-Henrik Heldin and Mary Purton. Published in 1996 by Chapman & Hall. ISBN 0 412 70810 8

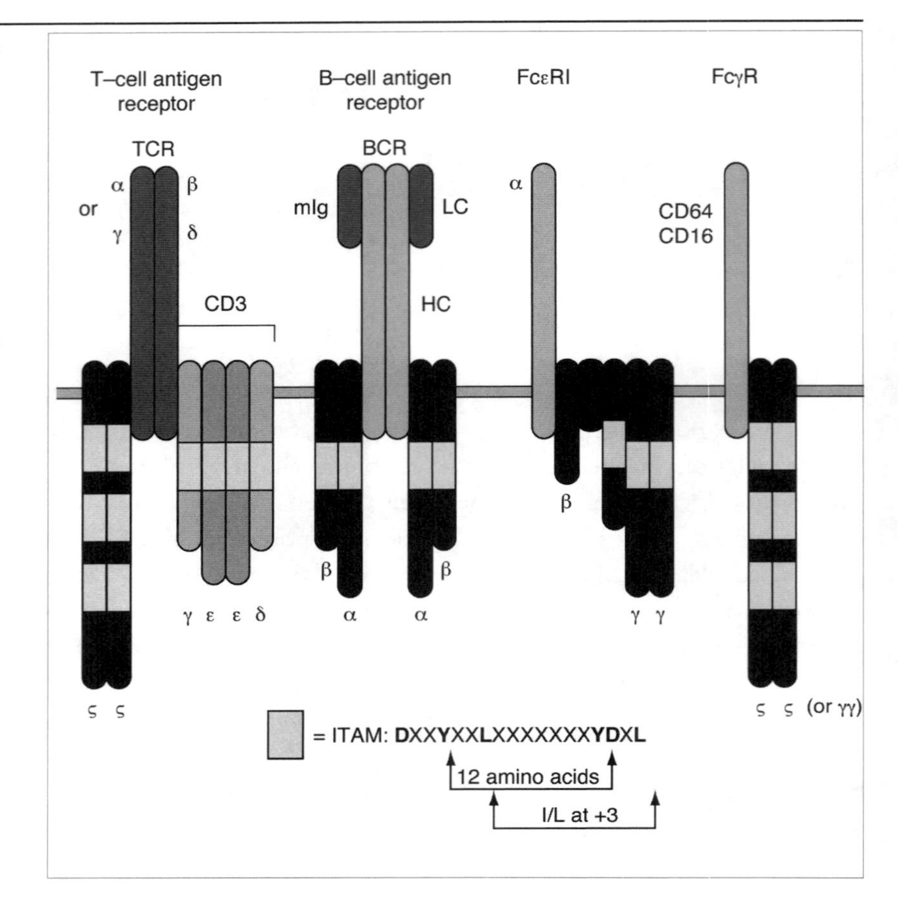

Figure 4.1 The major subunits of representative hematopoietic antigen receptors, also known as multichain immune recognition receptors (MIRRs), are indicated. The cytoplasmic signal-transducing subunits of each receptor contain immunoreceptor tyrosine activation motifs (ITAMs), the consensus protein sequence of which is shown below. D = aspartic acid; E = glutamic acid; Y = tyrosine; I = isoleucine; L = leucine; and X = any amino acid.

domain consisting of variable chains which confer antigen specificity upon the receptor. The extracellular portion of the complex is noncovalently associated with invariant subunits. Since these have no recognizable enzymatic activity, it was proposed that they serve to couple the receptor to cellular signaling molecules. A structural basis for the coupling mechanism was first identified within the cytoplasmic domain of the invariant subunits by Reth[5] and current convention labels the sequences as immunoreceptor tyrosine activation motifs (ITAMs). The importance of the relatively simple ITAM consensus sequence (see Figure 4.1) to the signal transduction capacity of the antigen receptor complex has been demonstrated in numerous experiments utilizing chimeric receptors and ITAM mutants.[6,7] Tyrosine phosphorylated ITAM sequences are thought to provide 'docking sites' for certain signaling proteins, thereby recruiting these molecules to the plasma membrane to propagate signals from activated MIRRs.

The T-cell antigen receptor

The extracellular domain of the T-cell antigen receptor (TCR) expressed on the surface of T lymphocytes is formed by the $\alpha\beta$ (or $\gamma\delta$) heterodimer which contains the variable domains responsible for recognition of small peptides bound to class I or class II major histocompatibility complex (MHC) antigens. The noncovalently associated, invariant proteins of the TCR include the CD3 complex (CD3δ, CD3γ, and two copies of CD3ε) and the ζ-homodimer. There is one copy of the ITAM sequence in each of CD3ε, γ and δ and three in each ζ-chain. (See ref. 8 for a more detailed description of the TCR complex and its variations.)

The B-cell antigen receptor

B-cell antigen receptors (BCR) are composed of membrane immunoglobulins [two transmembrane immunoglobulin heavy chains (HC) associated with two immunoglobulin light chains (LC)] noncovalently associated with heterodimers of the invariant Igα and Igβ subunits.[9] Igα and Igβ both contain one copy of the ITAM.

Fc receptors

A complex family of receptors that bind to soluble immunoglobulins and immune complexes via the Fc region of Ig are expressed on a wide variety of cells and display vast functional heterogeneity.[10] Even so, the Fc receptors are another example of receptors in which conserved domains are combined with variable sequences to yield signal transduction capability with diverse extracellular ligand-binding potential. The best studied Fc receptor is the high-affinity IgE receptor (FcεRI) which is expressed on the surface of mast cells and basophils (Figure 4.1). The α subunit of the FcεRI is responsible for binding the Fc region of IgE and is noncovalently associated with the β and γ subunits. The β subunit spans the membrane four times and contains a single ITAM in its carboxy-terminal cytoplasmic domain. The FcεRI γ exists as a disulfide-linked homodimer and each chain contains a single ITAM. There are three groups of FcγR (receptor for the Fc portion of IgG). FcγRI (CD64) has a relatively high affinity for monomeric IgG and is found predominantly on monocyte/macrophages. FcγRII and FcγRIII (CD16) have a lower affinity for monomeric IgG and are found on macrophages, natural killer cells and other types of lymphocytes. The ligand-binding chain of the FcγR receptor is found in association with either FcεRI γ or TCR ζ-homodimers.

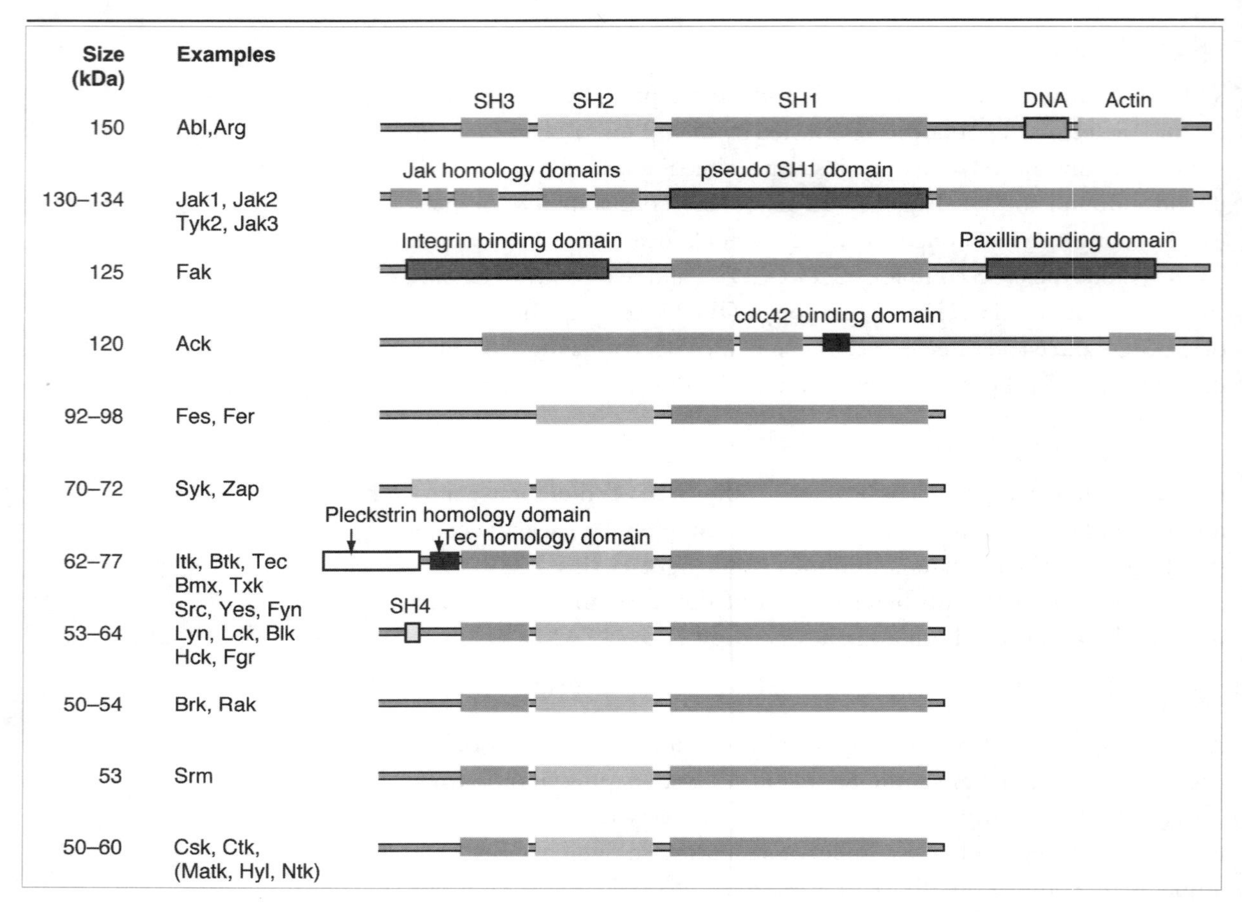

Figure 4.2 The non-transmembrane protein tyrosine kinases are arranged into 11 families based upon predicted protein sequences. The molecular mass range and major domains for each family are indicated.

Non-receptor tyrosine kinases

The participation of tyrosine kinases during antigen receptor signaling was suggested following the observation of rapid tyrosine phosphorylation of cellular proteins following engagement of MIRRs on the surface of hemopoietic cells (see reviews[1–3]). It was also observed that tyrosine kinase inhibitors retard signal generation from antigen receptors.[11,12] There are currently 30 identified mammalian nonreceptor tyrosine kinases which can be divided into 11 families based upon predicted amino acid sequences (see Figure 4.2). In addition to the catalytic domain, PTKs possess other conserved domains which are also found in other distinct signaling pathway components. It is thought that these conserved modules regulate signal transduction through their ability to mediate protein–protein interactions and, consequently, are used to build complex networks of interacting proteins.

SH1 domains

The major area of sequence homology and presumably the region of greatest structural identity of PTKs is their catalytic domain.[13] This SH1 (Src homology 1) domain shares the greatest sequence similarity with the c-Src catalytic domain. Most nonreceptor PTKs possess a site for tyrosine autophosphorylation within the SH1 domain that corresponds to Tyr416 of c-Src. The Csk class does not possess a comparable autophosphorylation site,[14] while members of the Syk/Zap class have two tyrosines at the predicted autophosphorylation site.[15] Members of the Janus kinase family (Jak1, Jak2, Jak3 and Tyk2) are unique in that they appear to possess two catalytic domains, although only the most carboxy-terminal domain has enzyme activity.[16]

SH2 domains

With the exception of the Jak, Fak and Ack groups, all the other nonreceptor PTKs contain one or more SH2 domains. SH2 domains are capable of high-affinity binding to selected phosphotyrosine-containing proteins and clearly play critical roles in the interactions between signaling components in tyrosine kinase pathways. A more detailed description of SH2 domain structure is provided in Chapter 9 and several reviews have been published recently which survey SH2 structure and function.[17,18]

SH3 domains

Another common feature amongst the families of PTKs is SH3 domains, which have been the subject of several recent reviews.[17,18] SH3 domains are made up of approximately 50 amino acid residues and bind short peptide motifs of approximately 10 amino acid residues containing key proline residues. Accumulated structural data indicate that nine-residue left-handed polyproline type II helices represent the preferred ligand for SH3 domains (see Chapter 9). Each SH3 domain has a distinct binding preference, as is demonstrated by studies using degenerative peptide libraries and phage display libraries.[19,20]

SH4 domains

Only members of the Src family and a single isoform of c-Abl/Arg (type IV) possess an SH4 domain. This sequence of 16 amino acid residues directs post-translational myristoylation of a common glycine residue and promotes association of PTKs with cell membranes.[21] With the exception of Blk and Src, other members of the Src family are also subject to post-translational palmitoylation of one or more cysteines located at residue three and either residue five or six. Evidence suggests that palmitoylated Src PTKs are located in regions of the plasma membrane that enhance their capacity to associate with glycosylphosphatidylinositol (GPI)-anchored proteins.[22] Thus, possession of an SH4 domain localizes the Src and c-Abl/Arg type IV PTKs to cell membranes where a significant portion of PTK-mediated signaling reactions are likely to take place. Important physiological substrates for Src PTKs are also likely to exist at these cellular sites.

PH domains

Members of the Tec family possess a domain that was recently added to the domain catalogue of signaling proteins: the pleckstrin homology or PH domain[23] (see Chapter 9). The precise functions of PH domains are still unknown, although they have been shown to bind the βγ subunits of heterotrimeric G proteins,[24] protein kinase C[25] and phospholipids.[26]

In addition to the above domains, additional regions have been identified that tend to be unique for a given PTK type. Undoubtedly, these unique domains will contribute to substrate specificity and specific functions of each enzyme.

Potential functions of nonreceptor PKTs in antigen receptor signaling

Lck and Fyn in TCR signaling

A great deal of insight into the roles of Src PTKs has been gained from *in vitro* studies with T-cell lines and genetic studies utilizing transgenic mice or mice rendered null for a specific PTK by homologous recombination (knock-out mice). TCR crosslinking results in rapid activation of Fyn and Lck in T lymphocytes.[27] Fyn can be found in association with the TCR and this interaction is between the unique domain of Fyn and the TCR ζ-chain.[28] Expression of constitutively activated Fyn in a T-cell hybridoma results in enhanced TCR responsiveness,[29] as does overexpression of Fyn in transgenic mice.[30] Overexpression of kinase-inactive Fyn in transgenic mice abrogated TCR-mediated proliferation.[31] Interestingly, TCR signaling was only partially hampered in T cells isolated from Fyn-null mice, suggesting compensatory activity by other Src PTKs.[32] Similar studies have indicated the importance of Lck in TCR signaling. A variant of the human Jurkat T-cell line that lacks Lck was found to be nonresponsive to TCR and reconstitution with wild-type Lck restored the TCR signaling capacity.[33] Arrested thymocyte development is observed in Lck-null mice and the small numbers of T cells that appear in the periphery are drastically impaired in their ability to proliferate in response to TCR crosslinking.[34] Thus, both Fyn and Lck appear to play important roles in TCR-mediated signal transduction.

Blk and Lyn in BCR signaling

In B cells, crosslinking of the BCR results in rapid stimulation of the specific activity of Blk, Lyn and Fyn.[35] Strong evidence that Src PTKs are critical to BCR-mediated signaling events is supplied by a Lyn-negative B-cell line that was generated by targeted disruption of the Lyn locus.[36] The induction of tyrosine phosphorylation of cellular proteins following BCR crosslinking that is normally observed in the parental B-cell line was abolished in the Lyn-negative line. Interestingly, B cells from Fyn-null mice and from Lck-null mice demonstrate normal BCR signaling capacity. The outcome of the studies of phenotypes of Lyn-null and Blk-null mice is awaited with interest.

Csk PTK

Csk appears to play a role in the regulation of Src PTKs. Negative regulation of the activity of Src PTKs is thought to occur, in part, through intramolecular interaction of the carboxy-terminal phosphotyrosine of Src PTKs with their own SH2 domain.[37] Csk appears to contribute to this process by phosphorylating the carboxy-terminal tyrosine residue conserved in all Src family members.[14] Overexpression of Csk in a mouse T-cell system negatively regulated TCR-induced tyrosine phosphorylation and lymphokine production, providing evidence for the involvement of Csk in TCR-initiated signal transduction events.[38] The role of Csk in B cells was implied by the demonstration that Lyn was highly phosphorylated at its autophosphorylation site and constitutively activated in Csk-negative chicken B cell clones.[39] Csk may also have other roles. Recent studies demonstrated that Csk phosphorylated CD45, a transmembrane protein tyrosine phosphatase expressed by most hemopoietic cells, and generated a binding site for the SH2 domain of Lck.[40] Binding of Src TPKs to Csk-phosphorylated CD45 may release the carboxy-terminal tyrosine–Src SH2 interaction, thereby facilitating the dephosphorylation of Src PTKs by CD45 and their subsequent activation.

Syk/Zap PTKs

Only the Syk/Zap family of PTKs is known to contain two tandem SH2 domains preceding the catalytic domain. Following TCR crosslinking, the specific activity of ZAP increases and ZAP is found in association with phosphorylated TCR ζ and ε subunits.[27] Similarly, crosslinking of Fc receptors or the BCR results in increased kinase activity of Syk and its association with phosphorylated receptor subunits.[41,42] Both Syk and Zap are found to associate with antigen receptors through interactions between their dual SH2 domains and tyrosine phosphorylated ITAMs in the receptor subunits.[43] Recent *in vitro* studies have demonstrated that binding of phosphorylated ITAM peptides to Syk resulted in increased enzymatic activity.[44] Relocalization of Syk/Zap from the cytoplasm to the cell membrane following receptor activation may also contribute to their regulation. Expression of Zap as a chimera with the extracellular domain of CD16 localized ZAP to the cell membrane, yet crosslinking of this chimera did not evoke calcium flux or cytolytic responses in this model system.[45] However, co-crosslinking of CD16-Zap with CD16-Fyn or CD16-Lck generated a strong response, implicating Src PTKs in the regulation of Syk/Zap.

The strongest evidence to date for the importance of Zap in TCR-mediated signal transduction is supplied by the recent discovery that a human immunodeficiency syndrome is correlated with mutations that abolish functional Zap expression.[46–48] Four families were identified in which children with severe combined immunodeficiency (SCID) had mutations at the Zap locus. The mutations were defined, mapped to the kinase domain and found to result in drastically reduced or undetectable amounts of Zap. The profile of peripheral T cells revealed the complete absence of a $CD8^+$ population, suggesting that Zap may play a role in the maturation of $CD8^+$ thymocytes. Peripheral T cells in these patients were able to respond normally to pharmacological stimulation of the signaling machinery, but failed to proliferate in response to TCR ligation. Studies of early responses to TCR crosslinking in these cells revealed a total absence of induced tyrosine phosphorylation and calcium flux.

Syk is likely to play a parallel role to that of Zap in B cells and FcR^+ hemopoietic cells. A Syk-negative chicken B-cell line derived by homologous recombination exhibits severely defective responses to BCR crosslinking.

Btk PTKs

The fourth family of PTKs that appears to be involved in MIRR signaling includes Tec, Itk (also referred to as Emt), Btk, Bmx and Txk. Members of this family possess SH1, SH2 and SH3 domains in addition to a long amino-terminal region containing a PH domain and a Tec homology (TH) domain. The TH domain is a conserved 27 amino acid stretch that is rich in proline residues and lies between the PH and SH3 domains.[49] Members of the Tec family share a high degree of homology in the amino terminus, and this portion of the protein may play an important role in associating with other signal transducers.

Following crosslinking of the BCR on B cells or the FcεRI on mast cells, Btk becomes enzymatically activated.[42,50] A small fraction of Btk relocates from the cytoplasm to the membrane in FcεRI-stimulated mast cells, yet no direct binding of Btk with the receptor has been detected. Evidence implicating Btk in B-cell function came from the discovery that mutations in Btk were responsible for X-linked agammaglobulinemia (XLA), a human disease resulting from a developmental block in the maturation of B cells.[51,52] XLA patients have a deficit of mature B cells and, consequently, lack circulating immunoglobulin. A number of deleterious mutations in Btk have been mapped that produce clinically recognizable XLA and include point mutations in the PH, SH2 and kinase domains, a deletion in the carboxy-terminal portion of the SH3 domain and mutations that destabilize the mRNA.[53] Interestingly, X-linked immunodeficiency (xid) in mice is the result of a mutation in the PH domain of Btk at the same amino acid as one of the mutations resulting in XLA in humans.[54] Mice with xid do have some mature B cells but these do not produce antibodies in response to polysaccharide antigens or proliferate upon stimulation through the BCR.

A model of early events in antigen receptor signaling

Signaling through antigen receptors can be initiated by the binding of either multivalent antigen or antibody. Figure 4.3 illustrates some of the intracellular events that are thought to occur in the initial stages of antigen receptor signaling.

Immediately following receptor crosslinking, members of the Src family of PTKs are activated.[27,35,42] The mechanism of activation of the Src PTKs is not known, but biochemical and genetic analyses of potential regulatory sites in Src PTKs has provided insight into possible mechanisms.[37] Within seconds of antigen receptor crosslinking, the phosphotyrosine content of Src PTKs increases, perhaps owing to intermolecular phosphorylation between Src PTK molecules at the SH1 autophosphorylation site. In resting lymphocytes, key Src PTKs are

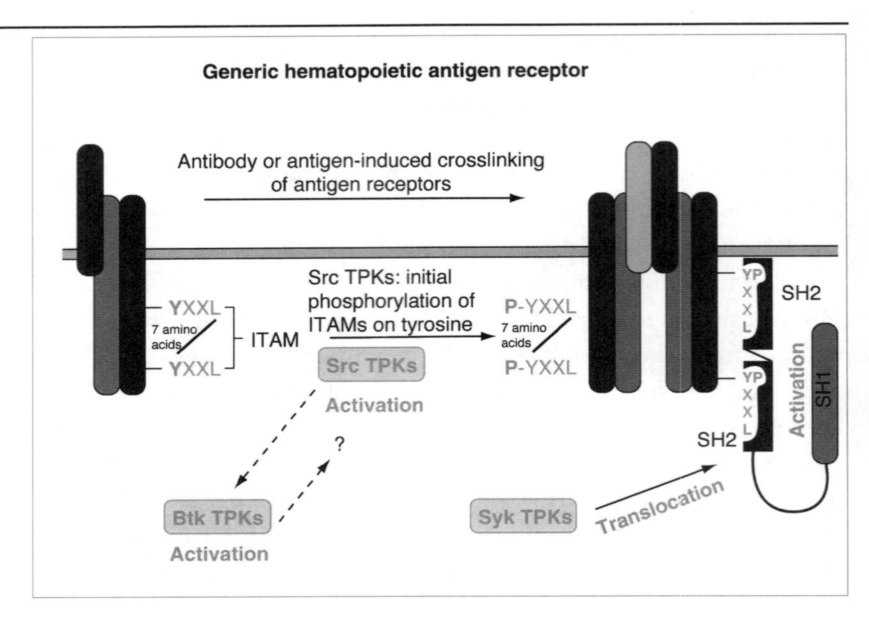

Figure 4.3 Model of the interactions of PTKs and receptor subunits following antigen receptor engagement. See text for discussion.

usually poorly phosphorylated at their regulatory tyrosine phosphorylation sites, perhaps owing to the function of phosphotyrosyl phosphatases. Receptor clustering may initiate the tyrosine kinase signaling cascade by providing an environment that excludes abundant membrane-associated phosphotyrosyl phosphatases such as CD45. Activation of Src PTKs coincides with the initial tyrosine phosphorylation of the receptor ITAMs, suggesting that the Src PTKs phosphorylate the ITAMs. Exclusion of phosphotyrosyl phosphatases would allow phosphorylation of the ITAMs to be maintained until bound by the appropriate SH2-containing proteins. Following the activation of Src PTKs, members of the Btk family are found to be enzymatically activated, although it is not clear yet if Src PTKs directly induce activation of Btk PTKs. The function and fate of activated Btk PTKs are not known. In B cells and mast cells, the activated Btk does not appear to associate with known antigen receptor components but may interact with such downstream signaling components. It has been well established that activation of Src PTKs and phosphorylation of ITAMs result in the translocation and association of Syk and Zap with tandemly phosphorylated ITAMs. In addition to the recruitment of Syk/Zap to the site of the activated receptor, both Syk and Zap become enzymatically activated. The mechanism of regulation of Syk/Zap PTKs is not known, but recent evidence indicates that tyrosine phosphorylation and association with ITAMs contribute to the enzymatic activation of this class of enzyme. It is clear that association of Syk/Zap with ITAMs is critical to antigen receptor signaling, since inhibition of this interaction abolishes downstream events.

Summary

The effector functions elicited as a consequence of antigen receptor engagement, signal transduction events and cellular commitment are extraordinarily diverse. A cell receiving signals through its MIRR may be stimulated to release stored effector molecules, synthesize new effector molecules, proliferate, differentiate or undergo programmed cell death. Results published recently have illuminated some of the processes involved in the complex responses of hemopoietic cells. Ligand binding to MIRRs activates PTKs and the phosphorylation of key substrates is likely to play important roles in most or all of the intracellular signaling events triggered by these receptors. It is conceivable that the path of signaling events that is taken following receptor crosslinking is determined by the choice of substrates of the PTKs and considerable insights have been gained into the mechanisms by which PTKs may choose their targets. It is also conceivable that the contribution of multiple PTKs in the kinase cascade may vary to evoke differentially the numerous possible responses to antigen receptor stimulation.

References

1 Weiss, A. and Littman, D.R. (1994) Signal transduction by lymphocyte antigen receptors. *Cell*, **76**, 263–74.
2 Sefton, B.M. and Taddie, J.A. (1994) Role of tyrosine kinases in lymphocyte activation. *Curr. Opinion Immunol.*, **6**, 372–9.
3 Mustelin, T. (1994) T-cell antigen receptor signaling: three families of tyrosine kinases and a phosphatase. *Immunity*, **1**, 351–6.
4 Keegan, A.D. and Paul, W.E. (1992) Multichain immune recognition receptors: similarities in structure and signaling pathways. *Immunol. Today*, **13**, 63–8.
5 Reth, M. (1992) Antigen receptor tail clue. *Nature*, **338**, 383–4.
6 Weiss, A. (1993) T-cell antigen receptor signal transduction: a tale of tails and cytoplasmic protein-tyrosine kinase. *Cell*, **73**, 209–12.
7 Burkhardt, A.L., Costa, T., Misulovin, Z. *et al.* (1994) Igα and Igβ are functionally homologous to the signaling proteins of the T-cell receptor. *Mol. Cell. Biol.*, **14**, 1095–103.
8 Ashwell, J.D. and Klausner, R.D. (1990) Genetic and mutational analysis of the T-cell antigen receptor. *Annu. Rev. Immunol.*, **8**, 139–67.
9 Reth, M. (1993) Antigen receptors on B lymphocytes. *Annu. Rev. Immunol.*, **10**, 98–121.

10 Ravetch, J.V. and Kinet, J.P. (1991) Fc receptors. *Annu. Rev. Immunol.*, **9**, 457–92.

11 June, C.H., Fletcher, M.C., Ledbetter, J.A. *et al.* (1990) Inhibition of tyrosine phosphorylation prevents T-cell receptor-mediated signal transduction. *Proc. Natl. Acad. Sci. USA*, **87**, 7722–6.

12 Lane, P.J.L., Ledbetter, J.A., McConnell, F.M. *et al.* (1991) The role of tyrosine phosphorylation in signal transduction through surface Ig in human B cells. *J. Immunol.*, **146**, 715–22.

13 Hanks, S.K., Quinn, A.M. and Hunter, T. (1988) The protein kinase family: conserved features and deduced phylogeny of the catalytic domains. *Science*, **241**, 42–52.

14 Nada, S., Okada, M., MacAuley, A. *et al.* (1991) Cloning of a complementary DNA for a protein-tyrosine kinase that specifically phosphorylates a negative regulatory site of $pp60^{c\text{-}Src}$. *Nature*, **351**, 69–72.

15 Chan, A.C., Iwashima, M., Turck, C.W. and Weiss, A. (1992) ZAP-70: a 70-kD protein tyrosine kinase that associates with the TCR ζ chain. *Cell*, **71**, 649–62.

16 Wilks, A.F., Harpur, A.G., Kurban, R.R. *et al.* (1991) Two novel protein-tyrosine kinases, each with a second phosphotransferase-related catalytic domain, define a new class of protein kinase. *Mol. Cell. Biol.*, **11**, 2057–65.

17 Cohen, G.B., Ren, R. and Baltimore, D. (1995) Modular binding domains in signal transduction proteins. *Cell*, **80**, 237–48.

18 Pawson, T. (1995) Protein modules and signaling networks. *Nature*, **373**, 579–8.

19 Yu, H., Chen, J., Feng, S. *et al.* (1994) Structural basis for the binding of proline-rich peptides to SH3 domains. *Cell*, **76**, 933–45.

20 Rickles, R.J., Botfield, M.C., Weng, Z. *et al.* (1994) Identification of Src, Fyn, Lyn, PI3K and Abl SH3 domain ligands using phage display libraries. *EMBO J.*, **13**, 5598–604.

21 Resh, M.D. (1994) Myristylation and palmitylation of Src family members: the fats of the matter. *Cell*, **76**, 411–13.

22 Rodgers, W., Crise, B. and Rose, J.K. (1994) Signals determining the protein tyrosine kinase and glycosyl-phosphatidylinositol-anchored protein targeting to a glycolipid-enriched membrane fraction. *Mol. Cell. Biol.*, **14**, 5384–91.

23 Gibson, T.J., Hyvonen, M., Musacchio, A. *et al.* (1994) PH domain: the first anniversary. *Trends Biochem. Sci.*, **19**, 349–53.

24 Touhara, K., Inglese, J., Pitcher, J.A. *et al.* (1994) Binding of G protein β–γ subunits to pleckstrin homology domains. *J. Biol. Chem.*, **269**, 10217–20.

25 Yao, L., Kawakami, Y. and Kawakami, T. (1994) The pleckstrin homology domain of Bruton tyrosine kinase interacts with protein kinase C. *Proc. Natl. Acad. Sci. USA*, **91**, 9175–9.

26 Harlan, R., Kolde, H.B. and Hemmings, B.A. (1994) Pleckstrin homology domains bind to phosophatidylinositol-4,5-bisphosphate. *Nature*, **371**, 168–70.

27 Burkhardt, A.L., Stealey, B., Rowley, R.B. *et al.* (1994) Temporal regulation of non-transmembrane protein tyrosine kinase enzyme activity following T-cell antigen receptor engagement. *J. Biol. Chem.*, **269**, 23642–7.

28 Timson Gauen, L.K., Kong, A.N., Samelson, L.E. and Shaw, A.S. (1992) $p59^{fyn}$ tyrosine kinase associates with multiple T-cell receptor subunits through its unique amino-terminal domain. *Mol. Cell. Biol.*, **12**, 5438–46.

29 Davidson, D., Chow, L.M., Fournel, M. and Veillette, A. (1992) Differential regulation of T-cell antigen responsiveness by isoforms of the Src-related tyrosine protein kinase $p59^{fyn}$. *J. Exp. Med.*, **175**, 1483–92.

30 Cooke, M.P., Abraham, K.M., Forbush, K.A. and Permutter, R.M. (1991) Regulation of T-cell receptor signaling by the Src family protein-tyrosine kinase ($p59^{fyn}$). *Cell*, **65**, 281–91.

31 Appleby, M.W., Gross, J.A., Cook, M.P. *et al.* (1992) Defective T-cell receptor signaling in mice lacking the thymic isoform of $p59^{fyn}$. *Cell*, **70**, 751–63.

32 Stein, P.L., Lee, H.-M., Rich, S. and Soriano, P. (1992) $pp59^{fyn}$ mutant mice display differential signaling in thymocytes and peripheral T cells. *Cell*, **70**, 741–50.

33 Strauss, D.B. and Weiss, A. (1992) Genetic evidence for the involvement of the lck tyrosine kinase in signal transduction through the T-cell antigen receptor. *Cell*, **70**, 585–93.

34 Molina, T.J., Kishihara, K., Siderowski, D.P. *et al.* (1992) Profound block in thymocyte development in mice lacking $p56^{lck}$. *Nature*, **357**, 161–4.

35 Burkhardt, A.L., Brunswick, M., Bolen, J.B. and Mond, J.J. (1991) Anti-immunoglobulin stimulation of B-lymphocytes activates src-related protein tyrosine kinases. *Proc. Natl. Acad. Sci. USA*, **88**, 7410–4.

36 Takata, M., Sabe, H., Hata, A. *et al.* (1994) Tyrosine kinase Lyn and Syk regulate B cell receptor-coupled Ca^{2+} mobilization through distinct pathways. *EMBO J.*, **13**, 1341–49.

37 Cooper, J.A. and Howell, B. (1993) The when and how of src regulation. *Cell*, **73**, 1051–4.

38 Chow, L.M.L., Fournel, M., Davidson, D. and Veillette, A. (1993) Negative regulation of T-cell receptor signaling by tyrosine protein kinase $p50^{csk}$. *Nature*, **365**, 156–60.

39 Hata, A., Sabe, H., Kurosaki, T. *et al.* (1994) Functional analysis of Csk in signal transduction through the B-cell antigen receptor. *Mol. Cell. Biol.*, **14**, 7306–13.

40 Autero, M., Saharinen, J., Pessa-Morikwa, T. *et al.* (1994) Tyrosine phosphorylation of CD45 phosphotyrosine phosphatase by $p50^{csk}$ kinase creates a binding site for $p56^{lck}$ tyrosine kinase and activates the phosphatase. *Mol. Cell. Biol.*, **14**, 1308–21.

41 Benhamou, M., Ryba, N.P.J., Kihara, H. *et al.* (1993) Protein-tyrosine kinase $p72^{syk}$ in high affinity IgE receptor signaling. *J. Biol. Chem.*, **268**, 23318–24.

42 Sauoaf, S.J., Mahajan, S., Rowley, R.B. *et al.* (1994) Temporal differences in the activation of three classes of non-transmembrane protein tyrosine kinases following B-cell antigen receptor surface engagement. *Proc. Natl. Acad. Sci. USA*, **91**, 9524–8.

43 Wange, R.L., Malek, S.N., Desiderio, S. and Samelson, L.E. (1993) Tandem SH2

domains of ZAP-70 bind to T-cell antigen receptor ζ and CD3 ε from activated Jurkat T cells. *J. Biol. Chem.*, **268**, 19797–801.

44 Rowley, R.B., Burkhardt, A.L., Chao, H.-G. *et al.* (1995) Syk protein tyrosine kinase is regulated by tyrosine phosphorylated Igα/Igβ ITAM binding and autophosphorylation. *J. Biol. Chem.*, **270**, 11590–4.

45 Kolanus, W., Romeo, C. and Seed, B. (1993) T-cell activation by clustered tyrosine kinases. *Cell*, **74**, 171–83.

46 Arpaia, E., Shahar, M., Dadi, H. *et al.* (1994) Defective T-cell receptor signaling and $CD8^+$ thymic selection in humans lacking ZAP-70 kinase. *Cell*, **76**, 947–58.

47 Elder, M.E., Lin, D., Clever, J. *et al.* (1994) Human severe combined immunodeficiency due to a defect in ZAP-70, a T-cell kinase. *Science*, **264**, 1596–9.

48 Chan, A.C., Kadlecek, T.A., Elder, M.E. *et al.* (1994) ZAP-70 deficiency in an autosomal recessive form of severe combined immunodeficiency. *Science*, **264**, 1599–601.

49 Vihinen, M., Nilsson, L. and Smith, C.I.E. (1994) Tec homology (TH) adjacent to the PH domain. *FEBS Lett.*, **350**, 263–5.

50 Kawakami, Y., Yao, L., Miura, T. *et al.* (1995) Tyrosine phosphorylation and activation of Bruton's tyrosine kinase (Btk) upon FcεRI crosslinking. *Mol. Cell. Biol.*, in press.

51 Vetrie, D., Vorechovsky, I., Sideras, P. *et al.* (1993) The gene involved in X-linked agammaglobulinemia is a member of the src family of protein tyrosine kinases. *Nature*, **361**, 226–33.

52 Tsukada, S., Saffran, D.C., Rawlings, D.J. *et al.* (1993) Deficient expression of a B cell cytoplasmic tyrosine kinase in human X-linked agammaglobulinemia. *Cell*, **72**, 279–90.

53 Ohta, Y., Haire, R.N., Litman, R.T. *et al.* (1994) Genomic organization and structure of Bruton agammaglobulinemia tyrosine kinase: localization of mutations associated with varied clinical presentations and course in X chromosome-linked agammaglobulinemia. *Proc. Natl. Acad. Sci. USA*, **91**, 9062–6.

54 Thomas, J.D., Sideras, P., Smith, C.I.E. *et al.* (1993) Colocalization of X-linked agammaglobulinemia and X-linked immunodeficiency genes. *Science*, **261**, 355–8.

5 Signaling through protein serine/threonine kinase receptors

Kohei Miyazono

The first example of an animal-cell receptor with a protein serine/threonine kinase domain was identified in *Caenorhabditis elegans* in 1990.[1] This worm forms a dauer larval stage to survive environmental stress. During the cloning of genes that control the development of dauer larva, a protein serine/threonine kinase receptor, DAF-1, was identified.

A second member of the serine/threonine kinase receptor family was found when a receptor for activins was obtained by an expression cloning strategy in 1991.[2] Activins are multifunctional proteins that belong to the transforming growth factor (TGF)-β superfamily. Receptors for the TGF-β superfamily are classified as types I, II, III, etc. on the basis of size and ligand-binding ability. The novel activin type II receptor clone had a serine/threonine kinase domain in its intracellular region. In 1992, a type II receptor for TGF-β was cloned by the same strategy[3] and also identified as a serine/threonine kinase.

Since then, more than a dozen serine/threonine kinase receptors have been cloned on the basis of amino acid sequence similarities, all of which are either type I or type II receptors for members of the TGF-β superfamily. In the TGF-β and activin systems, type II receptors bind ligand independently but type I receptors only bind ligand when type II receptors are also present.[4–7] Ligands induce the formation of heteromeric complexes of type II and type I receptors and transduce various signals.[8–10]

TGF-β superfamily

The TGF-β superfamily consists of more than 30 multifunctional proteins that are structurally related to each other. They can be

Signal Transduction. Edited by Carl-Henrik Heldin and Mary Purton. Published in 1996 by Chapman & Hall. ISBN 0 412 70810 8

classified into subgroups on the basis of similarities in their structures and biological activities[9,11] (Table 5.1). Most of them act as disulfide-bonded dimers, in which each monomer is composed of 110–140 amino acid residues. They have seven invariant cysteine residues which are conserved in most members. The crystal structure of TGF-β2 revealed that only one of the cysteine residues is used for intermolecular disulfide bonding, and the others form intramolecular disulfide bonds. It is likely that most members have similar three-dimensional structures to TGF-β2.

Proteins of the TGF-β family regulate growth, differentiation, migration and adhesion of various cell types. There are three mammalian TGF-β isoforms (TGF-β1, -β2 and -β3) with similar but not identical bioactivities. TGF-β is a potent growth inhibitor for most cell types.

Table 5.1 Members of the TGF-β superfamily[a]

Type	Mammals	Other vertebrates	Invertebrates
TGF-βs	TGF-β1	TGF-β5 [X]	
	TGF-β2		
	TGF-β3		
Inhibins/activins	Inhibin-α		
	Inhibin-βA		
	Inhibin-βB		
	Inhibin-βC		
Bone morphogenetic proteins (BMPs)	DPP/BMP-2 group:		
	BMP-2		DPP [D]
	BMP-4		
	60A/OP-1 group:		
	BMP-5		60A [D]
	BMP-6/Vgr-1		
	OP-1/BMP-7		
	OP-2/BMP-8		
	GDF-5 group:		
	GDF-5/CDMP-1		
	GDF-6/CDMP-2		
	GDF-7		
	BMP-3 group:		
	BMP-3		
	GDF-10		
	Other BMPs:		
	GDF-1	Vg1 [X]	Univin [e]
	GDF-3/Vgr-2		
	BMP-9	Dorsalin-1 [c]	Screw [D]
	Nodal		
Others	MIS/AMH		
	GDF-9		
	GDNF		

[a] In the BMP subgroup, proteins with high sequence similarities (more than 60%) have been grouped together. Species are indicated in brackets as follows: c, chicken; X, *Xenopus*; D, *Drosophila*; e, echinoderms. Other abbreviations are: Vgr, Vg-related; OP, osteogenic protein; GDF, growth/differentiation factor; CDMP, cartilage-derived morphogenetic protein; MIS, Müllerian inhibiting substance; AMH, anti-Müllerian hormone; GDNF, glial cell line-derived neurotrophic factor.

It also stimulates the deposition of extracellular matrix proteins by stimulating their production and inhibiting their degradation. Null mutation of the TGF-β1 gene resulted in multifocal inflammation in various organs after birth. This suggests that TGF-β1 is important for immune regulation *in vivo*.

Inhibins are heterodimers of one α-chain and one β-chain while activins are homodimers of two inhibin β-chains. Activins stimulate the secretion of follicle-stimulating hormone from pituitary cells and inhibins have an opposite effect. Activins were also shown to stimulate the differentiation of hematopoietic cells and induce the dorsal mesoderm in *Xenopus* embryos.

Bone morphogenetic proteins (BMPs) were originally identified as factors that stimulate bone and cartilage formation at extraskeletal sites *in vivo*, but they also act on many other cell types, including epithelial cells and neural cells. Proteins that are structurally similar to BMPs have been identified in invertebrates. In *Drosophila*, three BMP-like proteins, Decapentaplegic (DPP), Screw and 60A, have been found which play important roles in pattern formation during embryogenesis.

Three different types of TGF-β receptors

Among the receptors for the members in the TGF-β superfamily, the receptors for TGF-β have been best characterized.[8–10] Affinity cross-linking studies using radioiodinated TGF-β1 revealed three types of TGF-β receptors with different sizes. TGF-β type I receptor (TβR-I) is 53 kDa, type II receptor (TβR-II) is 75 kDa and type III receptor (also termed betaglycan) is more than 200 kDa.

Betaglycan is a proteoglycan in which glycosaminoglycan chains are attached to the 100 kDa core protein.[12,13] It is expressed in many cell types, but is absent in certain cells, including myoblasts, endothelial cells and hematopoietic cells. Some of these cells instead express endoglin, a 170 kDa disulfide-linked dimeric protein which is related to betaglycan.[14] Although betaglycan and endoglin are not directly involved in signal transduction, they play a role in facilitating the interaction between ligand and the types I and II signaling receptors.[15]

TβR-II and TβR-I are ubiquitously expressed in TGF-β responsive cells, and both are essential for signal transduction. Certain cell types, including hematopoietic cells, appear to express only TβR-I, but this may be because detection of TβR-II is obscured by poor efficiency in cross-linking of TGF-β to TβR-II in these cell types.[6]

TβR-II and TβR-I are signaling receptors

That both TβR-II and TβR-I are important in signal transduction has been shown using chemically mutagenized mink lung epithelial cells (Mv1Lu). Mv1Lu cells are highly sensitive to TGF-β and widely used for the studies on TGF-β. After chemical mutagenesis by ethyl methanesulfonate and culture of the cells in the presence of TGF-β, mutant cells that are resistant to the growth inhibitory activity of TGF-β were obtained.[16] The TGF-β-resistant mutants can be classified into three major categories, depending on the expression of the receptors: R mutants, which lack binding of TGF-β to TβR-I; DR mutants, which lack TGF-β binding to both TβR-II and TβR-I; and S mutants, which express all three TGF-β receptors and therefore probably have defects in the intracellular signaling pathway (Figure 5.1). Cells that are defective in betaglycan binding were not obtained, indicating that betaglycan is not required for signal transduction of TGF-β.

Somatic hybrid cells obtained by fusion of the R and DR mutants express both TβR-II and TβR-I and fully respond to TGF-β (Figure

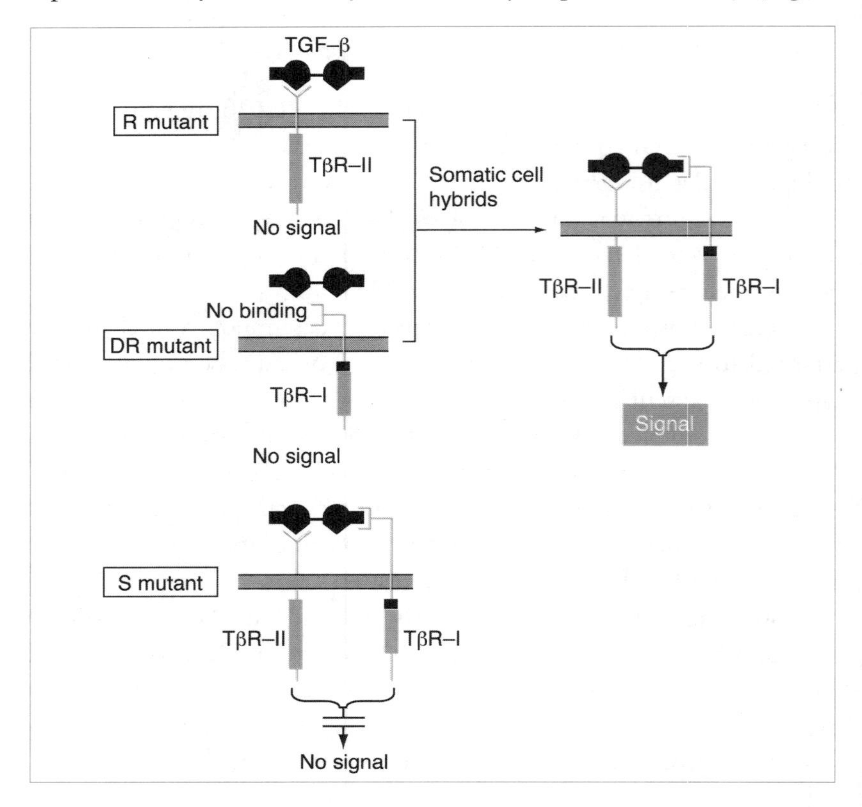

Figure 5.1 Receptor expression in chemically mutagenized Mv1Lu cells. Complex formation between single molecules of TβR-II and TβR-I and TGF-β are shown, but the ligand may induce hetero-oligomerization of receptors, most likely two molecules of each receptor (see text). In the S mutants, binding of ligand to TβR-II and TβR-I is intact, but signal transduction by intracellular signaling components including intracellular domains of TβR-II or TβR-I may be impaired by chemical mutagenesis.

5.1).[17] These data indicate that both TβR-II and TβR-I are required for signal transduction and that DR mutants have intact TβR-I but have defects in TβR-II. TβR-I cannot bind the ligands in the absence of TβR-II, and therefore neither binding to TβR-II nor to TβR-I were seen in the DR mutants. This hypothesis was confirmed after cloning of the TβR-II cDNA. Mutations in the TβR-II gene were found in the DR mutants. After transfection of the TβR-II cDNA into the DR mutants, TGF-β bound to both TβR-II and TβR-I, and the responsiveness to TGF-β was recovered.[18]

TβR-II and TβR-I are serine/threonine kinase receptors

cDNA cloning of TβR-II and TβR-I revealed that they are both serine/threonine kinase receptors. TβR-II is a 567 amino acid peptide with a hydrophobic leader sequence, followed by a short extracellular domain, a single transmembrane domain and an intracellular domain composed mainly of a serine/threonine kinase region[3] (Figure 5.2).

TβR-I was cloned by a strategy based on the polymerase chain reaction (PCR) using the sequence similarity in the serine/threonine kinase regions.[6] TβR-I is synthesized as a 503 amino acid peptide. Although smaller than TβR-II, the overall structure of TβR-I is similar. TβR-I does not bind ligand in the absence of TβR-II, but when the TβR-I cDNA is co-transfected with TβR-II cDNA, ligand is bound. Moreover, when the TβR-I cDNA was transfected into R mutant Mv1Lu cells, which lack functional TβR-I, it restored the responsiveness to TGF-β.

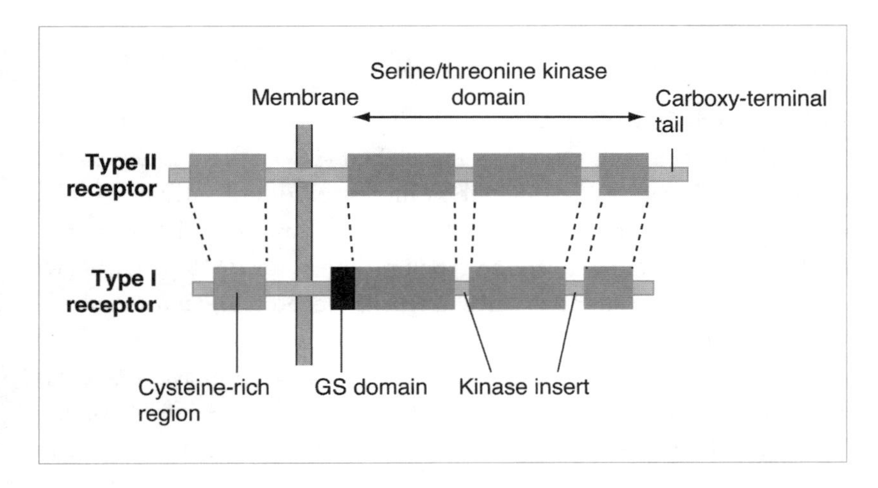

Figure 5.2 Schematic illustration of the structures of type II and type I receptors. Carboxy-terminal tails are short in type I receptors (less than 13 amino acid residues), but are longer in type II receptors (24–530 amino acid residues). GS domain is found only in type I receptors.

Table 5.2 Members of the serine/threonine kinase receptor family and their ligands[a]

Receptor	Ligand
Type II receptors:	
TβR-II	TGF-β
ActR-II	Activins, OP-1
ActR-IIB	Activins, OP-1
Atr-II/Punt [D]	Activins, BMP-2, DPP
BMPR-II	BMP-4, OP-1
DAF-4 [C]	BMP-2, -4, OP-1
C14/AMHR	MIS/AMH
Type I receptors:	
TβR-I/ALK-5	TGF-β
ActR-I/ALK-2/Tsk7L	Activins, OP-1
ActR-IB/ALK-4	Activins
Atr-I [D]	Activins
BMPR-IA/ALK-3	BMP-4, OP-1
BMPR-IB/ALK-6	BMP-4, OP-1
Saxophone [D]	BMP-2, DPP
Thick veins [D]	BMP-2, DPP
ALK-1/TSR-I	Unknown
DAF-1 [C]	Unknown

[a] Proteins from non-mammalian species are indicated as follows: D, *Drosophila*; C, *Caenorhabditis elegans*.

Multiple members in the serine/threonine kinase receptor family

Seventeen serine/threonine kinase receptors have thus far been cloned in different species, and the number of members in the family is still increasing (Table 5.2). The type I and type II receptors form subfamilies on the basis of differences in their structures.[8–10] Type II receptors in mammals include two activin type II receptors (ActR-II and ActR-IIB), a BMP type II receptor (BMPR-II) and a type II receptor which binds Müllerian inhibiting substance (MIS). DAF-4 in *Caenorhabditis elegans*, obtained as a protein responsible for dauer larva development, serves as a BMP type II receptor. In the type I receptor subfamily, there are two activin type I receptors (ActR-I and ActR-IB) and two BMP type I receptors (BMPR-IA and BMPR-IB) in mammals. The physiological ligand for ALK-1 (also termed TGF-β superfamily receptor-I or TSR-I) remains to be identified.

Most of the type I receptors are smaller than the corresponding type II receptors, but there are exceptions, e.g. Atr-I is larger than Atr-II.[19] The kinase domains are very similar to each other in each serine/threonine kinase receptor subfamily, but the other regions are less related. However, most of the cysteine residues in the extracellular domains can be well aligned, suggesting that they have similar three-dimensional structures. All type I receptors, but not type II receptors, have highly conserved glycine- and serine-rich domains (GS domains) preceding the kinase domains (Figure 5.2). There are two short insert sequences in the kinase domains. The carboxy-terminal tails are very short in type I receptors, whereas those in type II receptors are longer, up to 530 amino acid residues in BMPR-II.[20]

Ligand binding to serine/threonine kinase receptors

TβR-II binds TGF-β, which allows TβR-I to bind to the TGF-β–TβR-II complex. Thus, the binding of TGF-β to TβR-I is dependent on the binding of TβR-II. The binding properties of activins to their receptors are similar to those of TGF-β, but those of BMPs appear to be different. Binding of ligand to type I receptors for BMPs (BMPR-IA and -IB) does occur in the absence of type II receptors and is upregulated by type II receptors.[21] In part, this may be due to the presence of endogenous BMPR-II in the cells used for transfection assay, but it is also possible that the presence of BMPR-II is not required for the binding of BMPs to type I receptors. Interestingly, ligand binding to

BMPR-II alone is weak but is enhanced by the presence of BMP type I receptors.[20] Moreover, a *Drosophila* activin type II receptor (Atr-II/Punt) does not bind BMP-2 alone, but can bind it in the presence of *Drosophila* BMP type I receptors.[22]

The ligand binding to TβR-II is highly specific for TGF-β, but there is a redundancy in the binding of TGF-β to type I receptors. When overexpressed in mammalian cells by transfection, most of the mammalian type I receptors can bind TGF-β in the presence of TβR-II, and form heteromeric receptor complexes. However, only TβR-I has been shown to transduce intracellular signals.[6,7,23] Similarly, most mammalian type I receptors can bind activin in the presence of ActR-II, but only ActR-I and ActR-IB can transduce signals.[5,7,24]

The binding of activins to ActR-II is also relatively specific. However, osteogenic protein (OP)-1 (also termed BMP-7), but not BMP-4, can also bind ActR-II and transduce signals.[25] Moreover, OP-1/BMP-7 can bind to ActR-I, BMPR-IA and BMPR-IB, but not to ActR-IB, in the presence of different type II receptors (BMPR-II, ActR-II or DAF-4).[20,21,25] Thus, ligand binding to type I receptors is supported by type II receptors, but the specificity for ligand binding is determined by each of the type I receptors.

Ligands induce formation of hetero-oligomeric receptor complexes

TGF-β induces the formation of a heteromeric complex between TβR-II and TβR-I, but the complex does not appear to be a simple heterodimer. TβR-II forms homodimers in the absence or presence of ligand.[26,27] Homodimers of TβR-I have been observed in the presence of ligand, but it is not known whether they exist in its absence.[28] Thus, ligand binding to TβR-II and TβR-I may induce an oligomeric complex, most likely a hetero-tetramer composed of two molecules each of TβR-II and TβR-I (Figure 5.3).

Ligand-independent homo-oligomerization (homodimerization) of TβR-II was shown to occur at multiple contact points at the extracellular/transmembrane domains and at the cytoplasmic domain in mammalian cells.[27] Interactions between TβR-II and TβR-I also appear to occur via both extracellular/transmembrane domains and intracellular domains. Truncated TβR-II, which lacks its intracellular domain, can form a complex with TβR-I via extracellular/transmembrane domains after the addition of ligand.[29] Moreover, studies using

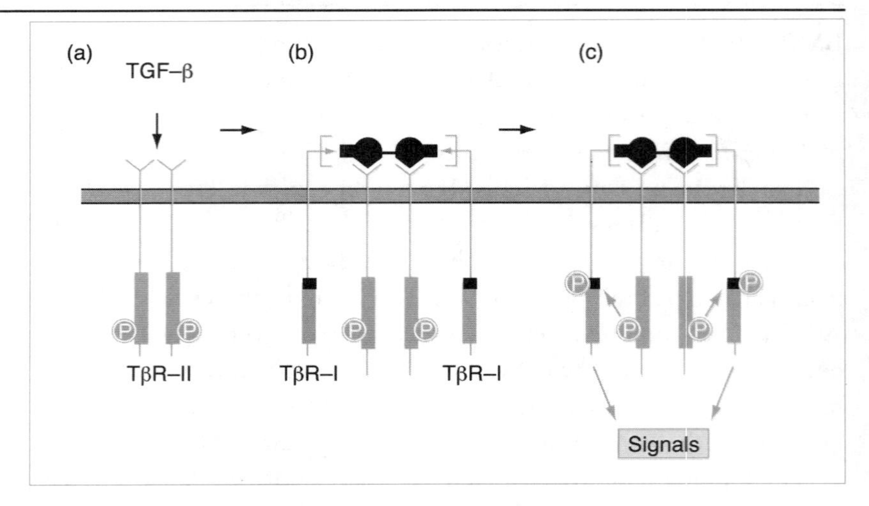

Figure 5.3 Activation of TGF-β receptors and signal transduction. TGF-β is shown to induce heterotetramerization of TβR-II and TβR-I in this figure, but the number of the receptor molecules in the ligand-induced receptor complex and their binding stoichiometry have not yet been elucidated.

the yeast two-hybrid system revealed that the TβR-I intracellular domain interacts with the TβR-II intracellular domain, although whether such an interaction occurs in mammalian cells remains to be elucidated.[30,31]

Activation of serine/threonine kinase receptors

TβR-II is a constitutively active kinase, in which serine residues are autophosphorylated. Whether TβR-II is transactivated by ligand-independent homodimerization, or by other enzyme(s) in the cells, is not fully understood. After binding ligand, TβR-II interacts with TβR-I and transphosphorylates its GS domain[32,33] (Figure 5.3). This triggers the activation of the TβR-I kinase, followed by transduction of the intracellular signals. Serine and threonine residues in the GS domain and possibly other parts in TβR-I are phosphorylated by TβR-II. The TβR-II intracellular domain cannot be replaced by the corresponding region of TβR-I, and vice versa, suggesting that they have distinct roles in signal transduction.[34]

In cells overexpressing TβR-II and TβR-I at very high amounts, the two receptors form a complex and the TβR-I kinase is activated in the absence of the ligand.[31] Moreover, by mutation of Thr204 to Asp, TβR-I kinase is constitutively activated and transduces signals in the absence of TβR-II and ligand.[35] Thus, type I receptors act downstream of type II receptors and specify the signals. Supporting this notion, TβR-I and ActR-IB, which are highly similar to each other, inhibit the growth of Mv1Lu cells, whereas ActR-I, which is more distantly related to TβR-I and ActR-IB, does not.[24] Whether type II receptors transduce certain signals independently of type I receptors, or whether they serve only to activate type I receptors, remain to be determined.

Signals induced by serine/threonine kinase receptors

The intracellular signals transduced by serine/threonine kinase receptors are not fully understood. Most of the substrates activated by tyrosine kinase receptors do not appear to be involved in the signal transduction by serine/threonine kinase receptors. It is not known whether downstream signaling molecules interact with serine/threonine kinase receptors by binding to phosphorylated serine or threonine residues, or other structures in the receptors.

Using the yeast two-hybrid system, FKBP-12, a binding protein for the immunosuppressants FK506 and rapamycin, was shown to bind to several type I receptors, including ALK-1/TSR-I, TβR-I and ActR-I, but not to kinase-defective type I receptors or type II receptors.[36] However, the physiological significance of this interaction has not been determined. TGF-β prevents the cell cycle progression at the late G1 phase. Since rapamycin also inhibits the growth of lymphocytes during the same phase of the cell cycle, it is possible that FKBP12 is involved in the signaling pathway of TGF-β.

Progression through the cell cycle is regulated by a family of serine/threonine kinase complexes composed of cyclins and cyclin-dependent kinases (Cdks).[37] TGF-β regulates the activity of cyclin–Cdk complexes by several mechanisms (Figure 5.4). The synthesis and activity of Cdk4 are dramatically repressed by TGF-β in Mv1Lu cells.[38] A decrease in Cdk4 results in an increase in the available levels of p27/Kip1, an inhibitor for cyclin–Cdk complexes, which then inhibits the activity of cyclin E–Cdk2 complex.[39,40] In human keratinocytes, TGF-β inhibits the expression of p15/Ink4B, an inhibitor of Cdk4 and Cdk6.[41] Production of p21/Cip1/Waf1, an inhibitor of cyclin E–Cdk2 and cyclin D–Cdk4 complexes, is also increased by TGF-β in certain cell lines, including human keratinocytes. Expression of p21/Cip1/Waf1 is positively regulated by p53, but the induction by TGF-β appears to occur in a p53-independent manner.[42,43] Inhibition of cyclin–Cdk kinases by TGF-β represses the phosphorylation of the retinoblastoma (Rb) protein.[44,45] Phosphorylated pRb releases the transcription factor E2F, which then induces expression of a number of genes critical for G1–S transition

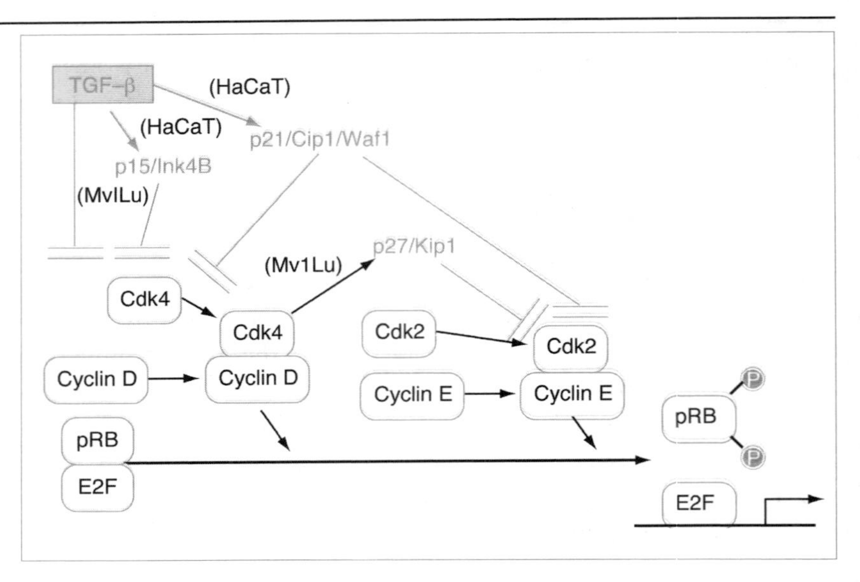

Figure 5.4 TGF-β inhibits progression in the cell cycle. In Mv1Lu cells, TGF-β inhibits the synthesis of Cdk4, which increases the available level of the cyclin–Cdk inhibitor, p27/Kip1. In HaCaT human keratinocytes, TGF-β induces the expression of p15/Ink4B, an inhibitor of Cdk4 and Cdk6, and p21/Cip1/Waf1, an inhibitor of cyclin–Cdk complexes. Inhibition of the cyclin–Cdk kinases by TGF-β prevents the phosphorylation of pRB, which results in cell cycle arrest late in the G1 phase.

and DNA replication. Thus, the growth inhibitory mechanism of TGF-β is linked to suppression of the phosphorylation of pRb.

TGF-β stimulates the growth of mesenchymal cells, but entry into S phase is delayed compared with that induced by other growth factors.[46] This is in part due to the autocrine activation of platelet-derived growth factors[46,47], but it is also possible that an increase in matrix formation facilitates the growth of certain cell types.

Certain cancer cells escape from negative growth control by TGF-β

Transformed cells often acquire resistance to the growth inhibitory activity of TGF-β, which results in the autonomous growth of cancer cells. Certain cells, including breast cancer cells, hepatoma cells, T-cell lymphoma cells and some gastric cancer cells, do not express TβR-II protein after transformation.[48–51] In these cells, transfection of the TβR-II cDNA restores the responsiveness to TGF-β and reduces tumorigenicity. Moreover, mutations in the TβR-II gene have been found in colorectal cancers with a defect in DNA repair.[52] Thus, loss of the TβR-II expression may be an important step in carcinogenesis in certain types of cancer.

Other cells have intact TβR-II and TβR-I, but they are not growth inhibited by TGF-β. Many of these cells respond to TGF-β with regard to the production of matrix proteins, suggesting that certain molecules involved in growth inhibitory pathway are perturbed. A potential such candidate may be p15/Ink4B (see above), which is often mutated in transformed cells.[41] Identification of such molecules may be useful for the understanding of the mechanisms of carcinogenesis and for the design of effective treatments for cancer.

Summary

Members of the TGF-β superfamily transduce their signals through binding to two different types of serine/threonine kinase receptors, type II and type I. These form heteromeric receptor complexes, most likely heterotetramers, after binding of ligand. TβR-II is a constitutively active kinase. After ligand binding, TβR-II interacts with TβR-I, and transphosphorylates the GS domain of TβR-I, which activates the kinase activity of TβR-I and thereby initiates downstream signal transduction pathways. The components of such pathways remain to be determined, but the downstream targets include molecules involved in cell cycle progression, including p27/Kip1, p15/Ink4B, p21/Cip1/Waf1 and Cdk4, which eventually inhibit the phosphorylation of pRb and prevent the progression of cell cycle at late G1 phase.

References

1 Georgi, L.L., Albert, P.S. and Riddle, D.L. (1990) *daf-1*, a *C. elegans* gene controlling dauer larva development, encodes a novel receptor kinase. *Cell*, **61**, 635–45.

2 Mathews, L.S. and Vale, W.W. (1991) Expression cloning of an activin receptor, a predicted transmembrane serine kinase. *Cell*, **65**, 973–82.

3 Lin, H.Y., Wang, X.-F., Ng-Eaton, E. *et al.* (1992) Expression cloning of the TGFβ type II receptor, a functional transmembrane serine/threonine kinase. *Cell*, **68**, 775–85.

4 Ebner, R., Chen, R.-H., Shum, L. *et al.* (1993) Cloning of a type I TGF-β receptor and its effect on TGF-β binding to the type II receptor. *Science*, **260**, 1344–8.

5 Attisano, L., Cárcamo, J., Ventura, F. *et al.* (1993) Identification of human activin and TGFβ type I receptors that form heteromeric kinase complexes with type II receptors. *Cell*, **75**, 671–80.

6 Franzén, P., ten Dijke, P., Ichijo, H. *et al.* (1993) Cloning of a TGFβ type I receptor that forms a heteromeric complex with the TGFβ type II receptor. *Cell*, **75**, 681–92.

7 ten Dijke, P., Yamashita, H., Ichijo, H. *et al.* (1994) Characterization of type I receptors for transforming growth factor-β and activin. *Science*, **264**, 101–4.

8 Miyazono, K., ten Dijke, P., Yamashita, H. and Heldin, C.-H. (1994) Signal transduction via serine/threonine kinase receptors. *Seminars Cell Biol.*, **5**, 389–98.

9 Massagué, J., Attisano, L. and Wrana, J.L. (1994) The TGF-β family and its composite receptors. *Trends Cell Biol.*, **4**, 172–8.

10 Derynck, R. (1994). TGF-β-receptor-mediated signaling. *Trends Biochem. Sci.*, **19**, 548–53.

11 Kingsley, D.M. (1994) The TGF-β superfamily: new members, new receptors, and new genetic tests of function in different organisms. *Genes Dev.*, **8**, 133–46.

12 López-Casillas, F., Cheifetz, S., Doody, J. *et al.* (1991) Structure and expression of the membrane proteoglycan betaglycan, a component of the TGF-β receptor system. *Cell*, **67**, 785–95.

13 Wang, X.-F., Lin, H.Y., Ng-Eaton, E. *et al.* (1991) Expression cloning and characterization of the TGF-β type III receptor. *Cell*, **67**, 797–805.

14 Cheifetz, S., Bellón, T., Calés, C. *et al.* (1992) Endoglin is a component of the transforming growth factor-β receptor system in human endothelial cells. *J. Biol. Chem.*, **267**, 19027–30.

15 López-Casillas, F., Wrana, J.L. and Massagué, J. (1993) Betaglycan presents ligand to the TGFβ signaling receptor. *Cell*, **73**, 1435–44.

16 Laiho, M., Weis, F.M.B. and Massagué, J. (1990) Concomitant loss of transforming growth factor TGFβ receptor types I and II in TGF-β-resistant cell mutants implicates both receptor types in signal transduction. *J. Biol. Chem.*, **265**, 18518–24.

17 Laiho, M., Weis, F.M.B., Boyd, F.T., Ignotz, R.A. and Massagué, J. (1991) Responsiveness to transforming growth factor-β (TGF-β) restored by a genetic complementation between cells defective in TGF-β receptors I and II. *J. Biol. Chem.*, **266**, 9108–12.

18 Wrana, J.L., Attisano, L., Cárcamo, J. *et al.* (1992) TGFβ signals through a heteromeric protein kinase receptor complex. *Cell*, **71**, 1003–14.

19 Wrana, J.L., Tran, H., Attisano, L. *et al.* (1994) Two distinct transmembrane serine/threonine kinases from *Drosophila melanogaster* form an activin receptor complex. *Mol. Cell. Biol.*, **14**, 944–50.

20 Rosenzweig, B.L., Imamura, T., Okadome, T. *et al.* (1995) Cloning and characterization of a human type II receptor for bone morphogenetic proteins. *Proc. Natl. Acad. Sci. USA*, **92**, 7632–6.

21 ten Dijke, P., Yamashita, H., Sampath, T.K. *et al.* (1994) Identification of type I receptors for osteogenic protein-1 and bone morphogenetic protein-4. *J. Biol. Chem.*, **269**, 16985–8.

22 Letsou, A., Arora, K., Wrana, J.L. *et al.* (1995) Drosophila Dpp signaling is mediated by the *punt* gene product: a dual ligand-binding type II receptor of the TGFβ receptor family. *Cell*, **80**, 899–908.

23 Bassing, C.H., Yingling, J.M., Howe, D.J. *et al.* (1994) A transforming growth factor β type I receptor that signals to activate gene expression. *Science*, **263**, 87–9.

24 Cárcamo, J., Weis, F.M.B., Ventura, F. *et al.* (1994) Type I receptors specify growth-inhibitory and transcriptional responses to transforming growth factor β and activin. *Mol. Cell. Biol.*, **14**, 3810–21.

25 Yamashita, H., ten Dijke, P., Huylebroeck, D. *et al.* (1995) Osteogenic protein-1 binds to activin type II receptor and induces activin-like effects. *J. Cell Biol.*, **130**, 217–26.

26 Henis, Y.I., Moustakas, A., Lin, H.Y. and Lodish, H.F. (1994) The types II and III transforming growth factor-β receptors form homo-oligomers. *J. Cell Biol.*, **126**, 139–54.

27 Chen, R.-H. and Derynck, R. (1994) Homomeric interactions between type II transforming growth factor-β receptors. *J. Biol. Chem.*, **269**, 22868–74.

28 Yamashita, H., ten Dijke, P., Franzén, P. *et al.* (1994) Formation of hetero-oligomeric complexes of type I and type II receptors for transforming growth factor-β. *J. Biol. Chem.*, **269**, 20172–8.

29 Chen, R.-H., Ebner, R. and Derynck, R. (1993) Inactivation of the type II receptor reveals two receptor pathways for the diverse TGF-β activities. *Science*, **260**, 1335–8.

30 Kawabata, M., Chytil, A. and Moses, H.L. (1995) Cloning of a novel type II serine/threonine kinase receptor through interaction with type I transforming growth factor-β receptor. *J. Biol. Chem.*, **270**, 5625–30.

31 Ventura, F., Doody, J., Liu, F. *et al.* (1994) Reconstitution and transphosphorylation of TGF-β receptor complex. *EMBO J.*, **13**, 5581–9.

32 Wrana, J.L., Attisano, L., Wieser, R. *et al.* (1994) Mechanism of activation of the TGF-β receptor. *Nature*, **370**, 341–7.

33 Chen, F. and Weinberg, R.A. (1995) Biochemical evidence for the autophosphorylation and transphosphorylation of transforming growth factor β receptor kinases. *Proc. Natl. Acad. Sci. USA*, **92**, 1565–9.

34 Okadome, T., Yamashita, H., Franzén, P. *et al.* (1994) Distinct roles of the intracellular domains of transforming growth factor-β type I and type II receptors in signal transduction. *J. Biol. Chem.*, **269**, 30753–6.

35 Wieser, R., Wrana, J.L. and Massagué, J. (1995) GS domain mutations that constitutively activate TβR-1, the downstream signaling component in the TGF-β receptor complex. *EMBO J.*, **14**, 2199–208.

36 Wang, T., Donahoe, P.K. and Zervos, A.S. (1994) Specific interaction of type I receptors of the TGF-β family with the immunophilin FKBP-12. *Science*, **265**, 674–6.

37 Sherr, C.J. (1995) G1 phase progression: cycling on cue. *Cell*, **79**, 551–5.

38 Ewen, M.E., Sluss, H.K., Whitehouse, L.L. and Livingston, D.M. (1993) TGFβ inhibition of Cdk4 synthesis is linked to cell cycle arrest. *Cell*, **74**, 1009–20.

39 Polyak, K., Lee, M.-H., Erdjument-Bromage, H. *et al.* (1994) Cloning of $p27^{Kip1}$, a cyclin-dependent kinase inhibitor and a potent mediator of extracellular antimitogenic signals. *Cell*, **78**, 59–66.

40 Toyoshima, H. and Hunter, T. (1994) p27, a novel inhibitor of G1 cyclin–Cdk protein kinase activity, is related to p21. *Cell*, **78**, 67–74.

41 Hannon, G.J. and Beach, D. (1994) $p15^{INK4B}$ is a potential effector of TGF-β-induced cell cycle arrest. *Nature*, **371**, 257–61.

42 Li, C.-Y., Suardet, L. and Little, J.B. (1995) Potential role of *WAF1*/Cip1/p21 as

a mediator of TGF-β cytoinhibitory effect. *J. Biol. Chem.*, **270**, 4971–4.

43 Datto, M.B., Li, Y., Panus, J.F. *et al.* (1995) Transforming growth factor β induces the cyclin-dependent kinase inhibitor p21 through a p53-independent mechanism. *Proc. Natl. Acad. Sci. USA*, **92**, 5545–9.

44 Pietenpol, J.A., Stein, R.W., Moran, E. *et al.* (1990) TGF-β1 inhibition of c-*myc* transcription and growth in keratinocytes is abrogated by viral transforming proteins with pRB binding domains. *Cell*, **61**, 777–85.

45 Laiho, M., DeCaprio, J.A., Ludlow, J.W. *et al.* (1990) Growth inhibition by TGF-β linked to suppression of retinoblastoma protein phosphorylation. *Cell*, **62**, 175–85.

46 Shipley, G.D., Tucker, R.F. and Moses, H.L. (1985) Type β transforming growth factor/growth inhibitor stimulates entry of monolayer cultures of AKR-2B cells into S phase after a prolonged prereplicative interval. *Proc. Natl. Acad. Sci. USA*, **82**, 4147–51.

47 Battegay, E.J., Raines, E.W., Seifert, R.A. *et al.* (1990) TGF-β induces bimodal proliferation of connective tissue cells via complex control of autocrine PDGF loop. *Cell*, **63**, 515–24.

48 Inagaki, M., Moustakas, A., Lin, H.Y. *et al.* (1993) Growth inhibition by transforming growth factor β (TGF-β) type I is restored in TGF-β-resistant hepatoma cells after expression of TGF-β receptor type II cDNA. *Proc. Natl. Acad. Sci. USA*, **90**, 5359–63.

49 Kadin, M.E., Cavaille-Coll, M.W., Gertz, R. *et al.* (1994) Loss of receptors for transforming growth factor β in human T-cell malignancies. *Proc. Natl. Acad. Sci. USA*, **91**, 6002–6.

50 Sun, L., Wu, G., Willson, J.K.V. *et al.* (1994) Expression of transforming growth factor β type II receptor leads to reduced malignancy in human breast cancer MCF-7 cells. *J. Biol. Chem.*, **269**, 26449–55.

51 Park, K., Kim, S.-J., Bang, Y.-J. *et al.* (1994) Genetic changes in the transforming growth factor β (TGF-β) type II receptor gene in human gastric cancer cells: correlation with sensitivity to growth inhibition. *Proc. Natl. Acad. Sci. USA*, **91**, 8772–6.

52 Markowitz, S., Wang, J. Myeroff, L. *et al.* (1995) Inactivation of the type II TGF-β receptor in colon cancer cells with microsatellite instability. *Science*, **268**, 1336–8.

6 Signaling through members of the tumor necrosis factor receptor family

Muneesh Tewari
and Vishva M. Dixit

The tumor necrosis factor (TNF) receptor family of cell-surface cytokine receptors contains more than a dozen members that signal a diverse array of responses (Table 6.1). Family members are defined by the presence of cysteine-rich repeats in their extracellular domains and are all type I transmembrane proteins, although in some cases soluble forms are generated by proteolytic cleavage. The activities they signal range from the induction of thymocyte proliferation (type 2 TNF receptor[1,2]) to apoptosis (programmed cell death) (Fas[3]) and to immunoglobulin class-switching (CD40[4,5]) (Table 6.1). Consistent with the observation that the receptors signal different functions, their cytoplasmic signaling domains show little sequence conservation.

The ligands for these receptors form a parallel gene family (Table 6.1). With the exception of lymphotoxin-α, each ligand appears to be specific for one receptor, and vice versa. The ligands are all thought to form trimers, and are defined by homology in their carboxy-terminal region to TNF, the prototypical family member. All the ligands are synthesized as type II transmembrane proteins although in many cases, for example TNF, soluble forms can exist following proteolytic cleavage. The notable exception to the transmembrane rule is lymphotoxin-α, which is exclusively a secreted protein. All the ligands appear to function as homo-oligomers, with the exception of lymphotoxin-α and lymphotoxin-β, which can hetero-oligomerize to form a hybrid ligand that binds the recently identified receptor TNFRrp (TNF receptor-related protein[6] (Table 6.1).

The receptors are activated by the clustering induced upon binding of their respective oligomeric ligands or, as has been taken advantage

Signal Transduction. Edited by Carl-Henrik Heldin and Mary Purton. Published in 1996 by Chapman & Hall. ISBN 0 412 70810 8

Table 6.1 Some members of the tumor necrosis factor receptor family and their ligands

Receptor	Ligand(s)	Major functions
TNF receptor 1	TNF, lymphotoxin-α	Apoptosis NF-κB activation
TNF receptor 2	TNF, lymphotoxin-α	Thymocyte proliferation NF-κB activation
Fas	Fas ligand	Apoptosis
CD40	CD40 ligand	Ig class-switching Blocks apoptosis
Low-affinity NGF receptor	Nerve growth factor (NGF)	Blocks apoptosis
TNFRrp (TNF receptor-related protein)	Lymphotoxin-$\alpha_2\beta$ heterotrimer	Involved in lymph node development
CD27	CD27 ligand	Role in T-cell proliferation
4-IBB	4-IBB ligand	Role in T-cell proliferation
OX40	OX40 ligand	Unknown
CD30	CD30 ligand	Unknown

of experimentally, by crosslinking with agonist antibodies. None of the receptors possess any obvious catalytic domains (e.g. kinase or phosphatase) in their cytoplasmic segments and they do not, in general, directly couple to any of the established signal transduction pathways discussed in the other chapters of this book. Instead, it is becoming apparent that signaling is accomplished via the binding of an emerging class of novel and diverse signaling molecules. In this chapter, we shall discuss recent advances in the understanding of signaling pathways initiated by four of the most intensively studied receptors in this family: the type 1 and type 2 TNF receptors, Fas and CD40.

Tumor necrosis factor receptors

A plethora of *in vivo* functions have been described for TNF, including roles in the pathophysiology of various disease states such as fever, cancer cachexia and septic shock.[7] Here we shall consider its more immediate, primary effects on cells. Depending upon the cellular context, TNF has been shown to initiate a wide variety of responses, including: (1) induction of gene transcription, in most cases mediated by the transcription factor NF-κB; (2) induction of apoptosis; (3) activation of a lipid signaling cascade involving ceramide and (4) activation of phospholipase A_2 (see review [8]).

The effects of TNF are mediated through two distinct receptors, TNFR1 and TNFR2, which have molecular masses of approximately 55 and 75 kDa, respectively (see review [9]). TNFR1 and TNFR2 can both independently signal NF-κB activation and gene induction, whereas the induction of apoptosis is largely mediated via TNFR1, although it has been suggested that under certain conditions TNFR2 may also be able to transmit a cytotoxic signal.[10] The cytoplasmic domains of the two receptors for TNF bear no homology, suggesting that they utilize different signaling molecules to couple to downstream responses.

TNF receptor type 1

TNFR1 has been reported to signal a variety of activities in different cell types and contexts. We shall focus on (1) its ability to initiate gene induction via the activation of the transcription factor NF-κB and (2) its ability to initiate a pathway of apoptosis. These are the most intensively studied effects of TNF and, more important, are ones for which recent advances have been made regarding their signaling.

Transcriptional activation of genes by TNFR1

Crosslinking of TNFR1 induces the transcription of a multitude of genes, many of which encode proteins involved in the cellular response to stress and infection. The major mode of gene induction appears to be the activation of the transcription factor NF-κB. NF-κB exists as a variety of dimeric complexes consisting of subunits from a family of sequence-specific DNA-binding proteins related to the proto-oncogene c-*rel* in their amino-terminal domains. Although initially characterized as a nuclear protein in B cells, NF-κB exists as a cytoplasmic complex in most cell types, bound to an inhibitory protein termed IκB (see reviews [11,12]). Activation of TNFR1 leads to the phosphorylation and degradation of IκB, following which the released NF-κB migrates to the nucleus, where it binds specific κB elements resulting in transcriptional activation[13] (Figure 6.1). The cytoplasmic pathway between the receptor and NF-κB had been enigmatic until the discovery of at least one candidate molecule that may be part of the pathway (see below).

Induction of apoptosis (programmed cell death) by TNFR1

In addition to the activation of NF-κB, engagement of an apoptotic pathway is an important activity signaled by TNFR1, particularly in certain tumor cell lines. Apoptosis, also known as programmed cell death, is a form of physiologic cell death defined morphologically by chromatin condensation, DNA fragmentation and cytoplasmic shrinkage and blebbing (see review [14]). Apoptosis can be triggered by a number of stimuli in addition to TNF, and occurs in multicellular organisms ranging from

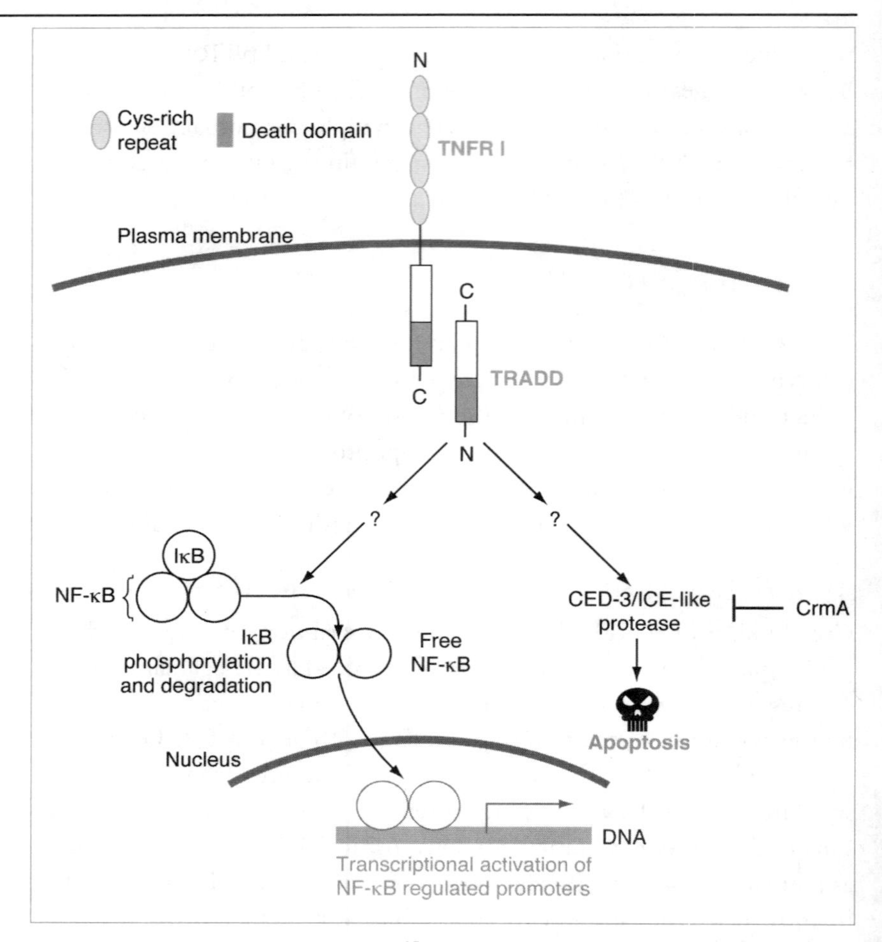

Figure 6.1 A model of signaling through the type 1 TNF receptor (TNFR1).

nematodes to insects to humans.[15] The mechanism of cell death has been largely unknown. Recently, however, evidence for the involvement of proteases has accumulated. Genetic analysis of programmed cell death in the nematode *Caenorhabditis elegans* first identified a gene, designated *ced-3*, that is obligatory for apoptosis.[16] The sequence of the CED-3 protein shows homology to a family of proteins related to the interleukin-1β-converting enzyme (ICE), all of which are cysteine proteases, indicating that CED-3 is also very likely a cysteine protease.[17]

Recent work indicates that the TNFR1-initiated apoptosis pathway involves a protease of the CED-3/ICE family. CrmA, a cowpox virus-encoded inhibitor of certain CED-3/ICE family members, potently inhibits TNF-induced apoptosis[18,19] (Figure 6.1). Furthermore, TNF-induced cell death is associated with CrmA-inhibitable proteolysis of certain substrates, including poly(ADP–ribose) polymerase (PARP)[20] and the 70 kDa protein component of the U1 small nuclear ribonucleoprotein.[21]

The PARP-cleaving enzyme has recently been identified to be a member of the CED-3/ICE family.[20,22] Given that the CED-3/ICE family cysteine proteases are themselves synthesized as inactive zymogens that require proteolytic activation,[20,22–24] it appears that TNFR1 may employ a proteolytic cascade to induce apoptosis.

Signaling molecules associated with TNFR1

Mutagenesis studies of the cytoplasmic segment of TNFR1 identified a protein domain of about 80 amino acids, designated the 'death domain', that is required for signaling apoptosis.[25] The same domain is also required for signaling NF-κB activation. Interestingly, this death domain shows homology to a similar region in Fas,[26], another receptor that signals apoptosis (see below).

The cytoplasmic segment of TNFR1 possesses no known enzymatic activity and no coupling proteins were known to engage it until the recent identification of TRADD, a 34 kDa protein that associates with the cytoplasmic domain of this receptor.[27] At the carboxyl terminus of TRADD (for TNF-receptor associated death domain) is a domain that has homology to the death domains found in TNFR1 and Fas. The large amino-terminal segment shows no significant homology to known proteins. TRADD can associate with the cytoplasmic domain of TNFR1, but not with that of Fas or TNFR2. Furthermore, this association is the result of a direct interaction between the death domain of TRADD and the death domain of TNFR1 (Figure 6.1).

Evidence for the involvement of TRADD in TNFR1 signaling comes from the demonstration that its overexpression can induce the two major responses to TNFR1 activation: NF-κB induction and apoptosis.[27] Furthermore, apoptosis induced by TRADD is inhibited by CrmA, indicating that TRADD activates a CED-3/ICE-like protease just as TNFR1 does. Although the death domain is required for both NF-κB and apoptosis induction, these two pathways appear to be divergent, as CrmA inhibits apoptosis, but not NF-κB induction.[27] Taken together, a model of the signaling in response to TNFR1 activation can be depicted (Figure 6.1).

Although the identification of TRADD is certainly a milestone in understanding TNFR1 signaling, there are many questions that remain to be answered, including whether TRADD is present constitutively bound to the receptor or whether it is recruited following TNF-mediated receptor coupling. Additionally, the mechanism of coupling of TRADD to downstream signaling events remains to be defined.

TNF receptor type 2

TNFR2 is thought to play a role in the immune system, as it has been shown to stimulate the proliferation of thymocytes.[1,2] Although signaling events induced by TNFR2 are not well characterized, it is known to activate NF-κB.[28] Rothe *et al.*[28] subjected the cytoplasmic domain of TNFR2 to a mutational analysis and identified a novel domain required for NF-κB signaling that had no similarity to sequences in TNFR1 that signal NF-κB activation. Using a combination of co-immunoprecipitation and yeast two-hybrid studies, they identified two proteins, designated TRAF1 and TRAF2 (for TNF receptor associated factors 1 and 2, respectively) that complex with the cytoplasmic segment of TNFR2. TRAF2 directly binds the TNFR2 cytoplasmic segment, whereas TRAF1 is indirectly linked to the receptor by its binding to TRAF2[28] (Figure 6.2).

TRAF1 and TRAF2 are structurally related molecules that contain a carboxy-terminal region of homology known as the TRAF domain and a central coiled-coil motif that presumably allows for oligomerization (Figure 6.2). In keeping with this, both TRAF1 and TRAF2 can form homodimers as well as heterodimers, although the heterodimer is the form thought to associate with TNFR2 in the cell. TRAF2 has additional motifs including an amino-terminal RING finger domain. In addition to TRAF2, another protein belonging to this

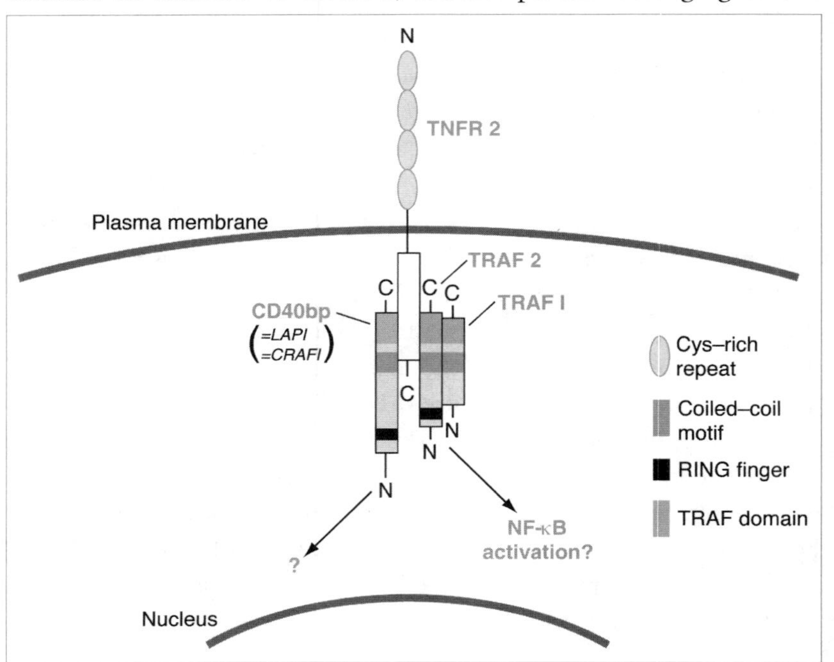

Figure 6.2 A model of signaling through the type 2 TNF receptor (TNFR2).

family has been identified that can bind to TNFR2 directly. This protein, designated CD40bp/LAP1/CRAF1, also has a carboxy-terminal TRAF homology domain and, like TRAF2, an amino-terminal RING finger motif.[29–31] This putative signal transducer will be discussed further in the section on CD40, as it is best known for its association with that receptor.

Although the identification of a new family of molecules that associate with TNFR2 is a significant advance, it represents only the beginning of our understanding of the mechanism of signaling initiated by this receptor. Indeed, it is not yet clear whether the molecules shown to bind the receptor are actually linked to downstream signaling events. Addressing this, in addition to identifying additional molecules involved in the signaling pathway, is the focus of current research in this area.

Fas/Apo-I/CD95

Fas, also known as Apo-I or CD95, is best recognized for its ability to signal programmed cell death.[3,32] It plays an important physiological role in the regulation of the immune system, as is evidenced by the fact that spontaneous mutations either in Fas itself or in its extracellular ligand result in a fatal lymphoproliferative syndrome in mice.[33–35] Furthermore, mutations in Fas have been found to be responsible for cases of a rare autoimmune condition found in children[36,37] and elevated concentrations of a soluble form of Fas have been reported in patients with the autoimmune condition, systemic lupus erythematosus.[38] Additionally, Fas is an important mediator of activation-induced death of T cells, a phenomenon thought to play a role in the pathogenesis of a variety of diseases, including AIDS.[39–42] Finally, activation of Fas on target cells by the Fas ligand expressed on cytotoxic T cells is part of the mechanism by which these cells deliver their lethal hit (see review [43]).

Signal transduction initiated by Fas

Mutational analysis of the cytoplasmic segment of Fas revealed a domain necessary for signaling cell death.[26] As discussed earlier, this domain has homology to the death domain in TNFR1, suggesting that both TNFR1 and Fas might signal apoptosis through the same or similar molecules. Experimental evidence for a common mechanism came from the demonstration that CrmA, which blocks TNF-induced apoptosis, also potently blocks Fas-induced cell death,[18,19,44] suggesting that a

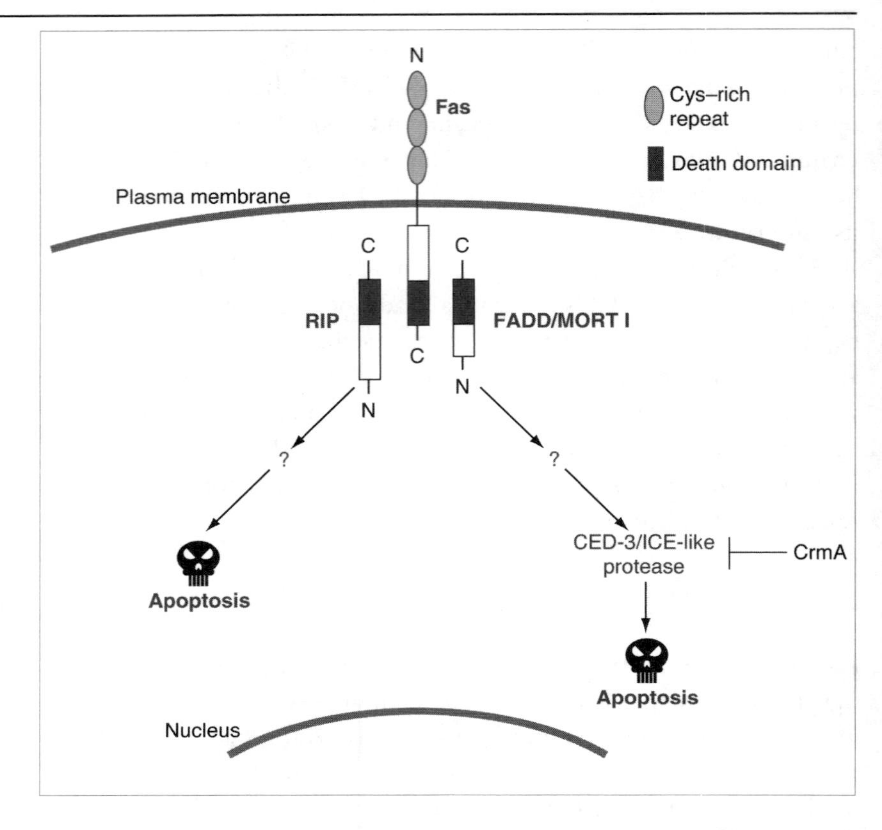

Figure 6.3 A model of signaling by Fas, a receptor that signals programmed cell death (apoptosis).

CED-3/ICE-like protease is an obligatory component of both death pathways (Figure 6.3).

Recently, two proteins have been identified, designated FADD (Fas-associated death domain, also known as MORT1) and RIP (receptor-interacting protein), that bind to the cytoplasmic segment of Fas and activate the death pathway.[45–47] Both of these putative signaling molecules have domains homologous to the death domains in TNFR1 and Fas, and in this respect they are similar to TRADD (Figure 6.3).

FADD/MORT1 is a 23 kDa protein that contains a death domain near its carboxyl terminus and an amino-terminal segment with no homology to known proteins. Overexpression of FADD/MORT1 is by itself able to initiate apoptosis that is CrmA-inhibitable, suggesting that the same downstream pathway is being utilized as in Fas-induced cell death.[45]

RIP is a 74 kDa protein containing a death domain at its carboxy-terminal end and an amino-terminal domain with homology to both protein tyrosine and serine/threonine protein kinases, although the presence of kinase activity has not been demonstrated. Overexpression of RIP, like FADD/MORT1, induces apoptosis. The kinase homology domain is not required for the induction of apoptosis, however.[47] Whether RIP-induced apoptosis can be inhibited by CrmA is not known at this time.

The discovery of FADD/MORT1 and RIP as putative signaling molecules associated with the cytoplasmic segment of Fas will hopefully allow for the identification of other molecules to delineate the pathway leading from the receptor to the CED-3/ICE-like protease and ultimately to cell death.

CD40

CD40 is a cell-surface transmembrane 45 kDa glycoprotein receptor expressed on B-lymphocytes and certain epithelial cells. The ligand for CD40, designated CD40L, is a type II transmembrane glycoprotein expressed on the surface of activated T cells.[48–50] Activation of CD40 induces B-cell proliferation and, in the concomitant presence of IL-4 or IL-13, allows for the generation of factor-dependent B-cell lines that secrete IgE following isotype switching.[51] Addition of IL-10 results in the induction of IgG3, IgG1 and IgA1 immunoglobulins.[52,53] Thus, the engagement of CD40 on B cells turns on their isotype switching machinery, the specificity of which is dictated by the cytokine environment. When the CD40–CD40L interaction is blocked *in vitro* with soluble CD40 or monoclonal antibodies to CD40L, B cells cannot proliferate or produce immunoglobulin in response to T-cell signals, indicating that this interaction is necessary for signaling.[54] *In vivo*, antibodies against CD40L block the development of both primary and secondary antibody responses and of collagen-induced arthritis, confirming the key role of this interaction in B-cell regulation.[55,56] The most compelling evidence for a critical role of CD40 and its ligand comes from the discovery that a fatal immunodeficiency (hyper-IgM syndrome), characterized by the lack of circulating IgG and IgA and the absence of germinal centers, is due to defective expression of CD40L.[57–60] In keeping with this, CD40-deficient mice generated by gene targeting display a phenotype similar to hyper-IgM syndrome patients with defective germinal center formation and defective immunoglobulin isotype switching.[5] Taken together, these studies support a key role for CD40 in T-cell dependent B-cell activation both *in vitro* and *in vivo*.

CD40 signal transduction

It has recently been demonstrated that cross-linking CD40 on resting human tonsillar B cells and on B-cell lines results in the activation of NF-κB.[61,62] Electrophoretic mobility shift assays and super-shift analysis revealed the activation of p50, p65 (RelA) and c-Rel NF-κB subunits.[62] Since a number of CD40-responsive genes, including IL-6 and ICAM-1, have NF-κB consensus sites in their 5′ regulatory regions, it is likely that CD40-dependent NF-κB activation is responsible for the induction of these genes.[63]

CD40 activation also results in increased protein tyrosine kinase (PTK) activity, which appears to play an important role in mediating the biological effects of CD40 since PTK inhibitors attenuate B-cell aggregation and isotype switching induced by CD40 crosslinking.[64] Within 1 min of CD40 crosslinking, there is an increase in phosphorylation of the Src-related kinase Lyn which persists for up to 20 min.[65] By contrast, the phosphorylation of related Src-like kinases including Fyn, Fgr and Lck remains unchanged.[65] Since it appears that CD40 does not directly associate with Lyn (as it is not possible to co-immunoprecipitate Lyn with CD40 from cell lysates), this raises the possibility that an intervening bridging molecule might function as an adaptor, bringing together the receptor (CD40) and effector (Lyn) portions of the signal transduction pathway.

Recently, a candidate for such an adaptor molecule was identified in a yeast two-hybrid screen employing the CD40 cytoplasmic domain as bait.[29,31] This molecule, designated CD40bp (for CD40-binding protein)[29] or CRAF1 (for CD40 receptor-associated factor 1),[31] contains an amino-terminal RING finger motif and a prominent central coiled-coil segment that may allow homo- or hetero-oligomerization. More important, the carboxyl terminus possesses substantial homology to the TRAF domain that, as discussed previously, is found in TRAF1 and TRAF2, two proteins that associate with the cytoplasmic domain of TNFR2. Although the exact function of CD40bp/CRAF1 remains to be defined, a recent report that the Epstein–Barr virus transforming protein LMP1 also associates with CD40bp/CRAF1 suggests an important role for this new molecule in signaling.[30]

Summary

The TNF receptor family of cytokine receptors is a group of proteins related by homology in their extracellular ligand-binding domains but demonstrating diversity in the activities signaled and in the structure of their cytoplasmic signaling domains. Recent discoveries indicate that signaling is likely achieved by an equally diverse class of novel proteins that can associate with the receptors. Delineation of the molecular components and steps of the signaling pathways engaged by these receptors is a major aim of current studies in this area.

References

1 Tartaglia, L.A., Weber, R.F., Figari, I.S. *et al.* (1991) The two different receptors for tumor necrosis factor mediate distinct cellular responses. *Proc. Natl. Acad. Sci. USA*, **88**, 9292–6.

2 Tartaglia, L.A., Goeddel, D.V., Reynolds, C. *et al.* (1993) Stimulation of human T-cell proliferation by specific activation of the 75-kDa tumor necrosis factor receptor. *J. Immunol.*, **151**, 4637–41.

3 Itoh, N., Yonehara, S., Ishii, A. *et al.* (1991) The polypeptide encoded by the cDNA for human cell surface antigen Fas can mediate apoptosis. *Cell*, **66**, 233–43.

4 Fuleihan, R., Ramesh, N. and Geha, R.S. (1993) Role of CD40–CD40-ligand interaction in Ig-isotype switching. *Curr. Opinion. Immunol.*, **5**, 963–7.

5 Kawabe, T., Naka, T., Yoshida, K. *et al.* (1994) The immune responses in CD40-deficient mice: impaired immunoglobulin class switching and germinal center formation. *Immunity*, **1**, 167–78.

6 Crowe, P.D., VanArsdale, T.L., Walter, B.N. *et al.* (1994) A lymphotoxin-beta-specific receptor. *Science*, **264**, 707–10.

7 Vassalli, P. (1992) The pathophysiology of tumor necrosis factors. *Annu. Rev. Immunol.*, **10**, 411–52.

8 Heller, R.A. and Kronke, M. (1994) Tumor necrosis factor receptor-mediated signaling pathways. *J. Cell. Biol.*, **126**, 5–9.

9 Tartaglia, L.A. and Goeddel, D.V. (1992) Two TNF receptors. *Immunol. Today*, **13**, 151–3.

10 Heller, R.A., Song, K., Fan, N. and Chang, D.J. (1992) The p70 tumor necrosis factor receptor mediates cytotoxicity. *Cell*, **70**, 47–56.

11 Siebenlist, U., Franzoso, G. and Brown, K. (1994) Structure, regulation and function of NF-kappa B. *Annu. Rev. Cell. Biol.*, **10**, 405–55.

12 Liou, H.C. and Baltimore, D. (1993) Regulation of the NF-κB/rel transcription factor and IκB inhibitor system. *Curr. Opinion Cell Biol.*, **5**, 477–87.

13 Baeuerle, P.A. and Henkel, T. (1994) Function and activation of NF-kappa B in the immune system. *Annu. Rev. Immunol.*, **12**, 141–79.

14 Duvall, E. and Wyllie, A.H. (1986) Death and the cell. *Immunol. Today*, **7**, 115–9.

15 Steller, H. (1995) Mechanisms and genes of cellular suicide. *Science*, **267**, 1445–9.

16 Ellis, H.M. and Horvitz, H.R. (1986) Genetic control of programmed cell death in the nematode *Caenorhabditis elegans*. *Cell*, **44**, 817–29.

17 Yuan, J., Shaham, S., Ledoux, S. *et al.* (1993) The *C. elegans* cell death gene *ced-3* encodes a protein similar to mammalian interleukin-1β-converting enzyme. *Cell*, **75**, 641–52.

18 Tewari, M. and Dixit, V.M. (1995) Fas- and tumor necrosis factor-induced apoptosis is inhibited by the poxvirus crmA gene product, *J. Biol. Chem.*, **270**, 3255–60.

19 Enari, M., Hug, H. and Nagata, S. (1995) Involvement of an ICE-like protease in Fas-mediated apoptosis. *Nature*, **375**, 78–81.

20 Tewari, M., Quan, L.T., O'Rourke, K. *et al.* (1995) Yama/CPP32β, a mammalian homolog of CED-3, is a CrmA-inhibitable protease that cleaves the death substrate poly(ADP-ribose) polymerase. *Cell*, **81**, 801–9.

21 Tewari, M., Beidler, D.R. and Dixit, V.M. (1995) CrmA-inhibitable cleavage of the 70 kDa protein component of the U1 small nuclear ribonucleoprotein during Fas- and tumor necrosis factor-induced apoptosis. *J. Biol. Chem.*, **270**, 18738–41.

22 Nicholson, D.W., Ali, A., Thornberry, N.A. *et al.* (1995) Identification and inhibition of the ICE/CED-3 protease necessary for mammalian apoptosis, *Nature*, **376**, 37–43.

23 Thornberry, N.A., Bull, H.G., Calaycay, J.R. *et al.* (1992) A novel heterodimeric cysteine protease is required for interleukin-1β processing in monocytes. *Nature*, **356**, 768–74.

24 Cerretti, D.P., Kozlosky, C.J., Mosley, B. *et al.* (1992) Molecular cloning of the interleukin-1β converting enzyme. *Science*, **256**, 97–100.

25 Tartaglia, L.A., Ayres, T.M., Wong, G.H.W. and Goeddel, D.V. (1993) A novel domain within the 55 kd TNF receptor signals cell death. *Cell*, **74**, 845–53.

26 Itoh, N. and Nagata, S. (1993) A novel protein domain required for apoptosis. Mutational analysis of human Fas antigen. *J. Biol. Chem.*, **268**, 10932–7.

27 Hsu, H., Xiong, J. and Goeddel, D.V. (1995) The TNF receptor 1-associated protein TRADD signals cell death and NF-kappa B activation. *Cell*, **81**, 495–504.

28 Rothe, M., Wong, S.C., Henzel, W.J. and Goeddel, D.V. (1994) A novel family of putative signal transducers associated with the cytoplasmic domain of the 75 kDa tumor necrosis factor receptor. *Cell*, **78**, 681–92.

29 Hu, H.M., O'Rourke, K., Boguski, M.S. and Dixit, V.M. (1994) A novel RING finger protein interacts with the cytoplasmic domain of CD40. *J. Biol. Chem.*, **269**, 30069–72.

30 Mosialos, G., Birkenbach, M., Yalamanchili, R. *et al.* (1995) The Epstein–Barr virus transforming protein LMP1 engages signaling proteins for the tumor necrosis factor receptor family. *Cell*, **80**, 389–99.

31 Cheng, G., Cleary, A.M., Ye, Z.S. *et al.* (1995) Involvement of CRAF1, a relative of TRAF, in CD40 signaling. *Science*, **267**, 1494–8.

32 Trauth, B.C., Klas, C., Peters, A.M.J. *et al.* (1989) Monoclonal antibody-mediated tumor regression by induction of apoptosis. *Science*, **245**, 301–5.

33 Watanabe-Fukunaga, R., Brannan, C.I., Copeland, N.G. *et al.* (1992) Lymphoproliferation disorder in mice explained by defects in Fas antigen that mediates apoptosis. *Nature*, **356**, 314–7.

34 Takahashi, T., Tanaka, M., Brannan, C.I. *et al.* (1994) Generalized lymphoproliferative disease in mice, caused by a point mutation in the Fas ligand. *Cell*, **76**, 969–76.

35 Lynch, D.H., Watson, M.L., Alderson, M.R. *et al.* (1994) The mouse Fas-ligand gene is mutated in gld mice and is part of a TNF family gene cluster. *Immunity*, **1**, 131–6.

36 Rieux-Laucat, F., Le Deist, F., Hivroz, C. *et al.* (1995) Mutations in Fas associated with human lymphoproliferative syndrome and autoimmunity. *Science*, **268**, 1347–9.

37 Fisher, G.H., Rosenberg, F.J., Straus, S.E. *et al.* (1995) Dominant interfering Fas gene mutations impair apoptosis in a human autoimmune lymphoproliferative syndrome. *Cell*, **81**, 935–46.

38 Cheng, J., Zhou, T., Liu, C. *et al.* (1994) Protection from Fas-mediated apoptosis by a soluble form of the Fas molecule. *Science*, **263**, 1759–62.

39 Alderson, M.R., Tough, T.W., Davis-Smith, T. *et al.* (1995) Fas ligand mediates activation-induced cell death in human T lymphocytes. *J. Exp. Med.*, **181**, 71–7.

40 Brunner, T., Mogil, R.J., LaFace, D. *et al.* (1995) Cell-autonomous Fas (CD95)/Fas-ligand interaction mediates activation-induced apoptosis in T-cell hybridomas. *Nature*, **373**, 441–4.

41 Ju, S.T., Panka, D.J., Cui, H. *et al.* (1995) Fas(CD95)/FasL interactions required for programmed cell death after T-cell activation, *Nature*, **373**, 444–8.

42 Yang, Y., Mercep, M., Ware, C.F. and Ashwell, J.D. (1995) Fas and activation-induced Fas ligand mediate apoptosis of T cell hybridomas: inhibition of Fas ligand expression by retinoic acid and glucocorticoids. *J. Exp. Med.*, **181**, 1673–82.

43 Henkart, P.A. and Sitkovsky, M.V. (1994) Cytotoxic lymphocytes. Two ways to kill target cells. *Curr. Biol.*, **4**, 923–5.

44 Los, M., Van de Craen, M., Penning, L.C. *et al.* (1995) Requirement of an ICE/CED-3 protease for Fas/APO-1-mediated apoptosis. *Nature*, **375**, 81–3.

45 Chinnaiyan, A.M., O'Rourke, K., Tewari, M. and Dixit, V.M. (1995) FADD, a novel death domain-containing protein, interacts with the death domain of Fas and initiates apoptosis. *Cell*, **81**, 505–12.

46 Boldin, M.P., Varfolomeev, E.E., Pancer, Z. *et al.* (1995) A novel protein that interacts with the death domain of Fas/APO1 contains a sequence motif related to the death domain. *J. Biol. Chem.*, **270**, 7795–8.

47 Stanger, B.Z., Leder, P., Lee, T.H. *et al.* (1995) RIP: a novel protein containing a death domain that interacts with Fas/APO-1 (CD95) in yeast and causes cell death. *Cell*, **81**, 513–23.

48 Armitage, R.J., Fanslow, W.C., Strockbine, L. *et al.* (1992) Molecular and biological characterization of a murine ligand for CD40. *Nature*, **357**, 80–2.

49 Hollenbaugh, D., Grosmaire, L.S., Kullas, C.D. *et al.* (1992) The human T-cell antigen gp39, a member of the TNF gene family, is a ligand for the CD40 receptor: expression of a soluble form of gp39 with B-cell co-stimulatory activity. *EMBO J.*, **11**, 4313–21.

50 Graf, D., Korthauer, U., Mages, H.W. *et al.* (1992) Cloning of TRAP, a ligand for CD40 on human T cells. *Eur. J. Immunol.*, **22**, 3191–4.

51 Banchereau, J., de Paoli, P., Valle, A. *et al.* (1991) Long-term human B cell lines dependent on interleukin-4 and antibody to CD40. *Science*, **251**, 70–2.

52 Rousset, F., Garcia, E. and Banchereau, J. (1991) Cytokine-induced proliferation and immunoglobulin production of human B lymphocytes triggered through their CD40 antigen. *J. Exp. Med.*, **173**, 705–10.

53 Rousset, F., Garcia, E., Defrance, T. *et al.* (1992) Interleukin 10 is a potent growth and differentiation factor for activated human B lymphocytes. *Proc. Natl. Acad. Sci. USA*, **89**, 1890–3.

54 Fanslow, W.C. Anderson, D.M., Grabstein, K.H. *et al.* (1992) Soluble forms of CD40 inhibit biologic responses of human B cells. *J. Immunol.*, **149**, 655–60.

55 Foy, T.M., Shepherd, D.M., Durie, F.H. *et al.* (1993) In vivo CD40-gp39 interactions are essential for thymus-dependent humoral immunity. II. Prolonged suppression of the humoral immune response by an antibody to the ligand for CD40, gp39. *J. Exp. Med.*, **178**, 1567–75.

56 Durie, F.H., Fava, R.A., Foy, T.M. *et al.* (1993) Prevention of collagen-induced arthritis with an antibody to gp39, the ligand for CD40. *Science*, **261**, 1328–30.

57 Allen, R.C., Armitage, R.J., Conley, M.E. *et al.* (1993) CD40 ligand gene defects responsible for X-linked hyper-IgM syndrome. *Science*, **259**, 990–3.

58 DiSanto, J.P., Bonnefoy, J.Y., Gauchat, J.F. *et al.* (1993) CD40 ligand mutations in X-linked immunodeficiency with hyper-IgM. *Nature*, **361**, 541–3.

59 Fuleihan, R., Ramesh, N., Loh, R. *et al.* (1993) Defective expression of the CD40 ligand in X chromosome-linked immunoglobulin deficiency with normal or elevated IgM. *Proc. Natl. Acad. Sci. USA*, **90**, 2170–3.

60 Korthauer, U., Graf, D., Mages, H.W. *et al.* (1993) Defective expression of T-cell CD40 ligand causes X-linked immunodeficiency with hyper-IgM. *Nature*, **361**, 539–41.

61 Lalmanach-Girard, A.C., Chiles, T.C., Parker, D.C. and Rothstein, T.L. (1993) T-cell-dependent induction of NF-kappa B in B cells. *J. Exp. Med.*, **177**, 1215–9.

62 Berberich, I., Shu, G.L. and Clark, E.A. (1994) Cross-linking CD40 on B cells rapidly activates nuclear factor-kappa B. *J. Immunol.*, **153**, 4357–66.

63 Barrett, T.B., Shu, G. and Clark, E.A. (1991) CD40 signaling activates CD11a/CD18 (LFA-1)-mediated adhesion in B cells. *J. Immunol.*, **146**, 1722–9.

64 Kansas, G.S. and Tedder, T.F. (1991) Transmembrane signals generated through MHC class II, CD19, CD20, CD39, and CD40 antigens induce LFA-1-dependent and independent adhesion in human B cells through a tyrosine kinase-dependent pathway. *J. Immunol.* **147**, 4094–102.

65 Ren, C.L., Morio, T., Fu, S.M. and Geha, R.S. (1994) Signal transduction via CD40 involves activation of lyn kinase and phosphatidylinositol-3-kinase, and phosphorylation of phospholipase C gamma 2. *J. Exp. Med.*, **179**, 673–80.

7 Signaling through G-protein-coupled receptors

Margaret A. Cascieri,
Tung Ming Fong,
Michael P. Graziano,
Michael R. Tota,
Mari R. Candelore
and
Catherine D. Strader

The existence of G-protein-coupled receptors was first discovered in the 1970s by Rodbell and Gilman. Rodbell and co-workers[1] discovered that glucagon activation of rat liver adenyl cyclase was modulated by GTP, but not ATP or other nucleotide triphosphates, and that the binding of glucagon to the rat liver receptor was inhibited by guanine nucleotides. Gilman and co-workers[2] showed that the binding of agonists to the β-adrenergic receptor displayed similar pharmacology. Gilman and his colleagues subsequently purified the guanine nucleotide binding protein (G protein) required to reconstitute GTP-sensitive agonist binding and stimulation of adenylyl cyclase.[3]

Subsequent to these seminal findings, the binding of many ligands of various structural classes was shown to be sensitive to guanine nucleotides. These ligands included small biogenic amines such as epinephrine and acetylcholine, small peptides such as enkephalins and substance P and large glycoproteins such as luteinizing hormone (see Table 7.1). These ligands transduce signals by various mechanisms including stimulation or inhibition of adenylyl cyclase or stimulation of calcium release via the inositol-1,4,5-trisphosphate pathway. For all of these receptors, binding of agonists was inhibited by GTP, whereas binding of antagonists was unaffected.

Within the last 10 years, the sequences of the receptors for many of these ligands have been characterized by molecular cloning techniques. The cloning of the β-adrenergic receptor in 1986 showed that the sequence had seven hydrophobic segments predicted to form trans-

Signal Transduction. Edited by Carl-Henrik Heldin and Mary Purton. Published in 1996 by Chapman & Hall. ISBN 0 412 70810 8

Table 7.1 Endogenous ligands for G-protein-coupled receptors

Small molecules	Glycoproteins	Peptides
Acetylcholine	Lutropin	Angiotensin
Adenosine	Thyrotropin	Bombesin
Adrenaline	FSH	Bradykinin
Cannabinoids		C5a
Dopamine		Calcitonin
Histamine		Cholecystokinin
Leukotrienes		Endothelin
Prostaglandins		f-MetLeuPhe
Retinal		Glucagon
Serotonin		Neurokinins
		Neuropeptide Y
		Neurotensin
		Opioids
		Oxytocin
		Parathyroid hormone
		Somatostatin
		Thrombin (amino-terminal cleavage peptide)
		Vasopressin

membrane helices, suggesting that the protein crossed the plasma membrane seven times.[4] This seven-transmembrane domain motif had been identified previously in the bacterial protein bacteriorhodopsin and in the G-protein-coupled receptor (GPCR) rhodopsin,[5] and this homology suggested a model for the structure of the β-adrenergic receptor (Figure 7.1).

The subsequent cloning of several hundred G-protein-coupled receptors has shown that this protein family is both structurally and functionally related. All display the seven-transmembrane helices motif, and the homologies between family members range from 20–30% for receptors for unrelated ligands up to 50–80% for subtypes of receptor recognizing the same or similar ligands.[6,7]

These receptors and their endogenous agonists are important for regulating the physiology of every major organ and physiological system and, as such, the receptors for these agonists are important targets for therapeutic intervention. Therapeutic agents discovered using pharmacologically defined receptors include β-adrenergic receptor blockers, histamine antagonists, anticholinergics and opiates. The molecular characterization of these receptors has shown that many more subtypes of receptors exist than were recognized by pharmacological criteria, and the distinct tissue distribution of these receptor subtypes has engendered a search for even more specific and selective therapeutic agents than was possible previously.[6]

Box 7.1 Agonists

An agonist is defined as an agent, either endogenous or synthetic, whose binding to the receptor results in activation of the G protein and the effector system. Conversely, an **antagonist** is an agent whose binding to the receptor does not result in activation. The **intrinsic efficacy** of an agent is defined as the degree of activation produced on binding, and can vary from zero (i.e. an antagonist) to 100% (i.e. a full agonist).

The guanine nucleotide binding protein family

As is reviewed in detail in Chapter 19, G proteins are heterotrimeric proteins consisting of α, β and γ subunits. They serve to transduce a signal initiated by the binding of a hormone or neurotransmitter to its receptor at the cell surface to an intracellular effector molecule. The receptor regulates a particular effector by interacting with a specific heterotrimeric G protein.[8] Currently, twenty α (i.e. α_s, α_i, α_q, etc.), five β and six γ subunits have been described. This gives rise to dozens of potential $\alpha\beta\gamma$ combinations, several of which can be expressed in a given cell type or tissue.

In some cells, it is clear that certain GPCRs will functionally interact with only a specific subset of G proteins, whereas in other cases a given receptor can interact with multiple G protein subtypes.[9] Historically, the specificity of this interaction was thought to be due solely to the interaction of the receptor with the α subunit of the G protein. Recently, the ability to coexpress specific $\beta\gamma$ pairs suggests that these subunits also contribute to the specificity of the receptor–G-protein interaction.[10,11] Immunocytochemical localization studies of both receptors and G proteins suggest that these proteins are not randomly distributed but are found in clusters and may associate with cytoskeletal elements.[12]

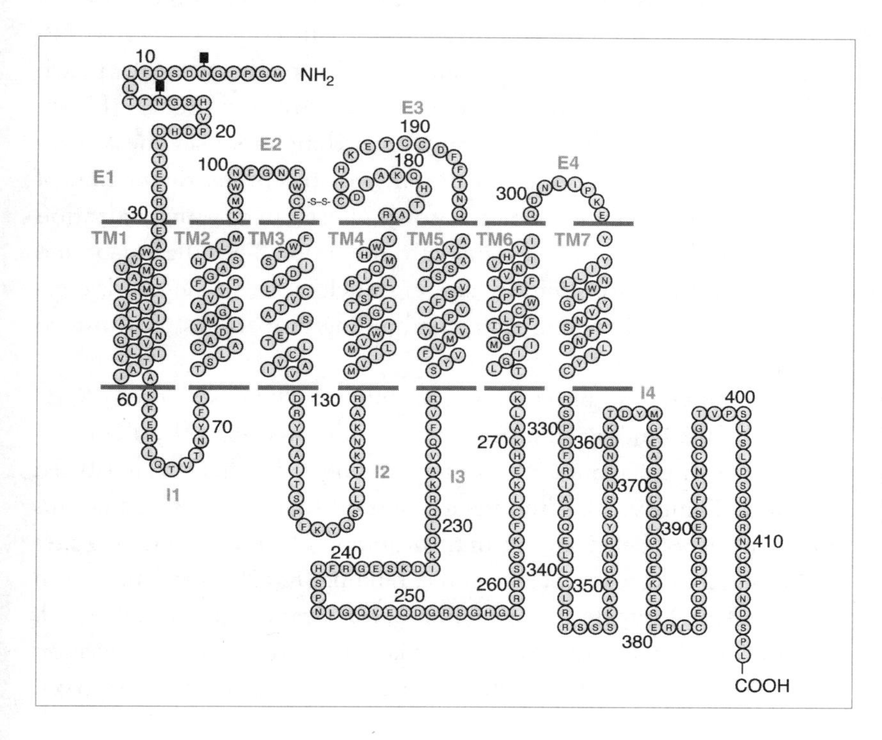

Figure 7.1 Model depicting the transmembrane topology of the β_2-adrenergic receptor. The positioning of the seven transmembrane domains was accomplished by hydropathy analysis of the primary amino acid sequence. Extended domains of hydrophobic amino acids were assigned as putative transmembrane spanning regions. Further examination of these sequences by Chou–Fasman analysis suggests that the transmembrane domains exist in an α-helical conformation. The putative transmembrane domains are labeled TM1–TM7, the extracellular domains E1–E4 and the intracellular domains I1–I4.

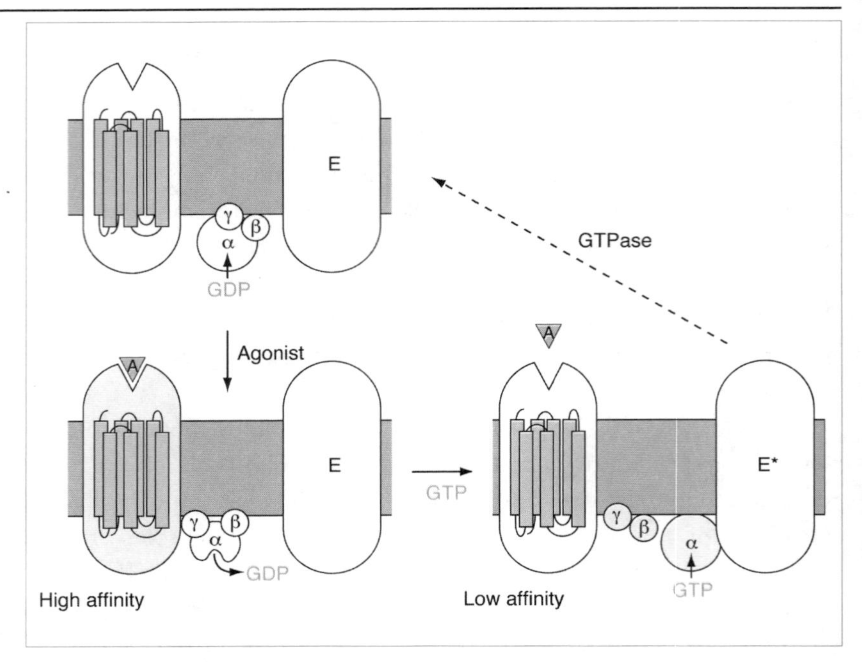

Figure 7.2 Model for the activation of G protein by GPCR. Agonist binding is hypothesized to catalyze the formation of a ternary complex between agonist, receptor and G protein that promotes the replacement of GDP with GTP in the nucleotide binding site of the α subunit. Upon GTP binding, the α subunit dissociates from the βγ subunits and from the receptor, resulting in the shift of the receptor conformation into the state with low affinity for agonist and in the activation of the effector system by the α or the βγ subunits. The GTPase activity of the α subunit returns the system to its basal state.

Activation of G-protein-coupled receptors

The receptor is activated by agonist binding and then interacts with and activates the G protein by increasing the rate of dissociation of a tightly bound molecule of GDP. Another guanine nucleotide (usually GTP) then binds and causes the G-protein subunits to dissociate from each other and the receptor, thus completing the activation process (Figure 7.2). The activated G-protein subunits bind to and modulate the activity of various effector enzymes, ion channels or receptors. Interaction of receptor and G protein has bidirectional effects, in that the activation of each protein affects ligand binding to the other. Thus, ligand-bound receptor activates G protein by altering nucleotide binding and, conversely, activated G protein alters the binding properties of agonists to the receptors.

The agonist may display two or three binding affinities for the GPCR. These multiple binding affinities are most often visualized by a competition curve where agonist inhibits the binding of a radioactively labeled antagonist (Figure 7.3a). Although it is possible to use increasing amounts of radioactively labeled agonist until saturation of all of the binding sites is achieved, the detection of low affinity binding (K_d>20 nM) is impractical by methods that rely on the separation of free ligand from bound ligand. Radioactively labeled agonists also display multiexponential dissociation rates, suggestive of multiple affinity states (Figure 7.3b). The observed

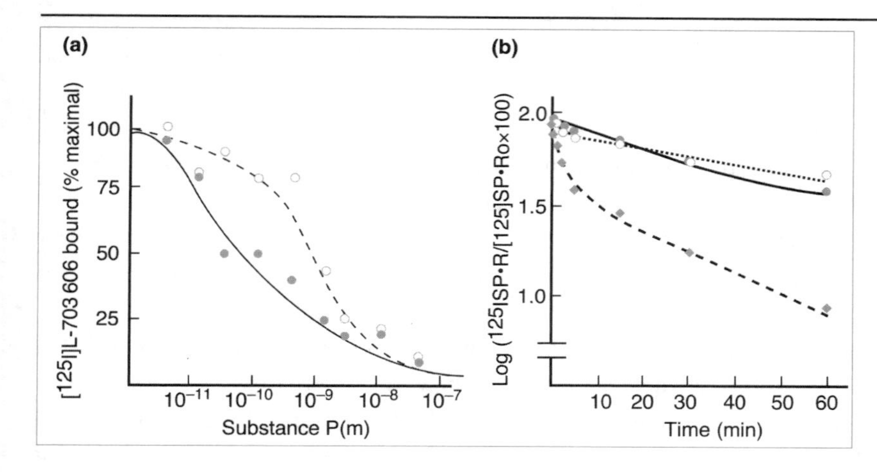

Figure 7.3 Demonstration of multiple affinity binding sites for substance P to the NK1 receptor. (a) Inhibition of binding of the radioactively labeled antagonist [^{125}I]L-703 606 to the human NK1 receptor by SP in the presence (open circles) and absence (closed circles) of the non-hydrolyzable GTP analog, guanylyl-5′-(β,γ-imido)diphosphate [Gpp(NH)p]. In the absence of Gpp(NH)p, the data are best described by a two-site fit, with SP having affinities of 40 pM and 1.5 nM for the high- and low-affinity sites, respectively. In the presence of Gpp(NH)p, which irreversibly binds to G protein, thereby locking the system in the state with low affinity for agonist, >90% of the sites are shifted to the low-affinity state. (b) Dissociation of [^{125}I]SP from the human NK1 receptor in the absence (closed circles) or presence of Gpp(NH)p (diamonds) or the competitive agonist L-703 606 (open circles). [^{125}I]SP dissociates from the receptor in a biphasic manner, with dissociation rates of 0.015 and 4 min^{-1} for the high- and low-affinity states of the receptor, respectively. Gpp(NH)p increases the percentage of the receptor in the low-affinity state from 10 to 45% of the receptor population. In contrast, L-703 606 has no effect on the dissociation of ligand from the receptor. [Reproduced with permission from Cascieri, M.A. *et al.* (1992) Characterization of the binding of a potent, selective radioiodinated antagonist to the human neurokinin-1 receptor. *Mol. Pharmacol.*, 42, 458–63.]

heterogeneity in agonist binding affinities has been well studied. Selective affinity labeling studies indicated that the high-affinity and low-affinity states were not interconvertible as a function of agonist concentration, suggesting that there were two (or more) subpopulations of receptors.[13,14] However, the cloning of the receptors revealed that a single subtype still displayed multiaffinity agonist binding.

It is now known that conversion of high- to low-affinity binding is mediated by guanine nucleotides. The state of the receptor with high-affinity for agonist is believed to be the form that is coupled to the G protein. Activation of the G protein by GTP (or other guanine nucleotides) not only dissociates the G-protein subunits, but also uncouples the G protein from the receptor, converting the latter into a state with low binding affinity for agonist. A kinetic model has been formulated to account for the guanine-nucleotide-mediated conversion of high-affinity to low-affinity binding in the β-adrenergic receptor[14,15] and the muscarinic receptor.[16] This concept was further tested using purified components. Receptor in the absence of G protein usually exists in a low-affinity form. Reconstitution of purified receptor and G protein restores guanine nucleotide-sensitive high-affinity agonist binding.[17–19]

The nature of the conversion from high to low affinity is fundamental to understanding how agonists function. Which regions of the receptor are involved in G-protein coupling? Site-directed mutagenesis of many GPCRs has shown that regions on the cytoplasmic face of the receptor are critical determinants of receptor–G protein interaction.[7] Removal of parts of the third intracellular loop (i.e. the region between transmembrane domains 5 and 6) has been shown to uncouple agonist binding from effector activation. In other cases, substitution of these sequences with those of another GPCR has been shown to alter the specificity of G protein coupling. While the specific amino acid sequences in these loop regions can vary, modeling studies suggest that the regions critical for G protein activation are amphipathic α helices. In addition, mutation in this region can induce constitutive activation of the receptor, indicating that disruption of local conformation can lead to G-protein activation.[20] This suggests that the specific secondary and tertiary structure of these loops may represent a common motif for regions of GPCRs that interact with G proteins.

Recent data suggest that antagonists are not always molecules that bind to the receptor with high affinity in a manner that will not activate the receptor. Antagonists of several GPCRs inhibit the basal activity of the receptor, suggesting that they actively promote the dissociation of GPCR from the G protein.[20,21] These so-called inverse agonists suggest that the model depicted in Figure 7.2 may be oversimplified. Many GPCRs may exist in a conformation capable of interaction with G protein in the absence of agonist, and the equilibrium between this conformation and the conformation that does not interact with G protein is altered by agonist or antagonist binding.

Structure–function analysis of G-protein-coupled receptors

One way to probe the ligand binding site on the receptor is through site-directed mutagenesis, although the data need to be interpreted with caution. Since the affinity of the receptor for agonist depends on both intrinsic binding affinity and the ability of the receptor to couple to G protein, the observed apparent affinity is a function of the intrinsic affinity and the equilibrium constant for receptor coupling. Mutations could affect the transition between the active and inactive states, which could be reflected as a change in the observable apparent affinity. Since the activation transition cannot be measured directly, this issue can be addressed qualitatively by determining the change in apparent binding affinity relative to the change in activation potency and maximal response, because any impairment of the receptor activation transition will be reflected in the magnitude of functional response.

After the effect of a given mutation on binding has been established, the second issue is how to determine whether the residues interact directly with the ligand or indirectly through other receptor residues. It is important to remember that mutational analysis is not structural analysis, and therefore it cannot prove unambiguously a direct ligand–residue contact. However, carefully designed experiments and multiple lines of evidence can often provide consistent data that strongly support a direct interaction. The best way to corroborate the receptor mutational analysis is to analyze various modified ligands in parallel.

Molecular pharmacology of the biogenic amine-binding site

The β_2-adrenergic receptor (βAR) was the first GPCR to be cloned and analyzed by these methods. This receptor binds catecholamine agonists, such as epinephrine (Figure 7.4a), and activates G_s to stimulate adenylyl cyclase. Scanning deletion mutagenesis, in which each of the transmembrane domains and connecting hydrophilic loops was deleted independently from the βAR, showed that the extracellular and intracellular loops were not required for ligand binding to the receptor.[22] From these studies, it seemed likely that the ligand binding site was located in the hydrophobic transmembrane core of the receptor in a pocket formed by the side chains of amino acids on various transmembrane helices. Medicinal chemistry efforts focused on the design of β-adrenergic agonists and antagonists[23] had shown that a protonatable amine group was critical for the activity of both agonists and antagonists at the βAR, suggesting that an acidic moiety in the binding site of the receptor provided a counterion for this amine. By replacing each of the acidic amino acids in the transmembrane

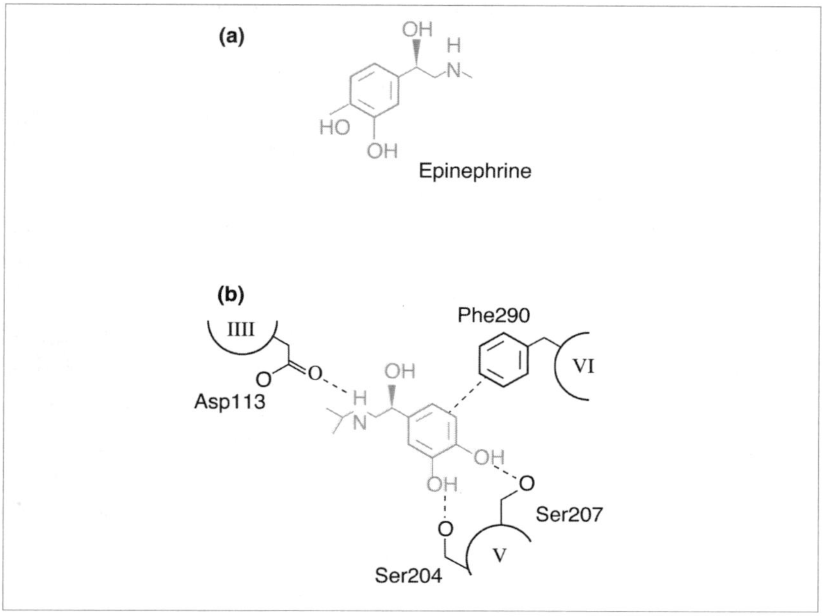

Figure 7.4 (a) Structure of the endogenous catecholamine agonist epinephrine. (b) Model of the ligand binding domain of the βAR, viewed from the extracellular face of the membrane, with the agonist isoproterenol docked in the binding site. Interactions with Asp113, Ser204, Ser207 and Phe290 are shown.

domain with neutral amino acids and analyzing the effects of such mutations on the ligand binding properties of the receptor, Asp113 in transmembrane helix 3 was identified as a likely candidate for this role.[24] To test this hypothesis, Asp113 was substituted with a Ser residue, which would be able to donate or accept a hydrogen bond from a ligand, but not participate in ion pairing. This mutant receptor was activated by analogs of epinephrine in which the ethanolamine side chain was replaced by hydrogen bond-accepting moieties, whereas these compounds failed to bind to the wild-type receptor.[25] Thus, simultaneously replacing Asp113 in the receptor and the amine group of the ligand generated a mutant receptor that could be activated by the modified neurotransmitter. This 'two-dimensional' mutagenesis approach, combining molecular biological modification of the receptor with chemical modification of the ligand, provides strong evidence for a specific ion pair linking the protonated amine group of epinephrine to the side chain of Asp113 in transmembrane helix 3 of the receptor.

Similar experiments have indicated that the catechol ring of the agonist binds to a pocket formed by residues in transmembrane helices 5 and 6 of the β-adrenergic receptor (Figure 7.4b). Two-dimensional mutagenesis experiments have suggested that the *meta*-hydroxyl group of the ligand forms a hydrogen bond with the hydroxyl side chain of Ser204 in transmembrane helix 5 of the receptor, whereas the *para*-hydroxyl group interacts with the side chain of Ser207.[26] The aromatic catechol-containing ring is held in place by an interaction with Phe290 in transmembrane helix 6.[27] Thus, the mutagenesis studies indicate that the binding site of the βAR is formed by residues from several transmembrane domains, buried approximately 30% into the membrane bilayer. This localization has been supported by biophysical studies using a fluorescent β-adrenergic ligand, in which the binding site was determined to be inaccessible to the solvent phase and buried at least 11 Å into the protein core.[28] Mutagenesis experiments on other GPCRs that bind biogenic amines have supported this model.[7]

The orientation of the catechol ring of the agonist in the ligand binding pocket of the receptor is constrained by the positions of three key residues: Ser204 and Ser207, located one helical turn apart in transmembrane helix 5 of the receptor, and Phe290, located at approximately the same distance from the surface in transmembrane helix 6. It is known that the aromatic catechol moiety of epinephrine is essential for its agonist activity,[23] and mutagenesis studies have shown that antagonists do not make these contacts with the helix 5–6 region of the receptor.[26] These data suggest that receptor activation by the agonist involves a conformational change, initiated by the contact of the catechol ring with these residues in helices 5 and 6, which functions to transfer the signal from the ligand binding site in the core of the receptor to the intracellular regions. This mechanism is in contrast to that defined for some other receptor classes where the ligand initiates a dimerization of the receptor, which then initiates signal transduction (see Chapters 1 and 5).

Molecular pharmacology of peptide receptors

Ligands that activate GPCRs can be divided into three groups: small molecules (molecular mass<300), peptides (3–40 amino acid residues) and glycoproteins (molecular mass 9000–40 000) (Table 7.1). Such a wide range of size distribution for ligands contrasts sharply with the relatively conserved size distribution of GPCRs that can be divided into two groups: one with a single structured domain of seven putative transmembrane helices (350–450 amino acid residues) and the other with an amino-terminal extended form of the heptahelical domain with a large amino-terminal domain of 400–600 amino acid residues. As discussed in the previous section, the binding site for most small-molecule ligands resides in the transmembrane region. However, the larger size of peptides and glycoprotein hormones raises the question of whether the transmembrane region alone would be sufficient for binding of large ligands.

For example, the neurokinin-1 receptor (NK1R) binds and is activated by the undecapeptide substance P (SP). The overall size of SP is about eight times that of epinephrine, and it seems unlikely that the cavity in the transmembrane region of the NK1 receptor is capable of accommodating the entire peptide. The structure–activity analysis of SP and related analogs[29] suggests that most of the peptide is involved in binding interactions with the NK1 receptor. Alanine substitution of selective residues in the extracellular region reduces SP affinity for the NK1 receptor without altering the affinities of several non-peptide antagonists.[30] These data suggest that part of the extracellular region is specifically required for the binding of peptides. Similarly, the large amino-terminal domains of glycoprotein hormone receptors are required for the binding of glycoprotein ligands.[31]

The carboxy-terminal amide of SP has been proposed to bind in the vicinity of Asn85 of the second transmembrane helix of the NK1 receptor.[32] An analog of SP in which the carboxy-terminal amide has been replaced with a methyl ester has a lower affinity than SP, suggesting that the amide is favored in receptor binding. When Asn85 was replaced with Ala, however, SP and many analogs with carboxy-terminal amides exhibit reduced affinity, whereas SP methyl ester binds to the Asn85Ala mutant with a higher affinity than that for the wild-type receptor. Since SP methyl ester shows a higher affinity for only the Asn85Ala mutant, the replacement of Asn85 with Ala appears to create a larger cavity accommodating the larger methyl ester group.

A schematic model of SP binding to the NK1 receptor is shown in Figure 7.5. Recent photoaffinity labeling studies also support the proximity of certain residues in a bound SP molecule to the extracellular regions of the NK1 receptor.[33] In some cases (such as the thrombin and IL-8 receptors), the extracellular region of peptide receptors also determines ligand binding specificity.[34,35] On the other hand, some naturally occurring small ligands may bind to their receptors in a manner that is significantly different from biogenic amines. For example, the amino-terminal region of the metabotropic glutamate receptor (mGluR) appears to play a role in determining agonist binding specificity.[36] Furthermore, the amino-terminal region in the mGluR has been modeled after the bacterial periplasmic binding proteins,[37] suggesting that the glutamate binding site is contained within the amino-terminal region. These studies suggest that the mGluRs may have evolved via the combination of an ancient soluble ligand binding protein and an ancient GPCR. This speculation is also consistent with the primary structure of the calcium-sensing receptor that has significant

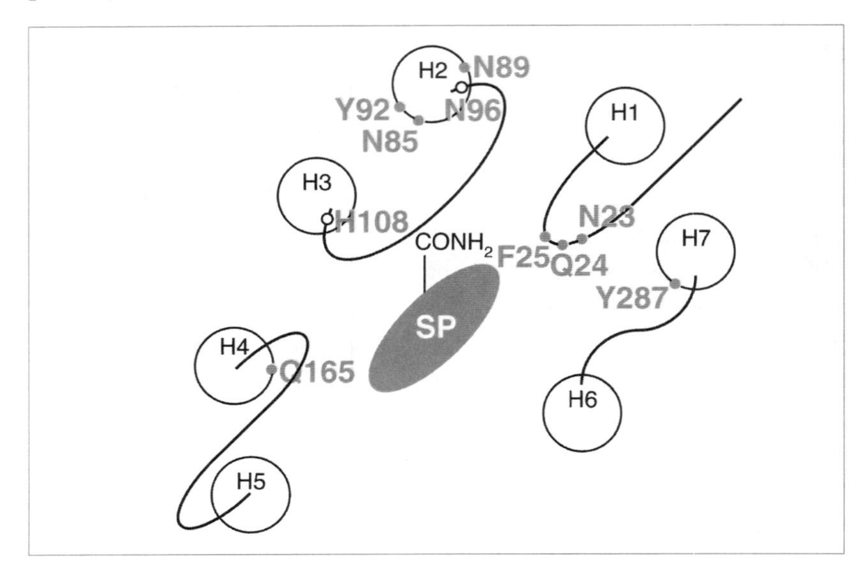

Figure 7.5 Model of the binding of SP with the NK1 receptor, as viewed from the extracellular domain. Residues whose substitution by alanine results in reduced affinity for substance P are shown. Most of these residues are predicted to be in the extracellular domain or the interface between the helical and extracellular domain. However, data suggest that the amidated carboxyl terminus of substance P penetrates the transmembrane domain and interacts in the vicinity of Asn85.

sequence similarity with mGluRs.[38] Therefore, the interaction between the large amino-terminal domain and the transmembrane domain in these receptors may be responsible for transmitting the conformational change from the ligand binding site in the amino-terminal domain to the transmembrane domain, and peptide–receptor interaction may serve as a prototype for such a mechanism of action.

Antagonist binding site for peptide receptors

In contrast to the data described above for the binding site for peptide agonists, non-peptide antagonists bind to the same receptors within the transmembrane domain. In the case of the NK1 receptor, five residues in transmembrane helices 4, 5, 6 and 7 have been identified that appear to interact directly with antagonists of various structural classes.[39] For example, the transmembrane residue His197 in the NK1 receptor has been proposed to interact with the antagonist CP-96 345 based on the observation that the His197Ala mutation has reduced binding affinity for CP-96 345.[40] Key functional groups in CP-96 345 were sequentially replaced with hydrogen and the affinities of the resulting compounds for the wild-type and the His197Ala mutant receptors were determined. If His197 interacts with a specific functional group, a compound lacking that group would be expected to have the same affinity for both the wild-type and the His197Ala mutant because the proposed interaction would no longer be present. Based on this lack of additivity principle, the benzhydryl group of CP-96 345 was proposed to maintain a direct interaction with His197.[40] Similar experimental strategies were utilized to identify several other contact points between CP-96 345 and the receptor (Figure 7.6). Further studies demonstrated that several non-peptide SP antagonists bind to the transmembrane region of the NK1 receptor in a manner that is topologically similar to ligands of the β_2 adrenergic receptor, although the exact molecular contacts for each small molecule can be very different.[41,42] In addition, recent mutagenesis studies on the AT_1 angiotensin receptor and the κ opioid receptor also support the localization of the binding site for non-peptide antagonists in the transmembrane region.[43,44] Although the currently available non-peptide antagonists appear to interact within the transmembrane domain, this does not preclude the possibility that novel, hydrophilic non-peptide antagonists that bind to the extracellular region of peptide receptors will be discovered.

It is important to note that the non-peptide antagonists and peptide

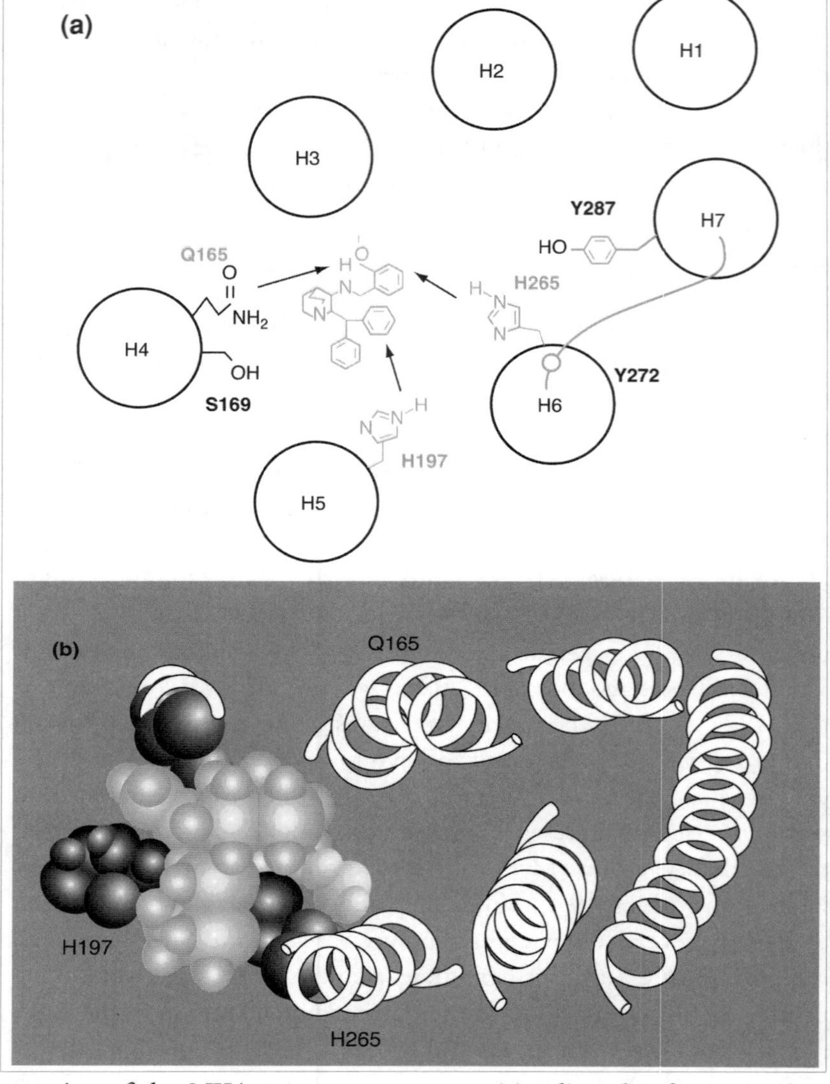

Figure 7.6 Model of the interaction of CP-96 345 with the NK1 receptor. (a) Schematic diagram of the interaction of the antagonist CP-96 345 with residues His197, His265 and Gln165 in transmembrane domains 5, 6 and 4 of the receptor. (b) A three-dimensional model of the putative binding site looking down into the membrane from the extracellular domain of the receptor. (Reproduced with permission from Fong, T.M. *et al.*, (1994) *J. Biol. Chem.*, 269, 14957–61.)

agonists of the NK1 receptor are competitive ligands of one another even though their binding sites are not identical. Thus, the effects of antagonist can be overcome by the addition of excess agonist, and the two ligands do not occupy the receptor at the same time.

Biophysical approaches to the study of receptor structure

As noted above, mutagenesis studies, although useful to support models of receptor structure and function, are not as definitive as structural studies.

To this end, several groups have recently employed fluorescently labeled agonists as tools for both structural and kinetic analysis.[45–47] These studies have used highly fluorescent fluorescein-labeled peptide agonists that can be used in membranes or cells, where the receptor makes up a small percentage of the total protein. The fluorescent probe is capable of reporting its local environment as a function of hydrophobicity, pH, local mobility and accessibility to solvent. Changes in fluorescence intensity and polarization can be used to monitor ligand binding for kinetic studies. In addition to mapping regions of agonist binding, more detailed ligand binding studies may be performed. The use of fluorescently labeled ligands (and fluorescently labeled receptor itself) may be merged with studies using fluorescently labeled G proteins[12] to correlate agonist binding, G-protein coupling and conformational changes.

The current potential to overexpress receptors in both mammalian and non-mammalian systems will hopefully lead to the ability to obtain structural information by the traditional X-ray crystallographic or NMR approaches. However, many technical problems relating to the crystallization of membrane proteins remain to be solved before these efforts reach fruition.

Summary

The molecular cloning of GPCRs and G proteins has led to a wealth of data regarding the structural and functional homologies of receptors for a wide variety of agonists. The distribution and specificity of these receptors has suggested many opportunities to utilize GPCRs as therapeutic targets with the potential to provide more selective and efficacious therapies for many diseases. The data from mutagenesis experiments support a model for a common binding site within the transmembrane domain of GPCRs that has the potential for interactions with a structurally diverse series of small molecule agonists and antagonists. Such models are useful both for understanding the nature of binding and receptor activation and for providing a framework for the design of more potent and selective agents with therapeutic potential.

References

1 Lin, M.C., Nicosia, S., Lad, P.M. and Rodbell, M. (1977) Effects of GTP on binding of tritiated glucagon to receptors in rat hepatic plasma membranes. *J. Biol. Chem.*, **252**, 2790–2.

2 Maguire, M.E., Van Arsdale, P.M. and Gilman, A.G. (1976) An agonist specific

effect of guanine nucleotides on binding to the beta adrenergic receptor. *Mol. Pharmacol.*, **12**, 335–9.

3 Northup, J.K., Sternweis, P.C., Smigel, M.D. *et al.* (1980) Purification of the regulatory component of adenylate cyclase. *Proc. Natl. Acad. Sci. USA*, **77**, 6516–20.

4 Dixon, R.A.F., Kobilka, B.K., Strader, D.J. *et al.* (1986) Cloning of the gene and cDNA for mammalian β-adrenergic receptor and homology with rhodopsin. *Nature*, **321**, 75–9.

5 Henderson, R., Baldwin, J.M., Ceska, T.A. *et al.* (1990) Model for the structure of bacteriorhopdopsin based on high resolution electron cryomicroscopy. *J. Mol. Biol.*, **213**, 899–929.

6 Gluchowski, C., Branchek, T.A., Weinshank, R.L. and Hartig, P.R. (1993) Molecular/cell biology of G-protein coupled CNS receptors. *Annu. Rep. Med. Chem.*, **28**, 29–38.

7 Strader, C.D., Fong, T.M., Tota, M.R. *et al.* (1994) Structure and function of G-protein coupled receptors. *Ann. Rev. Biochem.*, **63**, 101–32.

8 Neer, E.J. (1995) Heterotrimeric G proteins: organizers of transmembrane signals. *Cell*, **80**, 249–57.

9 Jelinek, L.J., Lok, S., Rosenberg, G.B. *et al.* (1993) Expression cloning and signaling properties of the rat glucagon receptor. *Science*, **259**, 1614–6.

10 Kleuss, C., Scherubl, H., Hescheler, J. *et al.* (1992) Different βγ subunits determine G-protein interaction with transmembrane receptors. *Nature*, **358**, 424–6.

11 Kleuss, C., Scherubl, H., Hescheler, J. *et al.* (1993). Selectivity in signal transduction determined by γ subunits of heterotrimeric G proteins. *Science*, **252**, 668–74.

12 Neubig, R.R. (1994) Membrane organization in G protein mechanisms. *FASEB J.*, **8**, 939–46.

13 Nathanson, N.M. (1987) Molecular properties of the muscarinic receptor. *Annu. Rev. Neurosci.*, **10**, 195–236.

14 De Lean A., Stadel, J.M. and Lefkowitz, R.J. (1980) A ternary complex model explains the agonist-specific binding properties of the adenylate cyclase-coupled β-adrenergic receptor. *J. Biol. Chem.*, **255**, 7108–17.

15 Samama, P., Cotecchia, S., Costa, T. and Lefkowitz, R.J., (1993) A mutation-induced activation state of the β2-adrenergic receptor: extending the ternary complex model. *J. Biol. Chem.*, **268**, 4625–36.

16 Ehlert, F.J. (1985) The relationship between muscarinic receptor occupancy and adenylate cyclase inhibition in the rabbit myocardium. *Mol. Pharmacol.*, **28**, 410–21.

17 Cerione R.A., Codina, J., Benovic, J.L. *et al.* (1984) The mammalian beta 2-adrenergic receptor: reconstitution of functional interactions between pure receptor and pure stimulatory nucleotide binding protein of the adenylate cyclase system. *Biochemistry*, **23**, 4519–25.

18 Haga, K., Haga, T. and Ichiyama, A. (1986) Reconstitution of the muscarinic acetylcholine receptor. Guanine nucleotide-sensitive high affinity binding of agonists to purified muscarinic receptors reconstituted with GTP-binding proteins (Gi and Go). *J. Biol. Chem.*, **261**, 10133–40.

19 Tota, M.R., Kahler, K.R. and Schimerlik, M.I. (1987) Reconstitution of the purified porcine atrial muscarinic acetylcholine receptor with purified porcine atrial guanine nucleotide binding protein. *Biochemistry*, **26**, 8175–82.

20 Lefkowitz, R.J., Cotecchia, S., Samama, P. and Costa, T. (1993) Constitutive activity of receptors coupled to guanine nucleotide regulatory proteins. *Trends Pharmacol. Sci.* **14**, 303–7.

21 Schutz, W. and Freissmuth, M. (1992) Reverse intrinsic activity of antagonists on G-protein coupled receptors. *Trends Pharmacol.* **13**, 376–80.

22 Dixon, R.A.F., Sigal, I.S., Rands, E. *et al.* (1987) Ligand binding to the β-adrenergic receptor involves its rhodopsin-like core. *Nature*, **326**, 73–7.

23 Main, B.G. and Tucker, H. (1985) Recent advances in β-adrenergic blocking agents. *Prog. Med. Chem.*, **22**, 122–64.

24 Strader, C.D., Sigal, I.S., Candelore, M.R. *et al.* (1988) Conserved aspartic acid residues 79 and 113 of the beta-adrenergic receptor have different roles in receptor function. *J. Biol. Chem.*, **263**, 10267–71.

25 Strader, C.D., Gaffney, T.G., Sugg, E.E. *et al.* (1991) Allele-specific activation of genetically engineered receptors. *J. Biol. Chem.*, **266**, 5–8.

26 Strader, C.D., Candelore, M.R., Hill, W.S. *et al.* (1989) Identification of two serine residues involved in agonist activation of the β-adrenergic receptor. *J. Biol. Chem.*, **264**, 13572–8.

27 Dixon, R.A.F., Sigal, I.S. and Strader, C.D. (1988) Structure–function analysis of the β-adrenergic receptor. *Cold Spring Harbor Symp. Quant. Biol.*, **53**, 487–97.

28 Tota, M.R. and Strader, C.D. (1990) Characterization of the binding domain of the β-adrenergic receptor with the fluorescent antagonist carazolol: evidence for a buried ligand binding site. *J. Biol. Chem.*, **28**, 16891–7.

29 Regoli, D., Escher, E. and Mizrahi, J. (1984) Substance P – structure–activity studies and the development of antagonists. *Pharmacology*, **28**, 301–320.

30 Fong, T.M., Huang, R.-R.C. and Strader, C.D. (1992) Localization of agonist and antagonist binding domains of the human neurokinin-1 receptor. *J. Biol. Chem.*, **267**, 25664–7.

31 Wadsworth, H.L., Chazenbalk, G.D., Nagayama, Y. *et al.* (1990) An insertion in the human thyrotropin receptor critical for the high affinity hormone binding. *Science*, **249**, 1423–5.

32 Huang, R.R.C., Yu, H., Strader, C.D. and Fong, T.M. (1994) Interaction of substance P with the second and seventh transmembrane domains of the neurokinin-1 receptor. *Biochemistry*, **33**, 3007–13.

33 Li, Y.-M., Marnerakis, M., Stimson, E.R. and Maggio, J.E. (1995) Mapping peptide-binding domains of the substance P (NK1) receptor from P388D1 cells with photolabile agonists. *J. Biol. Chem.*, **270**, 1213–20.

34 Gerszten, R.E., Chen, J., Ishli, M. *et al.* (1994) Specificity of the thrombin receptor for agonist peptide is defined by its extracellular surface. *Nature*, **368**, 648–51.

35 LaRosa, G.L., Thomas, K.M., Kaufmann, M.E. *et al.* (1992) Amino terminus of

the interleukin-8 receptor is a major determinant of receptor subtype specificity. *J. Biol. Chem.*, **267**, 25402–6.

36 Takahashi, K., Tsuchida, K., Tanabe, Y. *et al.* (1993) Role of the large extracellular domain of the metabotropic glutamate receptors in agonist selectivity determination. *J. Biol. Chem.*, **268**, 19341–5.

37 O'Hara, P.J., Sheppard, P.O., Thogersen, H. *et al.* (1993) The ligand-binding domain in metabotropic glutamate receptors is related to bacterial periplasmic binding proteins. *Neuron*, **11**, 45–52.

38 Brown, E.M., Gamba, G., Riccardi, D. *et al.* (1993) Cloning and characterization of an extracellular calcium sensing receptor from bovine parathyroid. *Nature*, **366**, 575–80.

39 Cascieri, M.A., Fong, T.M. and Strader, C.D. (1995) Molecular characterization of a common binding site for small molecules within the transmembrane domain of G-protein coupled receptors. *J. Pharmacol. Toxicol. Methods*, **33**, 179–85.

40 Fong, T.M., Cascieri, M.A., Yu, H. *et al.* (1993) Amino-aromatic interaction between histidine-197 of the human neurokinin-1 receptor and CP-96,345. *Nature*, **362**, 350–3.

41 Fong, T.M., Yu, H., Cascieri, M.A. *et al.* (1994) Interaction of glutamine-165 in the fourth transmembrane segment of the human neurokinin-1 receptor with quinuclidine antagonists. *J. Biol. Chem.*, **269**, 14957–61.

42 Cascieri, M.A., Macleod, A.M., Underwood, D. *et al.* (1994) Characterization of the interaction of *N*-acyl-L-tryptophan benzyl ester neurokinin antagonists with the human neurokinin-1 receptor. *J. Biol. Chem.*, **269**, 6587–91.

43 Ji, H., Leung, M., Zhang, Y. *et al.* (1994) Differential structural requirements for specific binding of nonpeptide and peptide antagonists to the AT_1 angiotensin receptor. *J. Biol. Chem.*, **269**, 16533–6.

44 Xue, J.-C., Chen, C., Zhu, J. *et al.* (1994) Differential binding domains of peptide and non-peptide ligands in the cloned rat κ opioid receptor. *J. Biol. Chem.*, **269**, 30195–9.

45 Sklar, L.A., Fay, S.P., Seligmann, B.E. *et al.* (1990) Fluorescence analysis of the size of a binding pocket of a peptide receptor at natural abundance. *Biochemistry*, **29**, 313–6.

46 Neubig, R.R. and Sklar, L.A. (1993) Subsecond modulation of formyl peptide-linked guanine nucleotide-binding proteins by guanosine 5′-O-(3-thio)triphosphate in permeabilized neutrophils. *Mol. Pharmacol.*, **43**, 734–40.

47 Tota, M.R., Daniel, S., Sirotina, A. *et al.* (1994) Characterization of a fluorescent substance P analog. *Biochemistry*, **33**, 13079–86.

8 Receptor-mediated endocytosis of growth factors

Alexander Sorkin

Many macromolecules gain entry to cells via receptor-mediated endocytosis. The macromolecule (or ligand) binds to specific receptors on the cell surface and this complex is carried into the cell by endocytosis, a process by which the plasma membrane pouches inward and pinches off vesicles. These endocytic vesicles fuse with specialized membrane organelles known as endosomes which deliver the receptors and their ligands to various intracellular destinations. Receptors can recycle back to the cell surface and participate in several rounds of endocytosis. Endocytosis, recycling, and therefore surface expression of many receptors are not affected by ligand binding. However, for receptors that transduce signals across the membrane, such as receptors for polypeptide growth factors (GFs), ligand binding results in a rapid loss of receptors from the surface and reduction of the total cellular receptor pool by as much as 70–90%[1–3] (see Figure 8.1). This 'down-regulation' has been found to be the consequence of accelerated endocytosis and lysosomal degradation of the receptors.[1–7] Endocytosis of GF receptors which possess intrinsic tyrosine kinase activity remains the classical model system in which to study down-regulation. Receptors for cytokines that become associated with cytoplasmic tyrosine kinases after ligand binding are also rapidly down-regulated in the presence of the ligand (for the most recent data, see refs 8 and 9).

The specific down-regulation of GF receptors, which have primarily mitogenic function, suggests that GF-induced endocytosis may play an important role in the control of mitogenic signaling. However, despite intensive study for almost 20 years, neither the molecular mechanisms by which GFs trigger receptor down-regulation nor the biological role of this phenomenon are fully understood. In this review, the general

Signal Transduction. Edited by Carl-Henrik Heldin and Mary Purton. Published in 1996 by Chapman & Hall. ISBN 0 412 70810 8

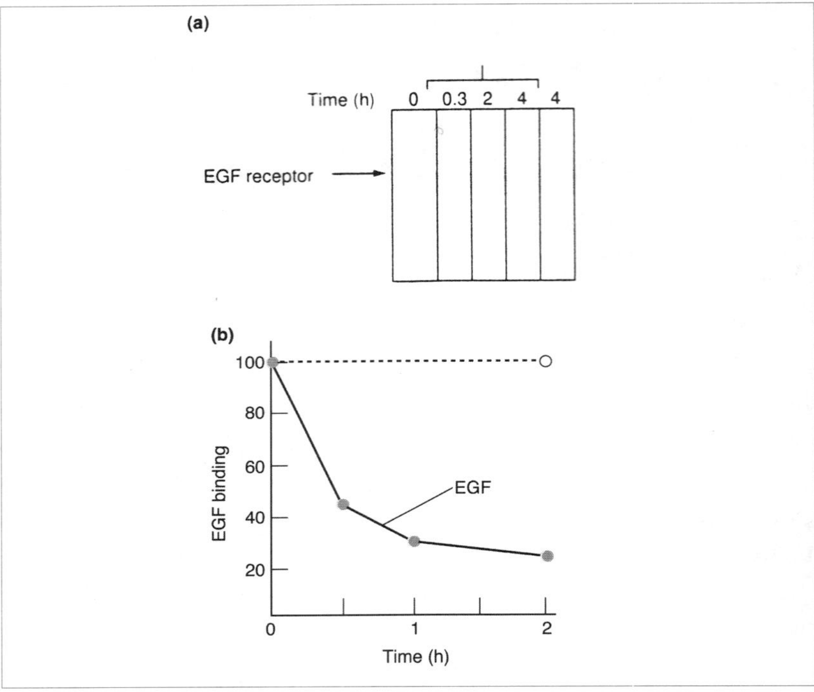

Figure 8.1 Down-regulation of EGF receptors by EGF. Mouse NR6 cells that express transfected human EGF receptors (10^5 per cell) were incubated with an excess of EGF at 37 °C. The total amount of EGF receptor protein (a), and the number of EGF binding sites on the cell surface (b) were measured at various time intervals using Western blotting and an ^{125}I-EGF binding assay, respectively, as described in ref. 25. Quantitation of the data shown in (a) revealed that the total cellular pool of EGF receptors was decreased by 70% after a 4 h incubation with EGF.

pathways of receptor-mediated endocytosis will be described, with the focus on the mechanisms that are unique for GF receptors. We shall build our discussion around endocytosis of the epidermal growth factor (EGF) receptor, since studies of this receptor have produced most of the original data on which the current model of GF receptor endocytosis is based.

Receptor recruitment into clathrin-coated pits

The formation of endocytic vesicles from the plasma membrane is a constitutive part of the recycling of cellular membranes. However, vesicle budding is significantly accelerated in specialized regions of the membrane which are coated from inside with a lattice of the protein clathrin and hence termed 'clathrin-coated pits'[10] (see Box 8.1). Although endocytosis of some receptors can occur in the absence of functional coated pits, clathrin-dependent endocytosis is the most efficient and fastest pathway. Coated pits recruit receptors with high efficiency and selectivity because they can recognize 'endocytic codes', sequence motifs in the cytoplasmic domains of the receptors that typically contain aromatic residues.[11] One group of receptors (type I) is clustered in coated pits and rapidly internalized even when no ligand

Box 8.1 Clathrin

Clathrin consists of three copies each of heavy chain (~190 kDa) and light chain (~23–27 kDa), forming a three-legged structure called a triskelion. Clathrin triskelions are the assembly units of the polygonal lattice on the surface of coated pits and coated vesicles. Clathrin assembles into coats on the cytoplasmic side of the plasma membrane by interacting with its adaptor protein complexes (AP-2s). AP-2 is a heterotetramer consisting of two large (~100 kDa), one medium (50 kDa) and one small (17 kDa) subunits which can be separated only under strongly denaturing conditions.

has bound. The well-known examples of this type are the receptors for transferrin, mannose-6-phosphate and low-density lipoprotein. Signaling receptors, for example epidermal growth factor (EGF) receptors, are type II receptors which are efficiently concentrated in coated pits and internalized only when occupied with the specific ligand.

In the absence of EGF, EGF receptors are diffusely distributed at the cell surface.[2,5,12] A small pool of unoccupied receptors can be detected in coated pits.[4,5] The relative size of this pool appears to be dependent on cell type and level of receptor expression.[2,4,13] Although quantitative studies of the distribution of EGF receptors in coated and uncoated regions of the plasma membrane are limited, several reports clearly demonstrated rapid aggregation and accumulation of EGF receptors in coated pits upon EGF binding[2,5,12–14] (Figure 8.2).

Coated pits can contain more than one type of receptor which can be internalized together. Conformational rearrangements within the clathrin lattice result in deep invagination of the coated pit and fission of the coated endocytic vesicle. Clathrin and associated proteins must then return back to the plasma membrane to form another coated pit (Figure 8.2). One or more stages of the coated pit cycle require energy and physiological temperature. Importantly, whereas late steps of internalization appear to be common for all types of receptors, receptor recruitment into coated pits is the rate-limiting and ligand-dependent step of the internalization of GF receptors.[14,15]

It has recently been proposed that EGF binding to the receptor results in a high-affinity interaction of receptors with clathrin adaptor protein complex AP-2[16] (see Box 8.1) AP-2 is essential for the assembly of the clathrin coat on the inner surface of the membrane[17] and has also been credited with the function of the selective recruitment of receptors bearing endocytic codes.[17] Although EGF receptor can bind AP-2 directly *in vivo* and *in vitro*,[18,19] it is not formally proven whether this interaction is important and sufficient for EGF-induced recruitment of receptors into coated pits.[14] The interactions of other GF receptors with AP-2 have not been demonstrated.

Kinetic analysis of EGF internalization reveals a saturable pathway

Receptor clustering in coated pits is a very dynamic and transient process which is technically difficult to monitor. Therefore, the molecular mechanisms of EGF receptor internalization have been mostly studied by measuring the uptake rates of radioactively labeled EGF. Analysis of EGF endocytosis revealed that, besides characteristics common for receptor-mediated endocytosis of all ligands (for instance, temperature- and ATP-dependence), the specific internalization rate of EGF was several-fold higher at low than at high EGF concentrations.[15,20] Mathematical modeling of these data led to the proposal that there are two pathways of internalization of EGF receptors: a rapid saturable pathway used by a limited number of EGF-activated receptors and a 5–10 times slower non-saturable endocytosis that is employed by the unoccupied or EGF-occupied receptors when the rapid pathway is saturated.[15] It has been postulated that whereas the saturable pathway involves clathrin-coated pits, the slow pathway is clathrin-independent.[15] It should be noted, however, that disruption of coated pit endocytosis diminishes internalization of high concentrations of EGF,[16,21] suggesting that the slow pathway may also require coated pits. This possibility is illustrated in the working model of EGF receptor internalization presented in Figure 8.2. In support of this model, unoccupied receptors can interact with AP-2, although to a lesser extent than occupied receptors[16,18] (A. Sorkin, unpublished results). It is possible that, because of the very high rate of recycling of unoccupied receptors and receptors saturated with EGF (see below), the internalization rate of the non-saturable pathway, calculated from the uptake curves without accounting for recycling, has been underestimated. In fact, direct measurements of the internalization rates of the unoccupied receptors and receptors saturated with EGF under conditions of blocked recycling revealed that these rates are similar and only two to three times lower than the maximal rate measured at low EGF concentration[22] (A. Sorkin, unpublished results).

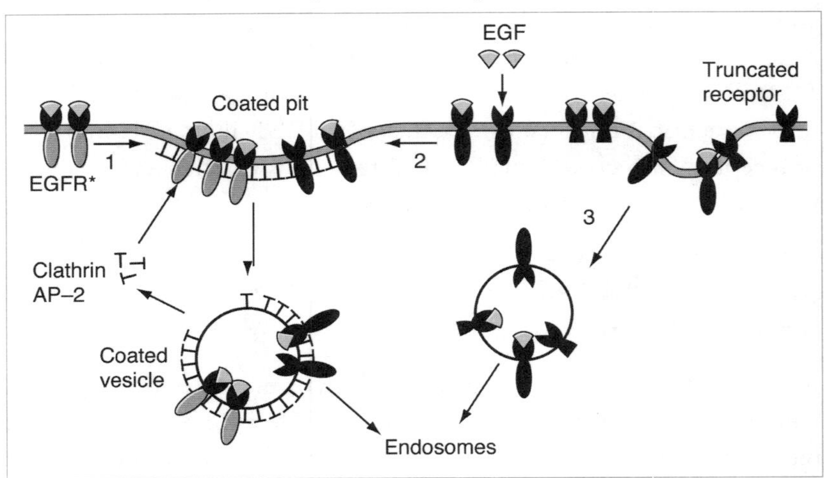

Figure 8.2 Internalization of EGF receptors. Three pathways of endocytosis of EGF receptors are indicated: (1) high-affinity saturable pathway via coated pits (internalization rate constant $k_e = 0.2–0.4$, capacity is typically less than 50 000 receptors) which is utilized by EGF-activated receptors; (2) non-saturable pathway with slow kinetics ($k_e = 0.08–0.12$) used by unoccupied and occupied EGF receptors that have internalization motifs (endocytic codes); and (3) clathrin-independent endocytosis ($k_e = 0.02–0.05$) which does not require endocytic codes. An EGF receptor mutant in which the carboxy-terminal domain had been truncated is shown as an example of the receptor internalized only through the third pathway. Numbered arrows indicate possible rate-limiting steps of the pathways.

Molecular mechanisms of internalization

Saturation of high-affinity EGF-induced receptor internalization suggested that unique mechanisms may regulate this pathway. It has been hypothesized that proteins other than AP-2, expressed in limited amounts in the cell, are involved in the rapid pathway.[15,23] although no attempt has been made so far to identify these proteins. At the same time, very intensive studies were directed to determine which regions of the receptor molecule are critical for internalization. Since the intracellular domain of the EGF receptor is essential for endocytosis, the importance of kinase activity and various regions of the carboxyl terminus were examined using receptor mutagenesis (for the structure of EGF receptor, see Box 8.2).

Several dozen EGF receptor mutants have been prepared and tested for their ability to internalize EGF. Mutational inactivation of the receptor kinase resulted in significant reduction of the internalization rate.[22,24–26] Kinase-negative receptors were capable of internalization at a moderate speed that did not depend on EGF concentration (non-saturable pathway), suggesting that kinase activity controls the specific saturable internalization. Kinase activity is also necessary for the maximum rapid internalization of platelet-derived growth factor (PDGF),[7] macrophage colony-stimulating factor (M-CSF),[27] fibroblast growth factor (FGF)[28] and c-Kit receptors.[29] Receptor tyrosine kinases are known to be autophosphorylated but it is unlikely that kinase activity is required only for receptor autophosphorylation, because rapid endocytosis of the EGF receptor mutant in which carboxy-terminal portion of the intracellular tail containing all autophosphorylation sites was truncated, preserved kinase dependence.[30] It is possible that tyrosine phosphorylation of an unidentified substrate of the receptor kinase is necessary for the receptor internalization through rapid saturable pathway.

Rapid internalization of EGF also requires multiple tyrosine phosphorylation of the receptor,[25] although the role of autophosphorylation is not clear. Experiments with partially truncated receptor mutants[30] suggested that tyrosine phosphorylation is not essential for the internalization of EGF receptors. This means that neither phosphorylated tyrosine residues nor proteins that contain a Src-homology 2 (SH2) domain and often interact with phosphotyrosines (see Chapter 1) directly mediate the association of EGF receptors with coated pits. However, tyrosine phosphorylation of the full-length receptor may be needed to support conformational changes that expose 'better'

Box 8.2 EGF Receptor

The EGF receptor is a glycoprotein (170 kDa) of which approximately 40 kDa is N-linked carbohydrate. The mature EGF receptor is composed of three major regions: an extracellular ligand-binding domain containing two cysteine-rich regions, a hydrophobic transmembrane region and a cytoplasmic domain. The cytoplasmic domain consists of the conserved kinase domain located between two regulatory regions: the juxtamembrane and carboxy-terminal domains. The carboxy-terminal domain contains at least five tyrosine and several serine residues that can be phosphorylated, whereas the juxtamembrane domain has two threonine phosphorylation sites. Phosphorylated residues are involved in regulation of the receptor kinase activity and the interaction of the receptors with other proteins.

endocytic codes than those on the unoccupied receptor. Data on the requirement of tyrosine phosphorylation sites of PDGF receptors are contradictory. Whereas one group reported that Tyr740 and Tyr751 of the PDGF β-receptor are critical for its trafficking,[31] in another study mutations of these tyrosine residues had no effect on PDGF endocytosis whereas Tyr579 was found to be involved in the control of internalization.[32] Phosphorylation of the m-CSF and c-Kit receptors is not important for internalization[27,29,33] whereas the phosphorylation site of the FGF receptor is crucial for its endocytosis.[28] It is possible that the effects on endocytosis seen when the autophosphorylation sites of FGF and PDGF receptors are mutated reflect the necessity of phosphorylation-dependent conformational changes rather than the direct involvement of SH2-containing proteins.

Besides kinase activity and tyrosine phosphorylation, several peptide sequences of the carboxyl terminus of the EGF receptor were found to be necessary for both saturable and non-saturable internalization.[30,34] These sequences are analogous to the coated pit localization motifs found in type I receptors and contain aromatic residues.[11,26,30] It is not clear, however, which of these endocytic codes function in native, full-length EGF receptor and whether different motifs can be involved in distinct pathways of internalization. Interestingly, the tyrosine-containing motif that is essential for internalization of the M-CSF receptor was found in its juxtamembrane region.[27]

In summary, the rapid saturable pathway of GF endocytosis is controlled by receptor kinase activity and, in some cases, by receptor phosphorylation. The non-saturable internalization of EGF does not require kinase activity, tyrosine phosphorylation or the kinase domain.[34] However, endocytic codes in the receptor's intracellular tail are required.[34] Receptors that lack endocytic codes, for instance EGF receptors in which the whole intracellular domain was truncated, do not interact with AP-2 and undergo internalization through constitutive clathrin-independent endocytosis, the third and slowest pathway of EGF receptor endocytosis (Figure 8.2).

Pathway through endosomes

After internalization, receptors and ligands begin their passage through the endosomal compartments (Figure 8.3). Internalized molecules can be either recycled back from endosomes to the plasma membrane, retained in endosomes for a long time or transported to lysosomes or other organelles. Classification of endosomes is based on their biogenesis and currently is the subject of much debate. Here, for simplicity, we shall use the terminology of early, intermediate and late endosomes according to the time of appearance of the endocytic markers in these compartments.

Endocytosed molecules are first delivered to early endosomes after uncoating of the coated vesicle and its fusion with the endosomal membranes. Early endosomes are the tubular and vesicular membrane structures often connected into networks and located close to the plasma membrane.[8] EGF and EGF receptors can be detected in this compartment within 2–5 min of EGF-induced internalization at 37 °C.[2,5,35,36] After 10–15 min, receptors begin to accumulate in large tubular-vesicular endosomes located mainly around the nucleus, often close to the centriole.[2,5,12,35] In electron microscopic sections, these endosomes frequently appear as multivesicular bodies (MVBs) because they contain internal vesicles.[2,5,12,35] In our classification, MVBs are likely to correspond to the intermediate (or 'carrier' from early to late endosomes) and late endosomes. Since the delivery of receptors from intermediate to late endosomes is highly temperature dependent, these two populations of organelles can be distinguished by lowering the temperature to 16–18 °C.[8,35,37] The late endosomal compartment has a complex morphology. As it serves as the last destination of molecules sorted to lysosomal pathway, it is also referred to as prelysosomal compartment.[10] EGF and EGF receptors become detectable in lysosomes after 30–60

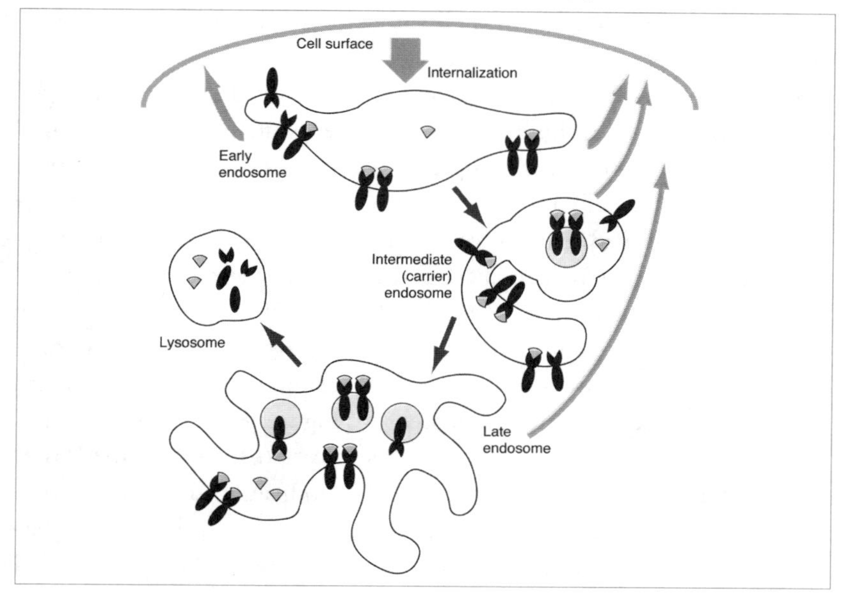

Figure 8.3 The transition of EGF receptors through the endosomal compartments. Both clathrin-dependent and -independent internalization pathways lead receptors to the same endosomal compartments. Recycling of occupied and unoccupied receptors from endosomes is shown by black arrows. The relative width of arrows indicates that the recycling is much slower from late than early endosomes.

min of internalization but can be seen in late endosomes for several hours.[35] Degradation of EGF and other GFs and also their receptors can be detected 20–30 min after initiation of endocytosis,[1,6,7] suggesting that the degradation begins in late endosomes which contain functionally active proteolytic enzymes. However, the complete degradation of EGF and its receptors is thought to occur in mature lysosomes. Both EGF and EGF receptors and the intermediate products of their proteolysis are difficult to detect in lysosomes, presumably because they are very rapidly degraded to low-molecular-mass peptides.[4,5,35]

The intravesicular pH drops along the endocytic pathway, from 6.0–6.5 in early endosomes to 4.5–5.5 in late endosomes and lysosomes, which causes dissociation of many ligand–receptor complexes.[10] However, direct measurements of the status of EGF–receptor complexes in endosomes and several sets of indirect data indicated that the release of EGF is insignificant until the late stages of endocytosis, and that there is a large pool of endosomal EGF–receptor complexes.[38–40] This is in agreement with the common localization of EGF and EGF receptor throughout the endocytic pathway[35] (Figure 8.3).

Kinetics of intracellular trafficking: recycling versus degradation

In the absence of EGF, the endosomal pool of EGF receptors is small compared with the surface pool, suggesting that unoccupied EGF receptors must recycle very rapidly after internalization[23] (Figure 8.4). Because EGF-accelerated receptor degradation is very rapid, it was assumed in early studies that recycling of internalized EGF receptors did not occur or was insignificant.[2,24] In later studies, however, rapid recycling of EGF–receptor complexes was demonstrated first in human carcinoma A-431 cells and then in all types of cells tested.[22,37,41] These observations showed that GF receptor sorting is not a simple, one-directional process of lysosomal targeting. In fact, after each round of endocytosis as much as 70–80% of EGF–receptor complexes can be recycled and then re-internalized, while only 20–30% are degraded.

The relative values of trafficking rate parameters of occupied and unoccupied receptors are compared in Figure 8.4. Recent studies suggested that unoccupied receptors recycle two to three times faster than receptors in the presence of EGF,[41,42] indicating that only 5–10% of unoccupied receptors are degraded after each round of internalization. Given the low internalization rate and small pool of endocytosed

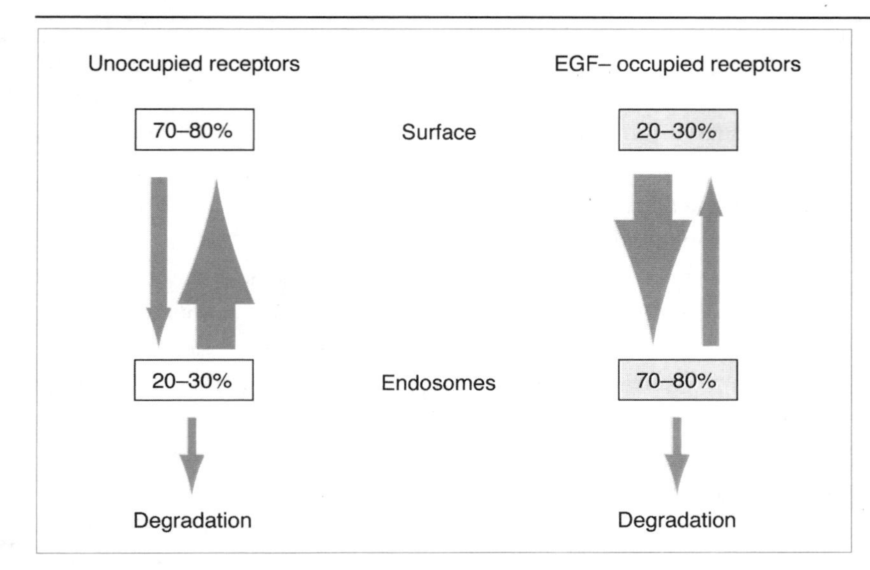

Figure 8.4 EGF-induced changes in the rate parameters of the trafficking and subcellular distribution of the receptors. The relative values of the specific rate constants of internalization, recycling and degradation were averaged from several studies[15,23,25,41–43] and expressed as the widths of arrows. The model is applicable to the cells expressing a moderate physiological amount of EGF receptors (not more than 100 000 per cell) and assumes no saturation of internalization and degradation systems. In this model we propose that EGF binding accelerates receptor internalization by activating the rapid pathway and reduces recycling of the receptors by retaining occupied receptors in endosomes and preventing them from recycling. The resulting accumulation of 70–80% of occupied receptors in endosomes leads to the increased degradation and down-regulation of receptors without any change in the specific rate of degradation.

receptors, the overall degradation rate of unoccupied receptors is very low, and the loss of receptors is compensated for by biosynthesis. EGF binding decreases the recycling rate and, therefore, increases the degradation/recycling ratio that determines the fate of receptors after endocytosis. This effect of EGF, together with EGF-accelerated internalization, results in the dramatic elevation of the overall degradation of receptors which causes receptor down-regulation.

The importance of ligand binding for the lysosomal targeting and down-regulation of internalized receptors has been shown by differential trafficking of EGF and transforming growth factor α (TGFα), which both bind to the EGF receptor. The complex of TGFα and the EGF receptor is much more sensitive to low pH than the EGF–receptor complex and, probably, most of the TGFα dissociates from the receptor at pH 5.5–6.0 early in the endosomal pathway.[44,45] This dissociation allows rapid recycling of unoccupied receptors and reduces receptor degradation. As a consequence, TGFα causes much less down-regulation of EGF receptors than EGF.[44,45]

Mechanism of receptor sorting in endosomes

Recycling of EGF–receptor complexes is partially inhibited at 18°C, suggesting that some recycling does occur from the late endosomes.[37] This recycling is, however, much slower than recycling from early endosomes. Therefore, the overall rate and extent of receptor recycling can be reduced by relocating receptors from early to late compartments.

The morphological pathways of intracellular sorting of EGF receptors have been compared with those of transferrin receptors which are targeted to lysosomes very insignificantly.[12,36] In early endosomes, EGF receptors tend to accumulate in vesicular parts of the compartments (Figure 8.3), whereas transferrin receptors are mostly located in tubular parts. It has been also noticed that in MVB-like endosomes, EGF receptors are preferentially associated with the internal vesicular structures, whereas transferrin receptors are distributed mainly in the outer membrane and tubular parts of endosomes. On the basis of these observations, the following model of intracellular sorting has been proposed.[12,46] Receptors located in the tubular portions of endosomes are constitutively recycled. In contrast, a pool of EGF receptors is trapped (retained) in the vesicular parts and they subsequently become incorporated into internal vesicles of MVBs. According to this model, sorting to the internal membranes prevents recycling of EGF receptors, and leads to their retention in late endosomes and subsequent delivery to lysosomes.

The molecular mechanisms of the EGF receptor retention in the vesicular compartments of endosomes is unclear. It has been hypothesized that EGF-induced activation of receptor kinase and tyrosine phosphorylation of annexin I in MVBs is important for inclusion of EGF receptors in internal vesicles of MVBs.[46,47] However, recent data indicating that receptor mutants lacking an ATP-binding site or entire kinase domain are degraded as fast as wild-type receptor have questioned the direct involvement of tyrosine kinase activity in the intracellular sorting of EGF receptors.[23,34,41,43] A recent report[34] suggested that carboxy-terminal regions of the receptor distinct from its internalization codes are necessary for receptor retention in endosomes. Interestingly, kinetic studies revealed that the lysosomal sorting (or retention) pathway of EGF receptors is also saturable,[43] indicating that either the proteins responsible for sorting or the pool of EGF receptors competent for lysosomal sorting are limited.

Biological function of down-regulation

It is currently accepted that GF receptor endocytosis is not essential for signal transduction, and that it attenuates mitogenic signaling by down-regulating the receptors and by depleting GFs from the extracellular fluid.[48] The key data in support to this notion have been obtained using an EGF receptor mutant that has an intact kinase but lacks the carboxyl terminus.[49] Although this receptor was internalized at a very slow rate and not down-regulated in response to EGF, its mitogenic and transforming activity was even higher than that of the wild-type receptor. The enhanced response to GFs in the cells expressing internalization-defective receptors is probably due to prolonged presence of GF-activated receptors in the cell and negligible depletion of GFs from the medium.[48]

In normal non-transformed cells that express a moderate number of receptors, GF-induced endocytosis is not saturated and down-regulation of receptors helps to prevent excessive signaling which may lead to cell transformation. In contrast, endocytosis does not influence signaling in transformed cells overexpressing GF receptors, because when internalization and degradation pathways are saturated there is insignificant down-regulation of the receptors.[20,25,42]

Since EGF remains associated with receptors after internalization, a pool of dimerized, tyrosine-phosphorylated and active receptors are present in endosomes.[39,40,50] The kinase domains of these receptors are oriented into the cytoplasm until they are included into internal vesicles of MVB-like endosomes. Therefore, endosomal receptors are capable of interacting with other proteins containing SH2 domains and phosphorylating endogenous substrates. A recent study by Guglielmo *et al.*[50] demonstrated that several proteins involved in the Ras signaling pathway are mainly associated with endosomal and not surface EGF receptors in rat liver. In A-431 cells, endosomal EGF receptor can phosphorylate annexin I, a tyrosine kinase substrate that does not have SH2 domain.[51] The involvement of internalized receptors in signaling depends on the relative amount of active and accessible receptors on the cell surface and endosomes. The subcellular distribution of receptors, in turn, depends on the parameters of receptor trafficking which may vary in different cells.

Summary

This chapter has discussed the morphology, kinetics and possible molecular mechanisms of receptor-mediated endocytosis of GFs. Two effects of GFs on receptor trafficking that cause down-regulation of receptors have been identified: (1) acceleration of internalization via coated pits and (2) reduction of recycling of internalized receptors through their retention in endosomes. Although mutagenesis of the EGF receptor revealed several putative internalization and endosome retention domains, further studies are required to determine which of these domains function in native forms of the EGF receptor and other GF receptors. The proteins that interact with these domains and the role of receptor kinase activity and phosphorylation in these interactions remain unknown. Because EGF-dependent internalization and lysosomal targeting are saturable pathways, they are probably regulated by specific sorting machineries. Elucidation of the mechanisms and identification of the principal players involved in GF receptor trafficking may provide new insights into general mechanisms of endocytosis and may unveil the possibility to control the expression of GF receptors and, therefore, the cellular responses to GFs.

Acknowledgements

The author is grateful to Drs Wiley, Lauffenberger, Gill, Nesterov and Reddy for communicating their unpublished results. This work was supported by NIH grant DK46817 and grant SA039 from the Council for Tobacco Research.

References

1 Carpenter, G. and Cohen, S. (1976) ^{125}I-labeled human epidermal growth factor: binding, internalization, and degradation in human fibroblasts. *J. Cell Biol.*, **71**, 159–71.

2 Beguinot, L., Liall, R.M., Willingham, M.C. and Pastan, I. (1984) Down-regulation of the epidermal growth factor receptor in KB cells is due to receptor internalization and subsequent degradation in lysosomes. *Proc. Natl. Acad. Sci. USA*, **81**, 2384–8.

3 Stoscheck, C.M. and Carpenter, G. (1984) 'Down-regulation' of EGF receptors: direct demonstration of receptor degradation in human fibroblasts. *J. Cell Biol.*, **98**, 1048–53.

4 Gorden, P., Carpentier, J.-L., Cohen, S. and Orci, L. (1978) Epidermal growth factor: morphological demonstration of binding, internalization, and lysosomal association in human fibroblasts. *Proc. Natl. Acad. Sci. USA*, **75**, 5025–9.

5 Haigler, H.T., McKanna, J.A. and Cohen, S. (1979) Direct visualization of the binding and internalization of a ferritin conjugate of epidermal growth factor in human carcinoma cells A-431. *J. Cell Biol.*, **81**, 382–95.

6 Heldin, C.-H., Wasteson, A. and Westermark, B. (1982) Interaction of platelet-derived growth factor with its fibroblast receptor. *J. Biol. Chem.*, **257**, 4216–21.

7 Sorkin, A., Westermark, B., Heldin, C.-H. and Claesson-Welsh, L. (1991) Effect of receptor kinase inactivation on the rate of internalization and degradation of PDGF and the PDGF β-receptor. *J. Cell Biol.*, **112**, 469–78.

8 Subtil, A., Hemar, A. and Dautry-Varsat, A. (1994) Rapid endocytosis of interleukin 2 receptors when clathrin-coated pit endocytosis is inhibited. *J. Cell Sci.*, **107**, 3461–3468.

9 Hemar, A., Subtil, A., Lieb, M. *et al.* (1995) Endocytosis of interleukin 2 receptors in human T lymphocytes: distinct intracellular localization and fate of the receptor α, β and γ chains. *J. Cell Biol.*, **129**, 55–64.

10 Smythe, E. and Warren, G. (1991) The mechanism of receptor-mediated endocytosis. *Eur. J. Biochem.*, **202**, 689–99.

11 Trowbridge, I.S., Collawn, J.F. and Hopkins, C.R. (1993) Signal-dependent membrane trafficking in the endocytic pathway. *Annu. Rev. Cell Biol.*, **9**, 129–61.

12 Hopkins, C.R. and Trowbridge, I.S. (1983) Internalization and processing of transferrin and the transferrin receptor in human carcinoma A-431 cells. *J. Cell Biol.*, **97**, 508–21.

13 Hanover, J.A., Willingham, M.C. and Pastan, I. (1984) Kinetics of transit of transferrin and epidermal growth factor through clathrin-coated membranes. *Cell*, **39**, 283–93.

14 Lamaze, C., Baba, T., Redelmeier, T.E. and Schmid, S.L. (1993) Recruitment of epidermal growth factor and transferrin receptors into coated pits *in vitro*. *Mol. Biol. Cell.*, **4**, 715–27.

15 Lund, K.A., Opresko, L.K., Starbuck, C. *et al.* (1990) Quantitative analysis of the endocytic system involved in hormone-induced receptor internalization. *J. Biol. Chem.*, **265**, 15713–23.

16 Sorkin, A. and Carpenter, G. (1993) Interaction of activated EGF receptor with coated pit adaptins. *Science*, **261**, 612–5.

17 Robinson, M.S. (1994) The role of clathrin, adaptors and dynamin in endocytosis. *Curr. Opinion Cell Biol.*, **6**, 538–44.

18 Nesterov, A., Kurten, R. and Gill, G.N. (1995) Association of epidermal growth factor receptors with coated pit adaptins via a tyrosine phosphorylation-regulated mechanism. *J. Biol. Chem.*, **270**, 6320–7.

19 Sorkin, A., McKinsey, T., Shih, W. *et al.* (1995) Stoichimetric interaction of the epidermal growth factor receptor with the clathrin-associated protein complex AP-2. *J. Biol. Chem.*, **270**, 619–5.

20 Wiley, H.S. (1988) Anomalous binding of epidermal growth factor to A431 cells is due to the effect of high receptor densities and a saturable endocytic system. *J. Cell Biol.*, **107**, 801–10.

21 Sandvig, K., Olsnes, S., Peterson, O.W. and van Deurs, B. (1987) Acidification of the cytosol inhibits endocytosis from coated pits. *J. Cell Biol.*, **105**, 679–689.

22 Felder, S., LaVin, J., Ullrich, A. and Schlessinger, J. (1992) Kinetics of binding, endocytosis and recycling of EGF receptor mutants. *J. Cell Biol.*, **117**, 203–12.

23 Wiley, H.S., Herbst, J.J., Walsh, B.J. *et al.* (1991) The role of tyrosine kinase activity in endocytosis, compartmentalization, and down-regulation of the epidermal growth factor receptor. *J. Biol. Chem.*, **266**, 11083–94.

24 Glenney, J.R., Jr, Chen, W.S., Lazar, C.S. *et al.* (1988) Ligand-induced endocytosis of the EGF receptor is blocked by mutational inactivation and by microinjection of anti-phosphotyrosine antibodies. *Cell*, **52**, 675–84.

25 Sorkin, A., Helin, K., Waters, C.M. *et al.* (1992) Multiple autophosphorylation sites of the epidermal growth factor receptor are essential for receptor kinase activity and internalization. *J. Biol. Chem.*, **267**, 8672–8.

26 Chen, W.S., Lazar, C.S., Lund, K.A. *et al.* (1989) Functional independence of the epidermal growth factor receptor from a domain required for ligand-induced internalization and calcium regulation. *Cell*, **59**, 33–43.

27 Myles, G.M., Brandt, C.S., Carlberg, K. and Rohrschneider, L. (1994) Tyrosine 569 in the c-fms juxtamembrane domain is essential for kinase activity and macrophage colony-stimulating factor-dependent internalization. *Mol. Cell. Biol.*, **14**, 4843–54.

28 Sorokin, A., Mohammadi, M., Huang, J. and Schlessinger, J. (1994) Internalization of fibroblast growth factor receptor is inhibited by point mutation at tyrosine 766. *J. Biol. Chem.*, **269**, 17056–61.

29 Yee, N.S., Hsiau, C.-W.M., Serve, H. *et al.* (1994) Mechanism of down-regulation of c-*kit* receptor. *J. Biol. Chem.*, **269**, 31991–8.

30 Chang, C.-P., Lazar, C.S., Walsh, B.J. *et al.* (1993) Ligand-induced internalization of epidermal growth factor receptor is mediated by multiple endocytic codes analogous to the tyrosine motif found in constitutively internalized receptors. *J. Biol. Chem.*, **268**, 19312–20.

31 Joly, M., Kazlauskas, A., Fay, F.S. and Corvera, S. (1994) Distribution of PDGF receptor trafficking by mutation of its PI-3 kinase binding sites. *Science*, **263**, 684–7.

32 Mori, S., Ronnstrand, L., Claesson-Welsh, L. and Heldin, C.-H. (1994) A tyrosine residue in the juxtamembrane segment of the platelet-derived growth factor β-receptor is critical for ligand-mediated endocytosis. *J. Biol. Chem.*, **269**, 4917–21.

33 Carlberg, K., Tapley, P., Haystead, C. and Rohrschneider, L. (1991) The role of kinase activity and the kinase insert region in ligand-induced internalization and degradation of the c-*fms* protein. *EMBO J.*, **4**, 877–83.

34 Opresko, L.K., Chang, C.-P., Will, B.H. *et al.* (1995) Endocytosis and lysosomal targeting of epidermal growth factor receptors are mediated by distinct sequences independent of the tyrosine kinase domain. *J. Biol. Chem.*, **270**, 4325–33.

35 Miller, K., Beardmore, J., Kanety, H. *et al.* (1986) Localization of epidermal growth factor (EGF) receptor within the endosome of EGF-stimulated epidermoid carcinoma (A431) cells. *J. Cell Biol.*, **102**, 500–9.

36 Hopkins, C.R., Gibson, A., Shipman, M. and Miller, K. (1990) Movement of internalized ligand–receptor complexes along a continuous endosomal reticulum. *Nature*, **346**, 335–9.

37 Sorkin, A., Krolenko, S., Kudrjavtceva, N. *et al.* (1991) Recycling of epidermal growth factor-receptor complexes in A431 cells: identification of dual pathways. *J. Cell Biol.*, **112**, 55–63.

38 Sorkin, A., Teslenko, L. and Nikolsky, N. (1988) The endocytosis of epidermal growth factor in A431 cells: a pH of microenvironment and the dynamics of receptor complexes dissociation. *Exp. Cell Res.*, **175**, 192–205.

39 Sorkin, A. and Carpenter, G. (1991) Dimerization of internalized epidermal growth factor receptors. *J. Biol. Chem.*, **266**, 8355–62.

40 Lai, W.H., Cameron, P.H., Doherty, J.-J., II, *et al.* (1989) Ligand-mediated autophosphorylation activity of the epidermal growth factor receptor during internalization. *J. Cell Biol.*, **109**, 2751–60.

41 Herbst, J.J., Opresko, L.K., Walsh, B.J. *et al.* (1994) Regulation of postendocytic

trafficking of the epidermal growth factor receptor through endosomal retention. *J. Biol. Chem.*, **269**, 12865–73.

42 Sorkin, A. and Waters, C.M. (1993) Endocytosis of growth factor receptors. *BioEssays*, **15**, 375–82.

43 French, A.R., Sudlow, G.P., Wiley, H.S. and Lauffenberger, D.A. (1994) Postendocytic trafficking of epidermal growth factor-receptor complexes is mediated through saturable and specific endosomal interactions. *J. Biol. Chem.*, **269**, 15749–55.

44 Ebner, R. and Derynck, R. (1991) Epidermal growth factor and transforming growth factor-α: differential intracellular routing and processing of ligand–receptor complexes. *Cell Regul.*, **2**, 599–612.

45 French, A.R., Tadaki, D.K., Niyogi, S.K. and Lauffenberger, D.A. (1995) Intracellular trafficking of epidermal growth factor family ligands is directly influenced by the pH sensitivity of the receptor/ligand interaction. *J. Biol. Chem.*, **270**, 4334–40.

46 Felder, S., Miller, K., Moehren, G. *et al.* (1990) Kinase activity controls the sorting epidermal growth factor receptor within the multivesicular body. *Cell*, **61**, 623–34.

47 Futter, C.E., Felder, S., Schlessinger, J. *et al.* (1993) Annexin I is phosphorylated in the multivesicular body during the processing of the epidermal growth factor receptor. *J. Cell Biol.*, **120**, 77–83.

48 Starbuck, C. and Lauffenburger, D.A. (1992) Mathematical model for the effects of epidermal growth factor receptor trafficking dynamics on fibroblast proliferation responses. *Biotechnol. Prog.*, **8**, 132–43.

49 Wells, A., Welsh, J.B., Lazar, C.S. *et al.* Ligand-induced transformation by a noninternalizing epidermal growth factor receptor (1990) *Science*, **247**, 962–4.

50 Di Guglielmo, G.M., Baass, P.C., Ou, W.-J. *et al.* (1994) Compartmentalization of SHC, GRB2 and mSOS, and hyperphosphorylation of Raf-1 by EGF but not insulin in liver parenchyma. *EMBO J.*, **13**, 4269–77.

51 Cohen, S. and Fava, R. (1985) Internalization of functional epidermal growth factor: receptor kinase complexes in A431 cells. *J. Biol. Chem.*, **260**, 12351–8.

Part Two

Cytoplasmic Signal Transduction

9 SH2, SH3 and PH domains

David Cowburn and John Kuriyan

The structures of complex eukaryotes – in which cells form organs which, in turn, make up the complete organism – are developed and maintained by subtle and diverse mechanisms. The positioning of cells, their differentiation, and maintenance relies on many interactions.[1,2] Growth factors and many other intercellular signals act at the cell surface by binding to specific receptors. Through this binding, complex signals are transmitted within the cells to control modulation of enzymatic activities, expression of genes and localization of cytoskeletal elements.

Despite its apparent complexity, signal transduction has apparently evolved using a small number of similar mechanisms. Recent advances in molecular cell biology have permitted the identification of the molecular players in these similar mechanisms. Many of the proteins involved in signal transduction contain recapitulations of polypeptide segments, commonly called domains. The first to be identified was found as a conserved region of about 100 residues in protein tyrosine kinases homologous to Src. This domain is different from the enzymatic (SH1) portion of the molecules and so was given the name Src homology 2 (SH2) domain.[3] Further analysis of the sequences of the Src family and related protein tyrosine kinases suggested another cluster of about 60 homologous residues, the Src homology 3 (SH3) domain.[3] More recently, by analysis of sequences of proteins binding to SH3 domains, a sequence block of about 110 amino acids has been identified as a conserved region homologous to the repeat within the protein pleckstrin and has hence been named the pleckstrin homology (PH) domain.[4,5] These domains, SH2, SH3 and PH, had been shown to exist in many other enzymes besides kinases, and in a few classes of transcription factors (Figure 9.1). In addition, new proteins containing these domains have been found which serve solely as adaptors, bringing two or more domain-containing signaling molecules into close proximity. Adaptors themselves have no enzymatic or gene-transcriptional activities.

Signal Transduction. Edited by Carl-Henrik Heldin and Mary Purton. Published in 1996 by Chapman & Hall. ISBN 0 412 70810 8

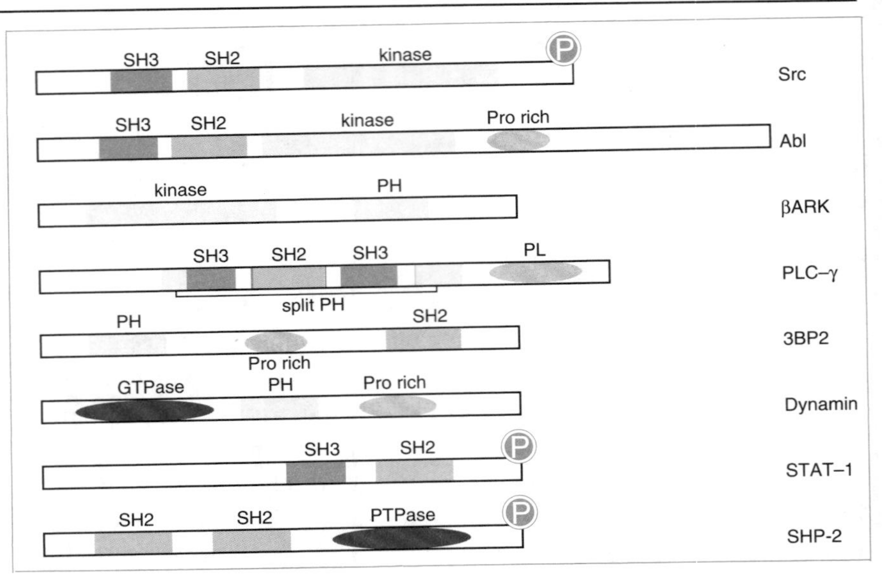

Figure 9.1 Schematic diagram of positions of domains in several molecules involved in intracellular signal transduction. Relative scales are approximate. The following abbreviations are used in this and other figures. P = phosphorylation site; Pro rich = proline-rich area; PL = phospholipase enzymatic sites; PTPase = protein tyrosine phosphatase site; Src = Rous asian sarcoma tyrosine kinase; Abl = Albeson tyrosyl kinase; βARK = β-adrenergic receptor kinase; PLC-γ = phospholipase C-γ; 3BP2 = binding protein for SH3, type 2; STAT = signal transducing and activating of transcription; SHP-2 = SH2-containing phosphatase, type 2.

Representatives of each of these three classes of domains have been isolated and studied structurally. The individual domains form compact, folded structures with their carboxyl and amino termini close together in space, suggesting that they can function as plug-in modules within larger proteins. Many domains contain well defined binding sites for interactions with their specific targets.[6] Chimeras containing domains from different proteins have properties consistent with the donor molecule. Thus, these domains, and others, appear to be used in a mix-and-match strategy to modify enzymatic activity by direct inhibition or allosteric control, and to recruit proteins containing these domains to specific complexes. SH2 domains bind to short amino acid sequences containing phosphotyrosine, and SH3 domains bind to proline-rich sequences. Their targets are apparently epitopic (a term borrowed from immunochemistry); that is, linear in sequence, without the use of extended tertiary structure. The tertiary structure of a macromolecular target may play a role in affinity and specificity to SH2 and SH3 domains, but this is not understood at present. Short target peptides apparently mimic the affinity and specificities of larger targets. Defined target ligands for the PH domains have not yet been fully identified.[5] One hypothesis is that phosphatidyl phospholipids with phosphorylated inositol head-groups[7] and/or the phosphorylated inositols can bind to PH domains. Some PH domains also have the capacity to bind to the WD40 motif of β subunits of G proteins.[8,9]

Along with the intrinsic biological interest of these signaling domains, they are also involved in pathways that provide potential targets for therapeutic agents.[10] The identification and characterization of SH2 and SH3 domains has involved studying their roles in proteins produced by oncogenes.[11] Uncontrolled signaling, a characteristic of cancer cells, is sometimes associated with abberant SH2 or SH3 interactions.

This chapter will describe highlights of the general characteristics, structure and ligand properties of each of these domains.

SH2 domains

Tyrosine side chains in specific proteins are used as on/off switches in intracellular signal transduction by eukaryotes. The two states of the switch are interconverted by their phosphorylation by kinases and by their dephosphorylation by phosphatases. SH2 domains recognize the state of the switch by their highly selective binding to phosphotyrosyl targets. They have essentially no affinity for the non-phosphorylated state. Individual switches are discriminated by the local residues surrounding the phosphorylated tyrosine. Sequence comparison and site directed mutagenesis[12] identified the highly conserved FLVR motif of the SH2 domain as essential to phosphotyrosine binding. The arginine residue βB5 (Figure 9.2) in this motif is highly conserved among the various sequences, along with the tryptophan residue at βA1 and leucine

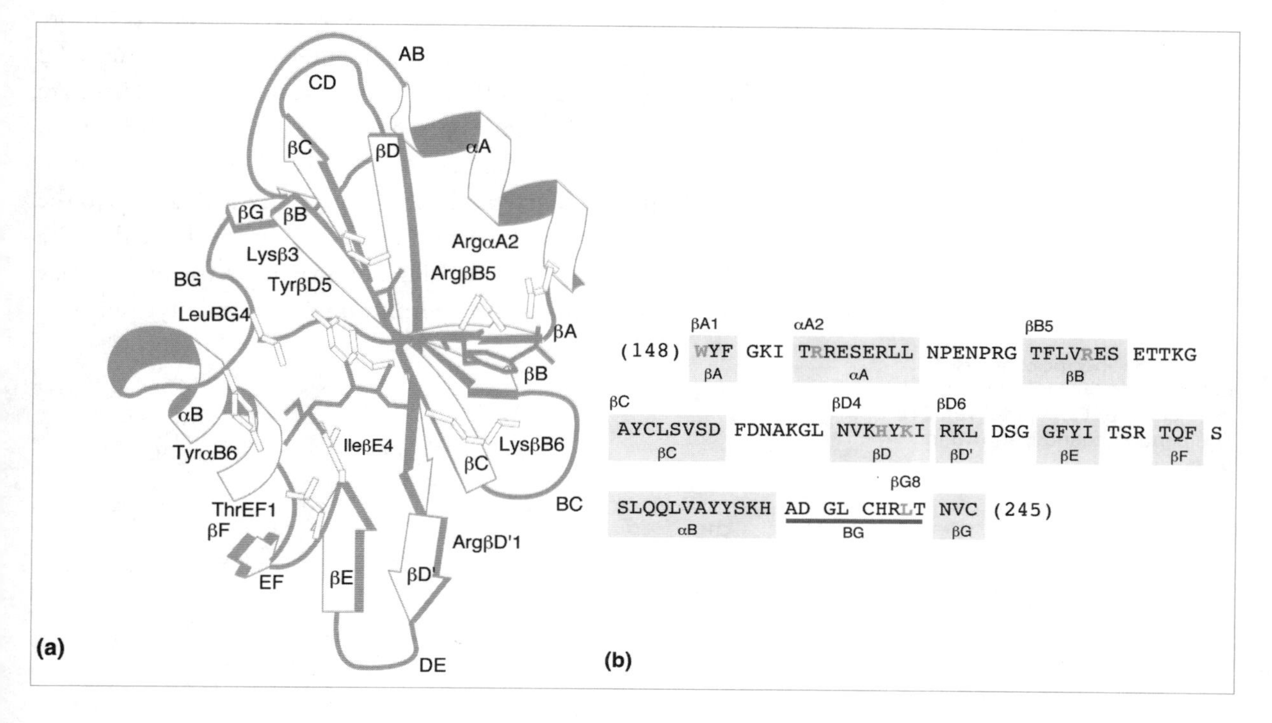

Figure 9.2 Structure of Src SH2 and its ligand. (a) Tertiary structure; reproduced with permission from Waksman, G. *et al.* (1993) *Cell*, 72, 779–90. (b) Amino acid sequence from residues 148–245. Elements of secondary structure are indicated by boxes (β-sheets and α-helices) and the BG loop is underlined. Key residues discussed in the text are shown in bold.

residue at BG8, which appear to be highly conserved as bookends, contributing their side chains to the protein's hydrophobic core.

The published SH2 structures show essentially the same protein fold and topology. Figure 9.2 shows an overall picture of the SH2 structural elements of Src bound to a peptide. At the center, a large three-stranded antiparallel β-sheet is extended by a highly twisted smaller β-sheet and flanked by two α-helices. The amino and carboxyl termini are defined precisely by mini β-strands. The conformations of the large β-sheet and the α-helices are highly conserved. The smaller sheet and the loops are more variable. The core conserved conformation is associated with the conserved function of phosphotyrosine recognition; the variable structure is correlated in part with the varying peptide specificity.

Phosphotyrosine binding site

The binding site for phosphotyrosine may be expected to be selective for the aromatic ring of this amino acid, discriminating phosphotyrosine from phosphoserine or phosphothreonine, and selective for the phosphate, since that is the key signaling switch. These two requirements are met by structures that are similar in all those so far published. The aromatic ring is discriminated by the deep insertion of the phosphotyrosyl side chain into the protein, and by specific contacts to the aromatic ring, although the latter may vary between proteins. In Src and Lck SH2 domains,[13,14] amino–aromatic interactions from Arg αA2 and Lys βD6 (Figures 9.2 and 9.3) flank the aromatic ring. These are not absolutely conserved; in the SypN-SH2 structure where there is no positively charged equivalent at the αA2 position, a close contact is provided by His βD4.[15] Since tight coordination of the Arg αA2 to the phosphotyrosine is not observed by NMR in solution for either Src SH2[16] or PLC-γ C-SH2,[17] this coordination may be transient, and may be part of the structural fluctuation permitting the phosphotyrosine side

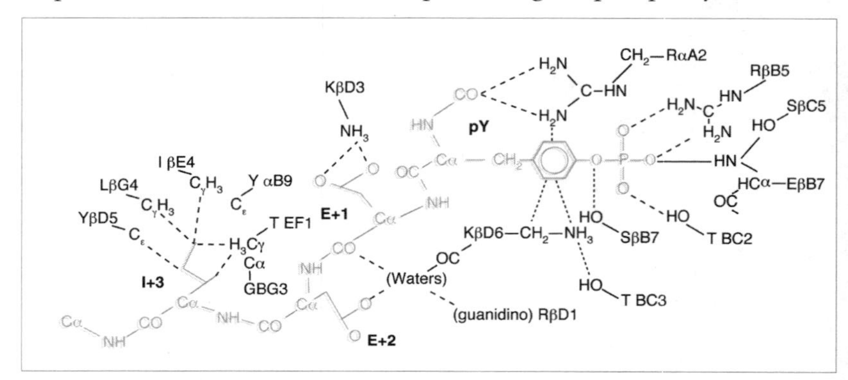

Figure 9.3 Specific contacts between the Src SH2 protein and its ligand – pYEEI – (after ref. 21). The peptide ligand is shown in blue.

chain to insert into the SH2 domain, or to leave it. A critical element of the discrimination between phosphotyrosine and phosphoserine/threonine appears to be steric: the conserved Arg βB5 is positioned at the base of the binding pocket in such a way that only the phosphotyrosine is long enough to reach it from the position of the tethered backbone of the peptide.

In high-resolution X-ray crystallographic structures, all of the oxygens of the phosphate group are observed to receive some hydrogen bond donation, particularly from the absolutely conserved Arg B5 guanidino amino groups (Figure 9.3). This very tight packing around the charged group is presumed to reflect the energy of the interaction and the specificity of this recognition.

Sequence specificity

Early experiments showed that the sequence specificities of phosphotyrosyl peptide–SH2 interactions lay in the portion of the peptide immediately carboxy-terminal of the phosphotyrosine. Using a limited combinatorial library (see Box 9.1), several classes of SH2 binding were identified.[18,19] There remains some controversy about the affinities of short phosphotyrosyl peptides, as least concerning those measurements using surface plasmon resonance (see Box 9.1).[20] The most likely range seems to be about 100 nM–1 μM for strong binders. There are so far no structural data on SH2 domains interacting with macromolecular targets. From the complexes with relatively short peptides, a clear picture emerges. The phosphotyrosine is bound essentially as described above, and the carboxy-terminal residues are selectively recognized on the surface, or by sockets, to provide a high degree of selectivity. Recognition is particularly specific for the side chain that is three residues toward the carboxyl terminus of the ligand peptides, (+3) for structures of Src,[21] Lck[14] and SypN.[15] For Src and Lck, the sequence specifically recognized is –pYEEI–, with the glutamic acid residues at + 1 and + 2 having extended side chains interacting with positive charges on the surface, and the isoleucine at + 3 largely buried into a hydrophobic pocket. For SypN SH2, the recognized sequences (–pYVNIDF– and –pYTAVQP–) form contacts principally at the + 1, + 3 and + 5 positions,[22] by complementation of hydrophobic surfaces. The solution structure of PLCγ SH2 with a 12-mer peptide provided a similar overview of a more extended hydrophobic groove.[17] Overall, the recognition of the carboxy-terminal sequence generally uses a relatively small area of the protein surface.

Box 9.1 Techniques

A **combinational library** is a collection of molecules in which some features are varied to provide all the possible combinations of selected fragments at specific sites. A binding assay can then be used to select those items of greater affinity from the library.

In **surface plasmon technology**, one of the binding partners is immobilized on a surface and a solution of the other flows over it. When the surface density of the immobilized ligand is high, self-association of the other ligand will result in apparently greater affinity than is actually the case.

Phosphotyrosine residues can also apparently be recognized by another domain type, PTB (phosphotyrosyl binder), which is clearly

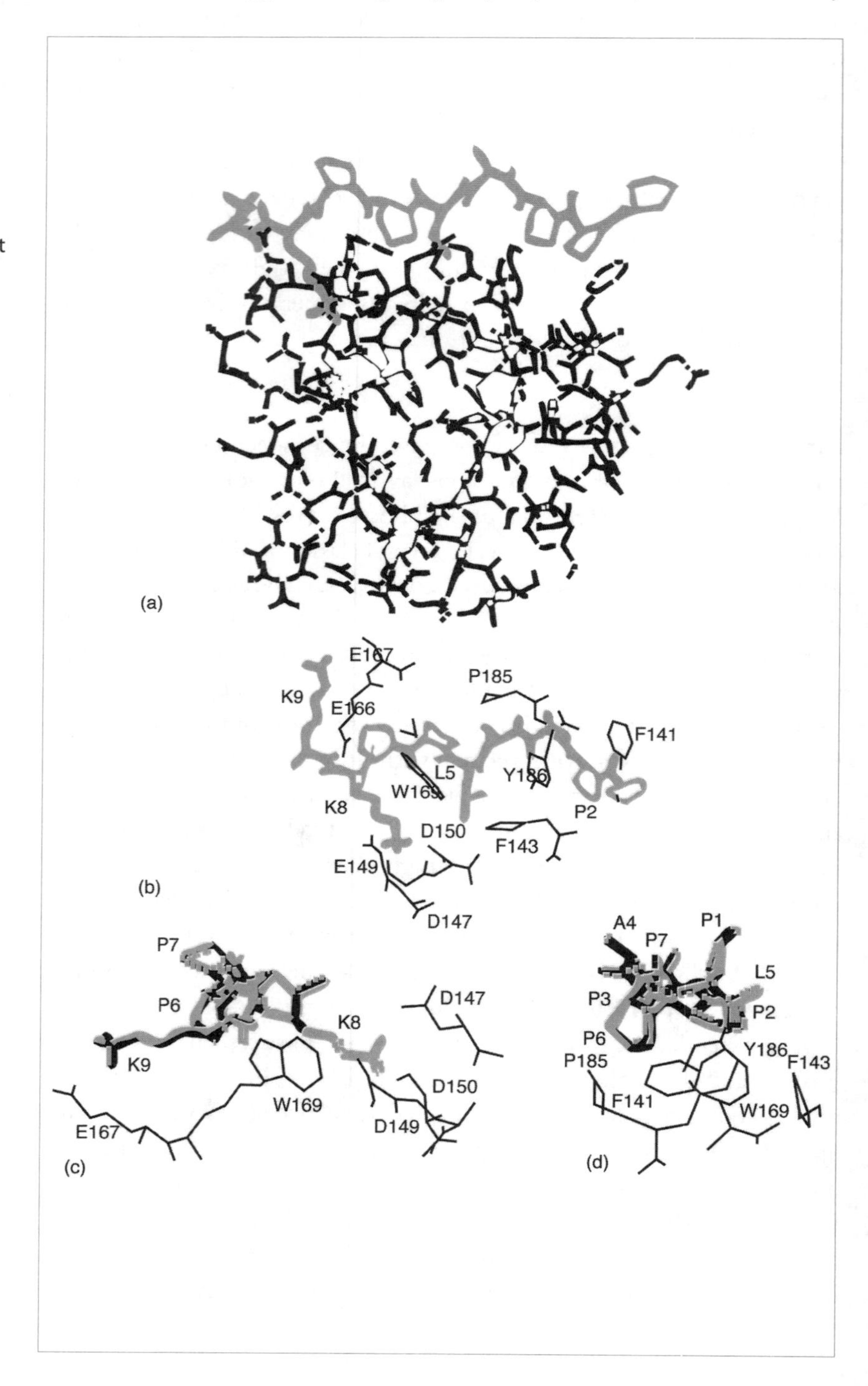

Figure 9.4 SH3 structure with ligand. The amino-terminal SH3 from Crk is shown[32] with ligand PPPALPPKKR (blue). (a) Face-on view of the binding site complexed. (b) Edge-on view of binding. Compare to lower left panel of Fig. 9.5. (c) End on view of binding. Compare to upper left panel of Fig. 9.5. The view is restricted to the amino-terminal prolyl-rich segment of the ligand. (d) End-on view of binding. The view is restricted to the carboxy-terminal positively charged region of the ligand.

distinct from SH2 domains (reviewed in refs 1 and 2; see Chapter 1). Very recently, the structure of a PTB domain complexed with a tightly bound tyrosine phosphorylated peptide has been published.[45]

SH3 domains

The overall fold of SH3 domains consists of two tightly packed anti-parallel β-sheets forming a partly opened barrel. The last strand is interrupted by a turn of a 3–10 helix. This structure is one of a set of similar folds from a variety of proteins and organisms.[23,24] SH3 domains specifically recognize proline-rich segments of 8–10 residues.[25,26] Studies with Abl SH3 permitted the design of a limited combinatorial library, which was used for identification of ligands to phosphoinositide 3′-kinase (PI3K) and Src SH3 domains.[27] Other ligands were identified by synthesis of proline-rich segments of likely biological target sequences. A solution structure of PI3K SH3 with a ligand[28] and crystal structures of ligands of Abl SH3 with Abl and Fyn SH3 domains[29] revealed an apparently common motif involving the left-handed type II polyproline helix binding to the shallow groove of conserved residues, many of which are aromatics, on the SH3 surface. Different residues to the amino-terminal side of the proline-rich motif provide different specificities.

The combinatorial library approach[27] and some mutational evidence suggested that specificity could also be supplied by residues to the

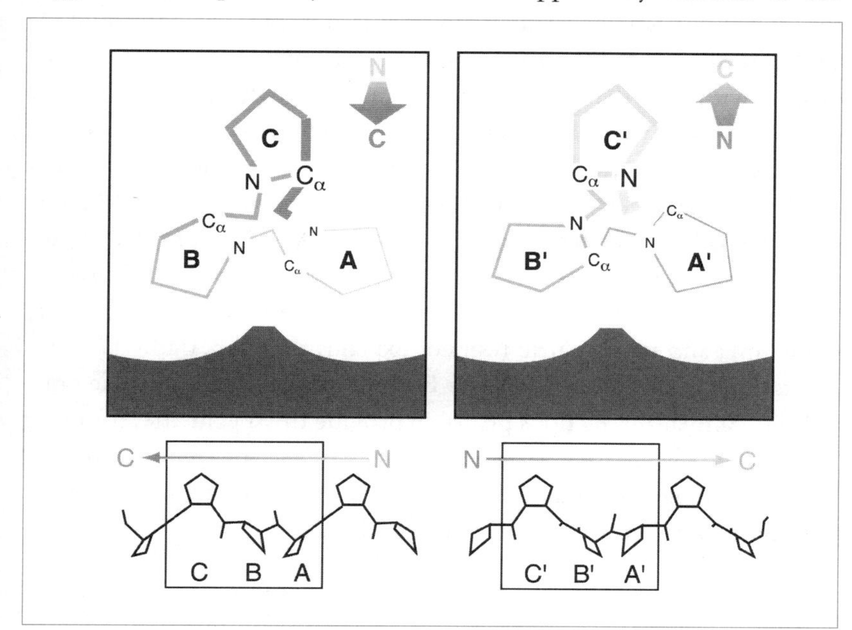

Figure 9.5 Alternate binding in SH3, after Ref. 31.

carboxy-terminal side of the proline-rich motif. Solution structures of two complexes to Src SH3[30] and a crystal structure of a Sem-5 complex[31] revealed a surprising story. For Src SH3 (Figure 9.4), the short proline helix in separate ligands could be bound in either of two directions at the surface, depending on the flanking residues for Src. The Sem-5 ligand structure showed the opposite orientation to that in Abl, Fyn and PI3K SH3 domains. Considerable further efforts (see reviews[1,2]) suggest that indeed both orientations may be naturally possible, with one selected by the specificity-determining residues outside the proline-rich segment (Figure 9.5). This directional ambivalence is intriguing, and apparently unique to the SH3 domains. The proline-rich helix lacks peptidic hydrogens to provide polarity and multiple hydrogen donors to the binding site, and the hydrophobic complementarity of the surfaces permits either of the two orientations to be accommodated in the highly conserved binding surface.

Whether any SH3 domain uses both directions of binding physiologically, at different times, is at present unknown. Specificity of ligands to SH3 domains then depends on the accommodation of the rest of the ligand by the SH3 domains and, as with SH2 domains, residues in loop regions are highly involved. Many SH3 domains (Abl is an exception) have a glutamine or asparagine residue at position 7 of the RT loop (Figure 9.4), and many of the structures determined (see review[1]) show that the positive residues in the ligand interact with the negatively charged side chain of this residue. The presence of an arginine or lysine residue will then orientate the ligand to complement the charge with the RT loop, but this provides little insight into the undoubted selectivity between ligands by different SH3 domains. In one case, the structural basis of selection of one ligand over another has been clearly demonstrated.[32] The peptide ligands PPPALPP**K**KR and PPPALPP**R**KR were co-crystallized with c-Crk SH3. The two differ in affinity by an order of magnitude, despite their similarity in sequence. Specific coordination of the lysine residue at position 8 by a glutamic and two aspartic residues explains the higher affinity, since an arginine at this position cannot fit the charged interface in the same fashion. Similar interactions probably provide the specificities for other SH3–ligand selectivities. Most SH3–ligand interactions reported to date are in the range of 1 μM and above. It is likely that there are higher affinity interactions that are physiologically significant.[2,22,33]

Interactions of SH2 and SH2 domains

Two crystal structures of proteins containing both SH3 and SH2 domains are available. The SH3 and SH2 segment of Lck was reported both with and without an SH2 ligand present.[34] The structure of the adapter protein Grb2 has also been published.[35] In both cases, the SH2 and SH3 domains have tenuous tertiary connections with each other, suggesting that considerable flexibility may be available in the short peptide segments linking them. It was suggested for Lck that occupancy of the SH2 binding site might influence a hypothetical dimerization in the complete kinase. Lck is closely analogous to Src, and its kinase is subtly regulated by SH2 and SH3 interactions with an intrachain tyrosine phosphorylation site.[2] Studies in solution of the similar segment, SH3 and SH2 from Abl, showed that the two domains have no apparent tertiary connections,[52] and established an upper limit for the separation of the two ligand binding sites using bivalent consolidated ligands.[53] Intramolecular SH2 phosphotyrosine interactions have been demonstrated in the (SH2 and SH3) construct from Crk.[54] SH2-mediated dimerization using homomolecular interchain phosphotyrosine–SH2 interaction has been demonstrated for the signal transducing and activating of transcription (STAT) pathway[36] (see Chapters 2 and 3).

Crystal structures of protein segments containing two SH2 domains have also been recently published.[46,47]

PH domains

PH domains were more recently identified and so less precise statements can be made about their structure and function. In particular, the lack of a single ligand type for the whole of the family remains troublesome. It is possible that new insights will resolve this issue in the near future, but it may be that the PH domain defines a class of proteins with only structural similarities and that different family members (or subclasses) serve different functions. In sequence comparisons, those residues that are most frequently conserved (see ref. 5) are hydrophobic. In the known structures of proteins containing PH domains [spectrin,[37] amino-terminal of pleckstrin,[38] dynamin[39,40] (Figure 9.6)], many of the side chains of these conserved residues are in the interior and do not provide possible sites of specific interaction.

The structures solved to date show a partially opened barrel of six β-strands capped with an α-helix[24] (Figure 9.7). These structures, and

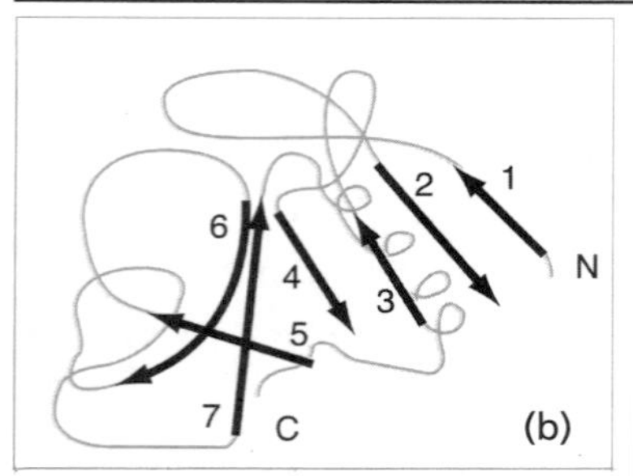

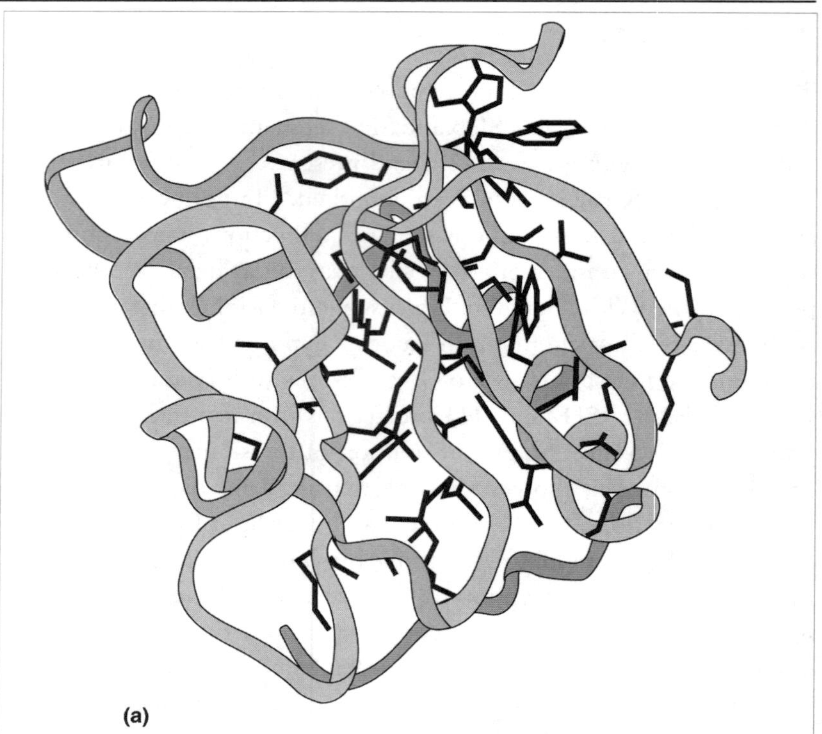

Figure 9.6 (a) Backbone fold for the PH domain of dynamin[39–42]. The residues in dynamin identified as strongly conserved in the family of PH domains[5] are shown as black side chains. Most of these are internal to the structure. (b) Cartoon trace of the secondary structure elements of the PH domain of dynamin. The tight turns between β strands (1 and 2) and (4 and 5) are omitted for clarity.

analysis of sequences of PH domains, suggest that large loops or inserts can be part of the structures of PH domains. For example, in PLC-γ the (2/3) loop contains an insert of several hundred residues including SH2 and SH3 domains.[5]

The overall topology of the PH domains has been compared with other proteins – the lipocalins,[38] streptavidin[39,40] and verotocin.[41,42] The analogy to lipocalins, and their interactions with lipids, led to the demonstration that phosphatidyl *myo*-D-inositol-4,5-bisphosphate [PtdIns(4,5)P_2] binds to N-pleckstrin PH, and several other PH domains.[7] Many proteins containing PH domains are membrane-associated, and/or in the inositol phosphate pathways, so such ligation

Figure 9.7 Alignment of PH domain sequences.

Peckstrin N	REGYLV**KKG**–	————**SVFNT**	**WK**PMWVVLL	ED	GIE**FY**	KKKSDNS————————	PK**G**MIP	LKG	STLTSPC
Dynamin									DVE
Spectrin	MEGFLNRKHE	WEAHNKKASSRS	WHNVYCVIN	NQ	EMGFY	KDAKSAASGIPY————H	SEVPVS	LKE	–AICEVA
Btk	ESIFLKRSQQ	——KKKKTSPLN	FKKRLFLLT	VH	KLSYY	EYDFERGRR—GS————	KKGSID	VE–	–KITCVE
PLC–δ	LKGSQLLKVK	S————SSW	RRERFYKLQ	ED	CKTIW	QESRKVMRTPESQLFSIEDIQ	EVRMGH	RT–	EGLEKFA
	β1	······1/2······	β2	2/3	β3	············3/4············	β4	4/5	β5

Peckstrin N	QDFGK————————————	RMFVF–KITT————TKQQDHFFQA	AF	LEERDAWVRDINKAI
Dynamin	KGFMS————————————	SKHIF–ALFNTEQRNVYKDYRQLELAC	ET	QEEVDSWKASFLRAG
Spectrin	LDYKK————————————	KKHVF–KLRL————SDGNEYLFQA	KD	DEEMNTWIQAISSA–
Btk	TVVPEKNPPPERQIPRRGEESSEMEQISIIER	FPYPF–QVVY————DEGP—LYVFS	PT	EELRKRWIHQLKNVI
PLC–δ	RDVPE————————————	DRCFSI–VFK————DQRNTLDLIA	PS	PADAQHWVLGLHKII
	·············5/6············	β6 ·····6/7····· β7		α1

is a reasonable hypothesis for a physiological function. An alternative set of ligands involves the binding of PH domains to β subunits of the heterotrimeric G proteins. Direct biological modulation by PH domains has been demonstrated by the transfection of transiently expressed PH domain minigenes and of individual G-protein-modulated receptors.[43] Physiological effects were specific for different PH domains. It has also been observed for βARK1, which contains a PH domain that can bind Gβ subunits, that there are co-ordinate interactions of lipids and G protein subunits.[44] It therefore seems likely that either PH domains can bind to both lipids and proteins, or that the identified class of PH domain is a structural family, and that subfamilies of PH domains have defined, separate ligand interactions. The structures of complexes of putative ligands have been published,[48,49] as well as NMR studies indicating[48,50,51] likely sites of ligand binding. While each of these individual studies casts light on the specific interactions studied, only a broad general conclusion seems presently likely for the PH domains, that phosphoinositol or phosphoinositide binding is associated with binding to several positive residues in the amino-terminal half of the molecule. For example, lysine residues identified as involved in binding to inositol phosphates for N-pleckstrin PH domain are not conserved as positively charged residues in other PH domains (Figure 9.7). This may again reflect the issue of different functional subfamilies.

Summary

The structural comments in this chapter are but a reflection of the richness of function in which these domains participate, as shown by many other chapters in this book. Structural biological studies of these domains have been enormously fruitful and are likely to play an essential role in deciphering transduction pathways. The individual folding motifs of these domains have been determined. For SH2s and SH3s, ligands have been clearly identified, and the principal recognition motifs are now understood. Key questions remain. How do these domains interact with each other, and achieve their specific functions? What is the predictive power of sequence analysis and homology modeling to identify correctly structure and function from sequence? What structural methods can be used to unravel the multiple combinatorial interactions stemming from these multiple domains? Answering these questions will provide important biological insights.

References

Complete referencing is not possible, so many of the following are reviews.

1 Pawson, T. (1995) Protein modules and signalling networks. *Nature*, **373**, 573–80.

2 Cohen, G.B., Ren, R. and Baltimore, D. (1995) Modular binding domains in signal transduction proteins. *Cell*, **80**, 237–48.

3 Koch, C.A., Anderson, D., Moran, M.F. *et al.* (1991) SH2 and SH3 domains: elements that control interactions of cytoplasmic signaling proteins. *Science*, **252**, 668–74.

4 Mayer, B.J., Ren, R., Clark, K.L. and Baltimore, D. (1993) A putative modular domain present in diverse signalling molecules. *Cell*, **73**, 629–30.

5 Gibson, T.J., Hyvonen, M., Musacchio, A. and Saraste, M. (1994) PH domain: the first anniversary. *Trends Biochem Sci.*, **19**, 349–53.

6 Kuriyan, J. and Cowburn, D. (1993) Structures of SH2 and SH3 domains. *Curr. Opinion Struct. Biol.*, **3**, 828–37.

7 Harlan, J.E., Hajduk, P.J., Yoon, H.S. and Fesik, S.W. (1994) Pleckstrin homology domains bind to phosphatidylinositol-4,5-bisphosphate. *Nature*, **371**, 168–70.

8 Touhara, K., Inglese, J., Pitcher, J.A. *et al.* (1994) Binding of G protein βγ-subunits to pleckstrin homology domains. *J. Biol. Chem.*, **269**, 10217–20.

9 Wang, D.S., Shaw, R., Winkelmann, J.C. and Shaw, G. (1994) Binding of PH domains of β-adrenergic receptor kinase and β-spectrin to WD40/β-transducin repeat containing regions of the β-subunit of trimeric G-proteins. *Biochem. Biophys. Res. Commun.*, **203**, 29–35.

10 Langdon, S.P. and Smyth, J.F. (1995) Inhibition of cell signalling pathways. *Cancer Treat. Rev.*, **21**, 65–89.

11 Cantley, L.C., Auger, K.R., Carpenter, C. *et al.* (1991) Oncogenes and signal transduction. *Cell*, **64**, 281–302.

12 Mayer, B.J., Jackson, P.K., Van Etten, R.A. and Baltimore, D. (1992) Point mutations in the Abl SH2 domain coordinately impair phosphotyrosine binding *in vitro* and transforming ability *in vivo*. *Mol. Cell. Biol.*, **12**, 609–18.

13 Waksman, G., Kominos, D., Robertson, S.C. *et al.* (1992) Crystal structure of the phosphotyrosine recognition domain of SH2 and v-Src complexed with tyrosine-phosphorylated peptides. *Nature*, **358**, 646–53.

14 Eck, M.J., Shoelson, S.E. and Harrison, S.C. (1993) Recognition of a high-affinity phosphotyrosyl peptide by the Src homology-2 domain of p56lck. *Nature*, **362**, 87–91.

15 Lee, C.-H., Kominos, D., Jacques, S. *et al.* (1994) Crystal structures of peptide complexes of the N-terminal SH2 domain of the Syp tyrosine phosphatase. *Structure*, **2**, 423–38.

16 Xu, R.X., Word, J.M., Davis, D.G. *et al.* (1995) Solution structure of the human pp60c-Src SH2 domain complexed with a phosphorylated tyrosine pentapeptide. *Biochemistry*, **34**, 2107–21.

17 Pascal, S.M., Singer, A.U., Gish, G. *et al.* (1994) Nuclear magnetic resonance structure of an SH2 domain of phospholipase C-γ1 complexed with a high affinity binding peptide. *Cell*, **77**, 461–72.

18 Songyang, Z., Shoelson, S.E., Chaudhuri, M. *et al.* (1993) SH2 domains recognize specific phosphopeptide sequences. *Cell*, **72**, 767–78.

19 Songyang, Z., Shoelson, S.E., McGlade, J. *et al.* (1994) Specific motifs recognized by the SH2 domains of Csk, 3BP2, fps/fes, GRB-2, HCP, SHC, Syk, and Vav. *Mol. Cell. Biol.*, **14**, 2777–85.

20 Ladbury, J.E., Lemmon, M.A., Zhou, M. *et al.* (1995) Measurement of the binding of tyrosyl phosphopeptides to SH2 domains: a reappraisal. *Proc. Natl. Acad. Sci. USA*, **92**, 3199–203.

21 Waksman, G., Shoelson, S.E., Pant, N. *et al.* (1993) Binding of a high affinity phosphotyrosyl peptide to the Src SH2 domain: crystal structures of the complexed and peptide-free forms. *Cell*, **72**, 779–90.

22 Lee, C.H., Leung, B., Lemmon, M.A., Zheng, J., Cowburn, D., Kuriyan, J. and Saksela, K. (1995) A single amino acid in the SH3 domain of Hck determines its high affinity and specificity in binding to HIV-1 Nef protein. *EMBO J.* **14**, 5006–15.

23 Musacchio, A., Wilmanns, M. and Saraste, M. (1994) Structure and function of the SH3 domain. *Prog. Biophys. Mol. Biol.*, **61**, 283–97.

24 Murzin, A.G., Brenner, S.E., Hubbard, T. and Chothia, C. (1995) SCOP: a structural classification of proteins database for the investigation of sequences and structures. *J. Mol. Biol.*, **274**, 536–40.

25 Cicchetti, P., Mayer, B.J., Thiel, G. and Baltimore, D. (1992) Identification of a protein that binds to the SH3 region of Abl, and is similar to Bcr and GAP-rho. *Science*, **257**, 803–6.

26 Ren, R., Mayer, B.K., Cicchetti, P. and Baltimore, D. (1993) Identification of a ten-amino acid proline rich SH3 binding site. *Science*, **259**, 1157–61.

27 Chen, J.K. and Schreiber, S.L. (1995) Combinatorial synthesis and multidimensional NMR spectroscopy: an approach to understanding peptide–ligand interactions. *Angew. Chem., Int. Ed. Engl.*, **34**, 953–69.

28 Yu, H., Chen, J.K., Feng, S. *et al.* (1994) Structural basis for the binding of proline-rich peptides to SH3 domains. *Cell*, **76**, 933–45.

29 Musacchio, A., Saraste, M. and Wilmanns, M. (1994) High-resolution crystal structures of tyrosine kinase SH3 domains complexed with proline-rich peptides. *Nature Struct. Biol.*, **1**, 546–51.

30 Feng, S., Chen, J.K., Yu, H. *et al.* (1994) Two binding orientations for peptides to the Src SH3 domain: development of a general model for SH3–ligand interactions. *Science*, **266**, 1241–7.

31 Lim, W.A., Richards, F.M. and Fox, R.O. (1994) Structural determinants of peptide-binding orientation and of sequence specificity in SH3 domains. *Nature*, **372**, 375–9.

32 Wu, X., Knudsen, N., Feller, S.M. *et al.* (1995) Structural basis for the specific interaction of lysine-containing proline-rich peptrides with the amino-terminal SH3 domain of c-Crk. *Structure*, **3**, 215–26.

33 Saksela, K., Cheng, G. and Baltimore, D. (1995) Proline-rich (PxxP) motifs in HIV-1 Nef bind to SH3 domains of a subset of Src kinases and are required for the enhanced growth of Nef + viruses but not for down-regulation of CD4. *EMBO J.*, **14**, 484–91.

34 Eck, M.J., Atwell, S.K., Shoelson, S.E. and Harrison, S.C. (1994) Structure of the regulatory domains of the Src-family tyrosine kinase Lck. *Nature*, **368**, 764–9.

35 Maignan, S., Guilloteau, J.-P., Becquart, J. and Ducruix, A. (1995) Crystal structure of the mammalian Grb2 adaptor. *Science*, **268**, 291–3.

36 Shuai, K., Horvath, C.M., Huang, L.H.T. *et al.* (1994) Interferon activation of the transcription factor Stat91 involves dimerization through SH2–phosphotyrosyl interactions. *Cell*, **76**, 821–8.

37 Macias, M.J., Musacchio, A., Ponstingl, H. *et al.* (1994) Structure of the pleckstrin homology domain from β-spectrin. *Nature*, **369**, 675–7.

38 Yoon, H.S., Hajduk, P.J., Petros, A.M. *et al.* (1994) Solution structure of a pleckstrin-homology domain. *Nature*, **369**, 672–5.

39 Ferguson, K.M., Lemmon, M.A., Schlessinger, J. and Sigler, P.B. (1994) Crystal structure at 2.2 Ang of the pleckstrin homology domain from human dynamin. *Cell*, **79**, 199–209.

40 Fushman, D., Cahill, S., Lemmon, M.A. *et al.* (1995) Solution structure of the pleckstrin homology domain of dynamin by heteronuclear NMR spectroscopy. *Proc. Natl. Acad. Sci. USA*, **92**, 816–20.

41 Downing, A.K., Driscoll, P.C., Gout, I. *et al.* (1994) Three-dimensional solution structure of the pleckstrin homology domain from dynamin. *Curr. Biol.*, **4**, 884–91.

42 Timm, D., Salim, K., Gout, I. *et al.* (1994) Crystal structure of the pleckstrin homology domain from dynamin. *Nature Struct. Biol.*, **1**, 782–8.

43 Luttrell, L.M., Hawkes, B.E., Touhara, K. *et al.* (1995) Effect of cellular expression of pleckstrin homology domains on G_i-coupled receptor signaling. *J. Biol. Chem.*, **270**, 12984–9.

44 Pitcher, J.A., Touhara, K., Payne, E.S. and Lefkowitz, R.J. (1995) Pleckstrin homology domain-mediated membrane association and activation of the β adrenergic receptor kinase requires coordinate interaction of Gβγ subunits and lipid. *J. Biol. Chem.*, **270**, 11707–10.

45 Zhou, M.M., Ravichandran, K.S., Olejniczak, E.T. *et al.* (1995) Structure and ligand recognition of the phosphotyrosine binding domain of SHC. *Nature*, **378**, 584–92.

46 Hatada, M.H., Lu, X., Laird, E.R. *et al.* (1995) Molecular basis for interaction of the protein tyrosine kinase ZAP-70 with the T-cell receptor. *Nature*, **376**, 32–8.

47 Eck, M.J., Pluskey, S., Trub, T. *et al.* (1995) Spatial constraints on the recognition of phosphoproteins by the tandem SH2 domains of the phosphatase SH-PTP2. *Nature*, **379**, 277–80.

48 Hyvonen, M., Macias, M.J., Nilges, M. *et al.* (1995) Structure of the binding site for inositol phosphates in a PH domain. *EMBO J.* **14**, 4676–85.

49 Ferguson, K.M., Lemmon, M.A., Schlessinger, J. and Sigler, P.B. (1995) Structure

of the high affinity complex of inositol trisphosphate with a phospholipase C pleckstrin homology domain. *Cell*, **83**, 1037–46.

50 Zhang, P., Talluri, S., Deng, H. *et al.* (1995) Solution structure of the pleckstrin homology domain of drosophila β-spectrin. *Structure*, **3**, 1185–95.

51 Zheng, J., Cahill, S.M., Lemmon, M.A. *et al.* (1996) Identification of the binding site for acidic phospholipids on the PH domain of dynamin – implications for stimulation of GTPase activity. *J. Mol. Biol.*, **255**, 14–21.

52 Gosser, Y.Q., Zheng, J., Overduin, M. *et al.* (1995) The solution structure of Abl SH3, and its relationship to SH2 in the SH(32) construct. *Structure*, **3**, 1075–86.

53 Cowburn, D., Zheng, J., Xu, Q.H. and Barany, G. (1995) Enhanced affinities and Specificities of consolidated ligands for the SRC homology (SH) 3 and SH2 domains of abelson protein-tyrosine kinase. *J. Biol. Chem.*, **270**, 26738–41.

54 Rosen, M.K., Yamazaki, T., Gish, G.D. *et al.* (1995) Direct demonstration of an intramolecular SH2-phosphotyrosine interaction in the Crk protein. *Nature*, **374**, 477–9.

10 Ras signaling

Sally J. Leevers

Ras is a small membrane-bound guanine nucleotide-binding protein that acts as a molecular switch linking receptor tyrosine kinase activation to downstream signaling events. In cells, Ras cycles between an active GTP-bound form and an inactive GDP-bound form. The activity of Ras is regulated by the opposing activities of guanine nucleotide exchange factors (GNEFs) and GTPase activating proteins (GAPs). GNEFs catalyze the activation of Ras via the exchange of Ras-bound GDP for GTP, whereas GAPs accelerate the normally slow intrinsic GTPase activity of Ras proteins, thereby inactivating them. The best characterized downstream effector molecule of RasGTP is Raf, a serine/threonine kinase which links Ras to the MAP kinase cascade. A striking feature of Ras signaling is its conservation in different organisms. The GAPs and GNEFs which regulate Ras are structurally conserved in higher and lower eukaryotes.

In this chapter, the biological and biochemical properties of Ras and the way in which it acts as a signaling molecule will be discussed. In addition, the molecules immediately upstream and downstream of Ras and their regulation will be described. More information on signaling through receptor tyrosine kinases and the MAP kinase cascade can be found in Chapters 1 and 11, respectively.

Oncogenic properties of Ras

The *ras* genes and the 21 kDa proteins they encode were first characterized as the transforming agents of highly oncogenic retroviruses. Then, when 'focus formation' assays were used to search for dominantly acting transforming genes (oncogenes) in tumor-derived DNA, three mammalian *ras* genes were identified as cellular oncogenes: Ha-*ras*, Ki-*ras* and N-*ras*. Oncogenically activated versions of all three genes, encoding proteins with point mutations at amino acid residues 12, 13 and 61, have been detected in numerous human malignancies, although the incidence and the specific *ras* gene that is mutated varies considerably with the type of malignancy.[1]

Signal Transduction. Edited by Carl-Henrik Heldin and Mary Purton. Published in 1996 by Chapman & Hall. ISBN 0 412 70810 8

Ras function in mammalian cells

The discovery that certain point mutations convert *ras* genes into oncogenes immediately implicated Ras in the control of cell growth. Subsequent experiments in which proteins were microinjected into cultured fibroblasts showed that activated Ras induced DNA synthesis and morphological transformation,[2,3] whereas Ras-neutralizing monoclonal antibodies inhibited serum-stimulated DNA synthesis and transformation by membrane-bound tyrosine kinases.[4,5] More recently, extensive analyses have shown that activated Ras initiates some of the intracellular signals normally seen in response to mitogens that signal via receptor tyrosine kinases. Consistent with this, inhibition of cellular Ras blocks activation of the same signals by ligands that activate receptor tyrosine kinases.

In contrast to the mitogenic effects observed in fibroblasts, activated Ras induces growth arrest and differentiation in other cell types. For example, in the rat phaeochromocytoma-derived PC12 cell line, activated Ras mimics the action of NGF and induces growth arrest and the extension of neurite-like processes,[6] whereas the inhibition of cellular Ras inhibits NGF-stimulated neurite extension.[7] The way in which Ras induces opposing effects in different cell types is not fully understood, although the initial state of the cell and differences in the duration of downstream signaling events are possible explanations.[8]

Ras function in other eukaryotes

Although this chapter will focus on the regulation and targets of Ras in mammalian cells, it should be remembered that Ras is found ubiquitously in eukaryotes. Ras is implicated in various developmental processes in amphibians, flies and nematodes, all of which are normally controlled by receptor tyrosine kinases. For example, insulin-induced maturation of *Xenopus laevis* oocytes is mimicked by activated Ras and abrogated by the inhibition of cellular Ras. In the nematode *Caenorhabditis elegans* the induction of vulval cell fate is dependent on the activity of the *let*-23 gene product, a receptor tyrosine kinase, and the *let*-60 gene product, a Ras protein.[9] Similarly, in the eye of the fruit fly *Drosophila melanogaster* the induction of the R7 photoreceptor requires the activity of the Sevenless receptor tyrosine kinase and the *Drosophila* Ras1 protein.[10]

Both fission yeast and budding yeast also contain Ras homologs, although these organisms lack receptor tyrosine kinases and the signaling pathways upstream and downstream of these proteins are somewhat diverged. However, yeast GAPs and GNEFs have catalytic domains that are highly homologous to those found in higher eukaryotes.

Biochemical and structural properties of Ras

Ras is a membrane-localized guanine nucleotide-binding protein with intrinsic GTPase activity. An alignment of the amino acid sequences of Ras proteins from various species reveals two regions of differing homology and function. The amino-terminal region (amino acid residues 1–165 of human Ras) is highly conserved and necessary for guanine nucleotide binding and GTPase activity. The carboxy-terminal portion (amino acid residues 166–189 for human Ras) contains sequences conferring membrane localization and is more variable, apart from the four carboxy-terminal amino acids that always encode a CAAX motif (C = cysteine, A = aliphatic amino acid, X = any amino acid).

Ras is structurally homologous to other members of the GTPase superfamily within the guanine nucleotide-binding and catalytic regions[11] and, like other GTPases, Ras cycles between an active GTP-bound form and an inactive GDP-bound form.[11,12] As stated above, Ras is inactivated by GAPs which accelerate the hydrolysis of Ras-bound GTP and activated by GNEFs which accelerate the conversion of RasGDP to RasGTP (see Figure 10.1). The way in which oncogenic point mutations increase the amount of GTP-bound Ras reflects these two modes of Ras regulation. The mutations either decrease the intrinsic GTPase activity of Ras and make it resistant to the action of GAPs or they increase the GDP-off rate of Ras so that GTP, which is more abundant *in vivo*, replaces it. Although this suggests that both GAPs and GNEFs co-operate to regulate RasGTP levels *in vivo*, only GTPase Ras mutants have been detected in human malignancy.[1]

The solution of the X-ray crystallographic structure of Ras bound to various guanine nucleotide homologs has provided tantalizing insights into the way in which Ras acts as a signaling molecule[13] (see Figure 10.2). Two so-called 'switch regions' on the surface of Ras adopt different conformations in the GDP- and GTP-bound forms. Of these 'switch region I' overlaps with loop 2 (L2) and the so-called 'effector' domain of Ras (amino acid residues 32–40). Point mutations in this region render Ras biologically inactive and unable to interact with downstream effector molecules such as Raf. Switch region II coincides with a region implicated in mediating Ras activation via nucleotide

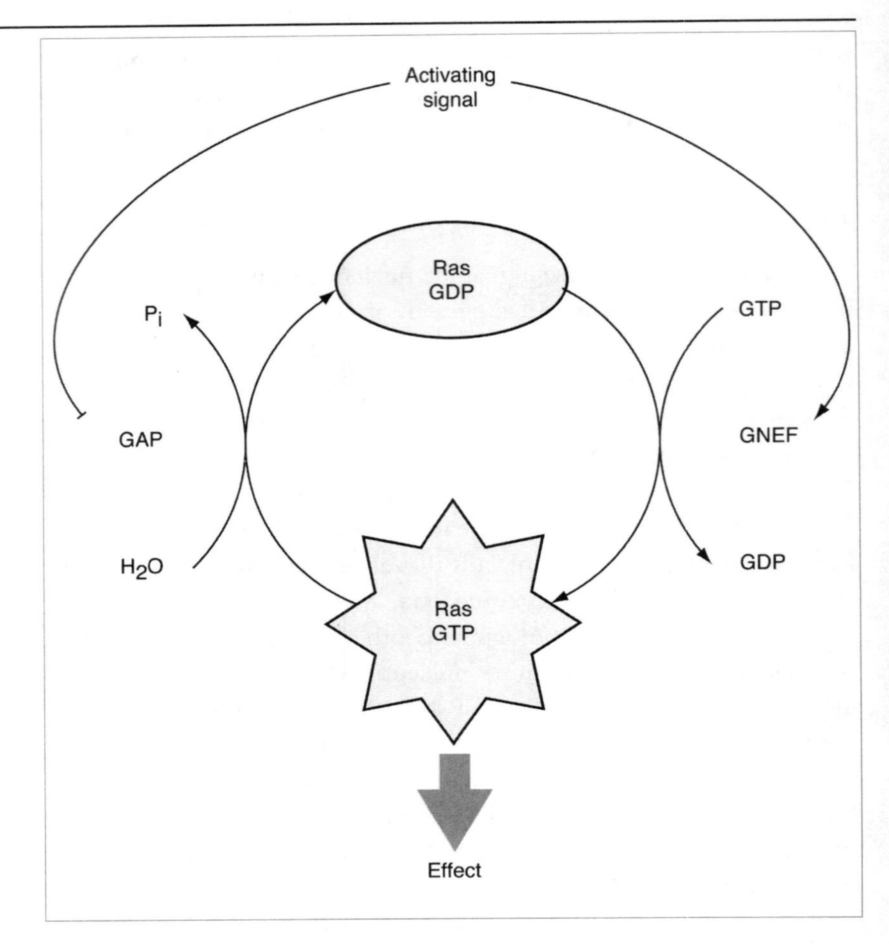

Figure 10.1 The Ras cycle. A model showing how Ras can be regulated *in vivo* by the opposing activities of GNEFs and GAPs.

exchange (amino acid residues 63–73). Hence, we can speculate that exchange factors responsible for Ras activation recognize inactive GDP-bound Ras via the conformation of switch region II, whereas Raf and other effectors of Ras recognize active GTP-bound Ras by selectively interacting with the GTP-bound conformation of switch region I.

Membrane localization of Ras

The carboxy-terminal hypervariable domain and CAAX box of Ras proteins specify their membrane localization – a property that is essential in order for Ras to induce the activation of Raf and the MAP kinase pathway.[14,15] A multistep post-translational modification pathway leads to the membrane localization of Ras.[16] The CAAX motif is modified by farnesylation on the cysteine residue, proteolysis

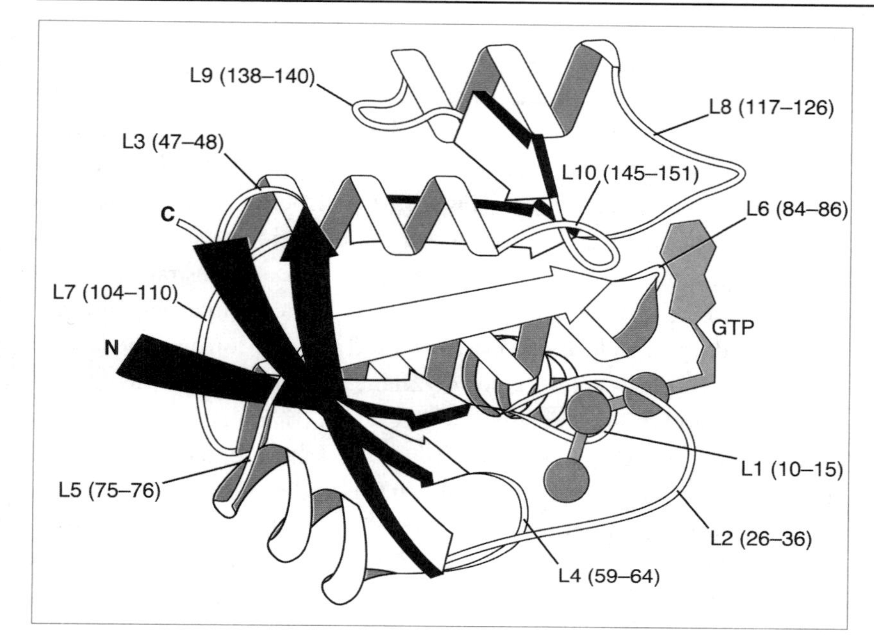

Figure 10.2 Cartoon of the three-dimensional structure of the guanine nucleotide binding domain of human Ha-Ras bound to the nucleotide analog GppNHp. The ten loops connecting six strands of β-sheet and five α-helices are labeled. [Reproduced with permission from Whittinghofer, A. and Pai, E.F. (1991) *Trends in Biochemical Sciences*, 16, 382–7.]

to remove the AAX amino acid residues and carboxymethylation, probably on the carboxy-terminal cysteine. Either palmitoylation on additional cysteines or the presence of a stretch of basic amino acids in the hypervariable region is also required to bring about full membrane localization.

The importance of Ras membrane localization has been demonstrated by recent research in two areas. First, drugs that inhibit the enzymes responsible for these posttranslational modifications are able to revert Ras-transformed cells and represent potentially powerful anti-tumorigenic agents.[17] Second, the requirement for Ras in the activation of its effector Raf and the MAP kinase pathway can be overcome by targeting Raf to the plasma membrane with the Ras hypervariable domain. Thus it seems that the role of Ras in Raf activation is to mediate the translocation of this normally cytosolic protein to a new microenvironment at the plasma membrane where additional signals lead to its activation[18,19] (see below).

Hence Ras represents an elegantly designed membrane-bound molecular 'switch' that transduces signals via conformational changes in portions of the molecule sensitive to the presence or absence of a γ-phosphate on the bound guanine nucleotide. Normally GNEFs and GAPs co-operate to control the switch, but oncogenic mutations disrupt that regulation and freeze or stabilize GTP-bound Ras.

Activation of Ras *in vivo*

Both cellular and oncogenic Ras are active when bound to GTP, and it is the relatively high abundance of the GTP-bound form of certain Ras mutants that confers their oncogenic potential.[12,20] The immunoprecipitation of cellular Ras from ^{32}P-labeled cells, followed by the resolution of Ras-bound guanine nucleotides using thin-layer chromatography, has been widely employed to demonstrate that numerous extracellular signals stimulate RasGTP accumulation. Interestingly, in addition to ligands known to bind to receptor tyrosine kinases or to activate non-receptor tyrosine kinases, G-protein-coupled receptor agonists and direct activators of PKC also activate Ras in certain cell types, although in most cases this activation is sensitive to tyrosine kinase inhibitors.[21,22] Recent data have also implied that ligation to integrins (receptors for components of the extracellular cell matrix) also activates Ras, possibly via the activation of the focal adhesion associated tyrosine kinase, p125FAK.[23] Typically, 2–7% of cellular Ras is in the GTP-bound state in unstimulated cells, and two- to threefold increases in RasGTP levels are observed following ligand treatment.

The measurement of Ras-specific GNEF and GAP activity in whole cell lysates implies that Ras regulation results from the close co-operation of these regulatory proteins. For example, NGF treatment of PC12 cells elicits slight (twofold) increases in GAP activity, and more marked increases in guanine nucleotide exchange factor activity.[24] In fibroblasts, it seems that GAP activity (and therefore the inactivation of Ras) increases with the density of cultured cells, peaking at confluence, whereas GNEF activity (and therefore Ras activation) increases following the ligand treatment of cells.[25,26] In T cells, however, ligand-induced Ras activation is mediated by the inhibition of GAP activity and not nucleotide exchange.[21]

Mammalian Ras GNEFs

To date, five potential mammalian Ras GNEFs containing a related domain encoding Ras GNEF activity have been described: Sos, C3G, p140Ras GRF, smgGDS and Vav (see Figure 10.3).

Sos

The best characterized Ras GNEF, Sos (Son of Sevenless), was first identified in *Drosophila*, although mouse and human homologs have subsequently been cloned. Sos is ubiquitiously expressed and genetic and

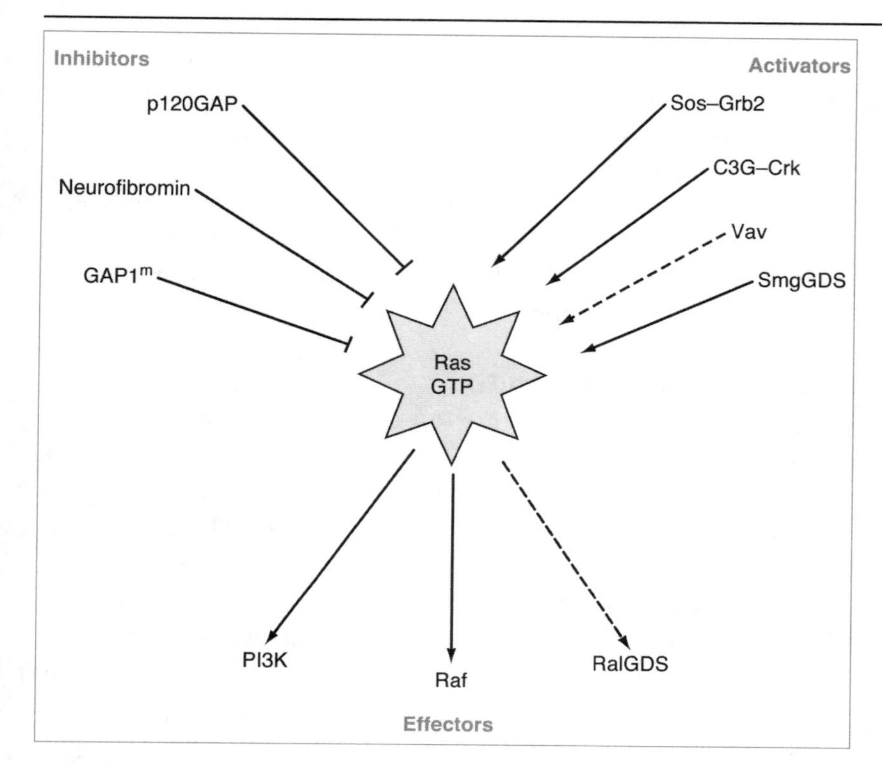

Figure 10.3 A diagram summarizing the proteins able to interact with Ras and their postulated function.

biochemical evidence strongly suggest that it functions as a Ras GNEF *in vivo*.[27] In addition to a central Ras GNEF domain, the Sos proteins have putative regulatory sequences at the amino and carboxyl termini. The carboxy-terminal portion of Sos contains proline-rich sequences typical of those which interact with SH3 domains (see Chapter 9). These proline-rich motifs mediate the interaction of Sos with Grb2, an adaptor molecule comprising of one SH2 domain flanked by two SH3 domains. Following receptor tyrosine kinase activation and autophosphorylation, the SH2 domain of Grb2 binds to specific phosphotyrosine docking sites on the autophosphorylated receptor and associated molecules such as Shc, Crk and Syp,[28–30] thus translocating the Sos–Grb2 complex to the plasma membrane and the vicinity of Ras (see Chapter 1). The amino-terminal portion of Sos contains a pleckstrin homology (PH) domain (see Chapter 9) that is essential for Sos function in *Drosophila* and might also participate in the membrane localization and activation of Sos by the Sevenless receptor tyrosine kinase.[31] Although it remains unclear whether subsequent regulation of Sos activity occurs, for example via protein phosphorylation,[32] a membrane-targeted version of Sos can activate Ras and transform cells, suggesting that the translocation of Sos to the plasma membrane alone is sufficient to induce Ras activation.[33]

C3G

C3G is another ubiquitously expressed Ras GNEF with many structural and functional similarities to Sos. It contains proline-rich sequences that enable it to interact with the SH3 domain of Crk, a Grb2-like SH2 and SH3 domain-containing adaptor molecule.[34] Furthermore, the C3G–Crk complex also associates with autophosphorylated receptor tyrosine kinases and their substrates and is implicated in Ras activation.[35]

p140RasGRF, Vav and SmgGDS

In contrast to Sos and C3G, other Ras GNEFs have a more restricted expression pattern and function *in vivo*. p140RasGRF is expressed exclusively in the brain,[36] while Vav may represent a GNEF for Ras that is only expressed in haematopoietic cells.[37] SmgGDS was originally described as an exchange factor for the Ras-related protein, Rap1, but has subsequently been shown to have exchange factor activity towards proteins in the Ras superfamily with polybasic regions in their carboxy-terminal hypervariable domains including K-Ras(B).[38] Consistent with its ability to discriminate substrates on the basis of their hypervariable domain, smgGDS only acts on the post-translationally modified versions of these proteins.

Mammalian Ras GAPs

The existence of factors able to enhance the GTPase activity of Ras was first implied when the rate of GTP hydrolysis of Ras-GTP injected into frog oocytes was found to exceed greatly that of Ras-GTP *in vitro*.[12] Since then, three mammalian Ras GAPs with homologous Ras GTPase activating domains have been described: p120GAP, neurofibromin [the product of the von Recklinghausen neurofibromatosis type I (*NF1*) gene] and $GAP1^{m}$.

p120GAP

p120GAP is a multi-domain protein containing two SH2, one SH3 and one PH domain, proline-rich sequences and and a region similar to the CalB domain of phospholipase A_2.[39] While numerous experiments have demonstrated that p120GAP inactivates cellular but not oncogenic Ras by virtue of its carboxy-terminal GTPase activating domain, many aspects of p120GAP regulation and function are unresolved. The demonstration that Ras has to be GTP-bound with an intact effector domain in order to interact with p120GAP led to numerous experiments designed to determine whether p120GAP might be an effector as well as downregulator of RasGTP.[40] Although some experiments did suggest that p120GAP might mediate certain signals emerging from Ras,[41] more likely Ras effectors have now been identified and p120GAP currently seems more likely to be an inhibitor of Ras function.

A number of possible regulatory mechanisms for p120GAP have been identified. p120GAP is one of several signaling proteins that associate with and are tyrosine phosphorylated by activated membrane-bound receptor tyrosine kinases. In addition, in ligand-treated cells p120RasGAP associates with two other phosphoproteins, p190 and p62, and translocates to the particulate fraction of fractionated cells.[42] These protein–protein interactions or the intracellular translocation might regulate p120GAP; *in vitro*, p120GAP complexed with p190[43] or the activated EGF receptor[44] has a reduced specific activity towards RasGTP.

There are also *in vitro* data showing that p120GAP activity is inhibited by certain lipids, including arachidonic acid and arachidonic acid-containing phosphatidic acid. However, the correlation between the concentration of a lipid required to inhibit GAP activity and the critical micellar concentration of that lipid[45] implies that GAP inhibition may arise simply because of the detergent action of these lipids, and not because of stoichiometric biochemical interactions.

Neurofibromin

Neurofibromin is a 250 kDa protein with a central GTPase activating domain which has Ras-specific GAP activity. Mutations that result in the absence of or a lower GAP activity of neurofibromin are associated with *NF1*, and there is evidence that the benign and sometimes malignant tumors associated with *NF1* may result from elevated levels of RasGTP.[46,47] *In vitro*, neurofibromin has a higher affinity than p120GAP for Ras but a lower specific activity,[48] suggesting that neurofibromin might be a more effective down-regulator than p120GAP when RasGTP is present at low levels. The GAP activity of the neurofibromin can also be inhibited by lipids *in vitro*.[48]

GAP1^{m}

The expression pattern of GAP1^{m} is more restricted than that of p120GAP and neurofibromin, with high levels in the brain, kidneys and placenta and lower levels elsewhere.[49] In addition to a conserved GTPase activating domain, GAP1^{m} also has a region homologous to the phospholipid binding domain of synaptotagmin, suggesting that this Ras GAP might also have the potential to be regulated by lipids.

Effectors of Ras

Raf

Once it had been established that the Raf serine/threonine kinase functions downstream of Ras both in mammalian cells and *Drosophila*,[14,50] several groups showed that Ras and Raf can interact directly.[51] Raf has a carboxy-terminal kinase domain and an amino-terminal regulatory domain which contains a zinc finger motif and serine/threonine-rich region. Ras binds to a region at the amino terminus of Raf overlapping with part of the zinc finger motif, but only when Ras itself has an intact effector domain and is bound to GTP. *In vitro*, this interaction with Ras is unable to increase the kinase activity of Raf. However, in Ras-transformed or ligand-treated cells Raf translocates from the cytoplasm to the plasma membrane, suggesting that RasGTP might activate Raf by inducing a redistribution of the protein.[52,53] This hypothesis was tested by examining the activity of a membrane-bound version of Raf, RafCAAX, generated by fusing the hypervariable domain of Ras on to the carboxyl terminus of Raf. RafCAAX is constitutively active, and can be further activated by epidermal growth factor (EGF). Importantly, both the constitutive and the ligand-stimulated activity of RafCAAX do not require cellular Ras, suggesting that the role of Ras in the activation of Raf is to translocate Raf to a microenvironment at the plasma membrane where subsequent modifications lead to its full activation.[18,19]

Precisely how membrane-localized Raf becomes activated is currently under intense investigation. Raf is heavily phosphorylated in activated cells and there is evidence that at least some of these phosphorylation events can increase its kinase activity.[54,55] Furthermore, Raf also interacts with the proteins hsp90, p50 and 14-3-3, although whether these interactions play a passive or active role in Raf activation remains to be determined.[56]

PI3 Kinase

Recent research has suggested that phosphoinositide 3′-kinase (PI3K) might also represent a target of active Ras. GTP-bound Ras interacts via its effector domain with the catalytic subunit of PI3K *in vitro*. Furthermore, levels of the phospholipid products of PI3K are elevated by the co-expression of active Ras and PI3K, while their production in response to ligands is reduced when cellular Ras is inhibited.[57] However, other experiments with a constitutively active form of PI3K suggest that PI3K might lie upstream of Ras.[58] While the precise relationship between PI3K and Ras is not yet resolved, the apparent involvement of PI3K in cytoskeletal reorganization[59] (see Chapter 13) and the ability of Ras to induce morphological transformation make the notion that PI3K acts as another effector of Ras an attractive hypothesis.

RalGDS, RGL

Other potential effectors of Ras, identified using the yeast two-hybrid system, are RalGDS, a GNEF for the Ras-related molecule Ral, and RGL, a RalGDS-related molecule.[60] Although these molecules are clearly able to interact with RasGTP via the effector domain, the physiological relevance of these interactions has yet to be demonstrated.

Antagonism of Ras signaling by cAMP

Recent evidence suggests that in, some cell types, elevated concentrations of intracellular cAMP and thus activation of protein kinase A (PKA) might antagonize Ras signaling[40,61] (see Chapter 15). This antagonism seems to occur at the level of Raf activation and, consistent with this, it has been shown that PKA can phosphorylate the amino terminus of Raf.

Perspectives

While our understanding of how Ras acts as a molecular switch in signal transduction is relatively good, our grasp of how the molecules that interact with Ras function is somewhat weaker. Precisely how the factors that regulate Ras do so in a co-ordinated manner *in vivo* is unclear. Similarly, which effector Ras interacts with when, and why is another issue to be addressed. The functional significance of the different Ras isoforms, and the reason for the association of different point-mutated *ras* genes with different carcinomas are other areas requiring further investigation.

Another likely area of future research into signaling which is clearly important for the Ras pathway is the intracellular movement of signaling molecules. While it is clear that both the Sos–Grb2 complex and Raf translocate from the cytoplasm to the vicinity of Ras at the cell membrane during cell activation (see above and Figure 10.4), the way in which these intracellular translocations occur is uncharacterized.

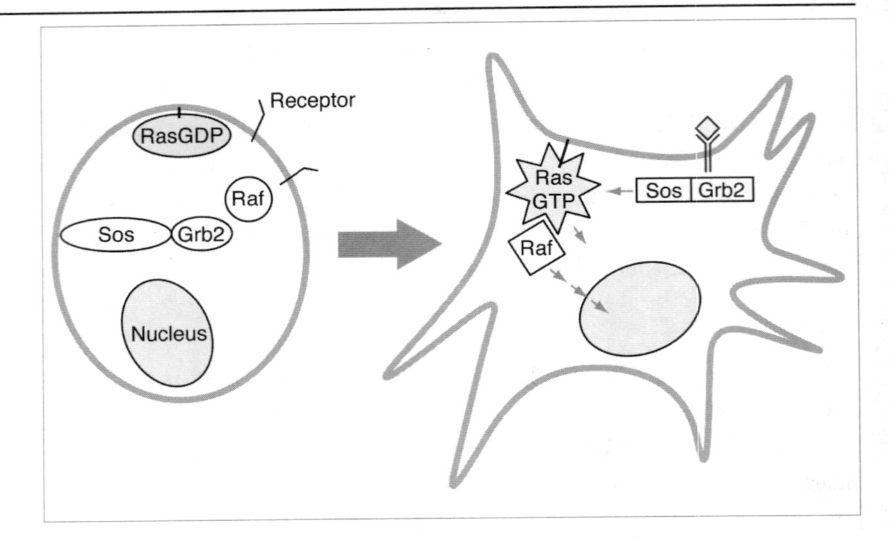

Figure 10.4 Ras signaling during cell activation. A model showing the redistribution of signaling molecules upon activation of the Ras signaling pathway.

Ras and its regulators and effectors clearly illustrate that intracellular signaling is the result of a network, as opposed to linear pathway of molecular interactions. Many of the molecules described above interact with other signaling molecules in addition to Ras, and there is the potential for numerous signaling pathways to be crosslinked, enabling cells to respond to changes in extracellular conditions with a plethora of responses.

References

Owing to space limitations, many important references could not be cited, and the author apologizes to the researchers involved.

1 Bos, J.L. (1989) Ras oncogenes in human cancer: a review. *Cancer Res.*, **49**, 4682–9.

2 Feramisco, J.R., Gross, M., Kamata, T. *et al.* (1984) Microinjection of the oncogene form of the human ras (T24) protein results in rapid proliferation of quiescent cells. *Cell*, **38**, 109–17.

3 Stacey, D.W. and Kung, H.-F. (1984) Transformation of NIH 3T3 cells by microinjection of Ha-ras p21 protein. *Nature*, **310**, 508–11.

4 Mulcahy, L.S., Smith, M.R. and Stacey, D.W. (1985) Requirement for *ras* proto-oncogene function during serum-stimulated growth of NIH3T3 cells. *Nature*, **313**, 241–3.

5 Smith, M.R., DeGubicus, S.J. and Stacey, D.W. (1986) Requirement for c-ras protein during viral oncogene transformation. *Nature*, **320**, 540–3.

6 Bar-Sagi, D. and Feramisco, J.R. (1985) Microinjection of the ras oncogene into PC12 cells induces morphological differentiation. *Cell*, **42**, 841–8.

7 Szeberényi, J., Cai, H. and Cooper, G.M. (1990) Effect of a dominant inhibitory

Ha-ras mutation on neuronal differentiation of PC12 cells. *Mol. Cell. Biol.*, **10**, 5324–32.

8 Marshall, C.J. (1995) Specificity of receptor tyrosine kinase signaling: transient versus sustained extracellular signal-regulated kinase activation. *Cell*, **80**, 179–85.

9 Han, M. and Sternberg, P.W. (1990) *let*-60, a gene that specifies cell fates during *C. elegans* vulval induction, encodes a *ras* protein. *Cell*, **63**, 921–31.

10 Dickson, B. and Hafen, E. (1993) Genetic dissection of eye development in *Drosophila*, in *The Development of Drosophila* (eds Martinez-Arias, A. and Bate, M.), Cold Spring Harbor Laboratory Press, Cold Spring Harbor, 1327–62.

11 Bourne, H.R., Sanders, D.A. and McCormick, F. (1991) The GTPase superfamily: conserved structure and molecular mechanism. *Nature*, **349**, 117–27.

12 Trahey, M. and McCormick, F. (1987) A cytoplasmic protein stimulates normal Nras p21 GTPase, but does not affect oncogenic mutants. *Science*, **238**, 542–5.

13 Wittinghofer, A. and Pai, E.F. (1991) The structure of Ras protein: a model for a universal molecular switch. *Trends Biochem. Sci.*, **16**, 382–7.

14 Howe, L.R., Leevers, S.J., Gómez, N. *et al.* (1992) Activation of the MAP kinase pathway by the protein kinase raf. *Cell*, **71**, 335–42.

15 Leevers, S.J. and Marshall, C.J. (1992) Activation of extracellular signal-regulated kinase, ERK2, by p21ras oncoprotein. *EMBO J.*, **11**, 569–74.

16 Hancock, J. and Marshall, C.J. (1993) Post-translational processing of Ras and Ras-related proteins in *The Ras Superfamily of GTPases* (eds Lacal, J. and McCormick, F.), CRC Press Inc., 65–84.

17 Travis, J. (1993) Novel anticancer agents move closer to reality. *Science*, **260**, 1877–8.

18 Leevers, S.J., Paterson, H.F. and Marshall, C.J. (1994) Requirement for Ras in Raf activation is overcome by targeting Raf to the plasma membrane. *Nature*, **369**, 411–4.

19 Stokoe, D., Macdonald, S.G., Cadwallader, M. *et al.* (1994) Activation of Raf as a result of recruitment to the plasma membrane. *Science*, **264**, 1463–7.

20 McCormick, F. (1989) Ras GTPase activating protein: signal transmitter and signal terminator. *Cell*, **56**, 5–8.

21 Downward, J., Graves, J.D., Warne, P.H. *et al.* (1990) Stimulation of p21ras upon T-cell activation. *Nature*, **346**, 719–23.

22 van Corven, E.J., Hordijk, P.L., Medema, R.H. *et al.* (1993) Pertussis toxin-sensitive activation of p21ras by G protein-coupled receptor agonists in fibroblasts. *Proc. Natl. Acad. Sci. USA*, **90**, 1257–61.

23 Schlaepfer, D.D., Hanks, S.K., Hunter, T. and van der Geer, P. (1994) Integrin-mediated signal transduction linked to Ras pathway by GRB2 binding to focal adhesion kinase. *Nature*, **372**, 786–91.

24 Li, B., Kaplan, D., Kung, H. and Kamata, T. (1992) Nerve growth factor stimulation of Ras guanine nucleotide exchange factor and GTPase activating protein activity. *Science*, **256**, 1456–9.

25 Hoshino, M., Kawakita, M. and Hattori, S. (1988) Characterisation of a factor that stimulates hydrolysis of GTP bound to p21Ras and correlation of its activity with cell density. *Mol. Cell. Biol.*, **8**, 4169–73.

26 Zhang, K., Papageorge, A.G. and Lowy, D.R. (1992) Mechanistic aspects of signaling through Ras in NIH 3T3 cells. *Science*, **257**, 671–4.

27 Bar-Sagi, D. (1994) The Sos (Son of Sevenless) protein. *Trends Endocrinol. Metab.*, **5**, 165–9.

28 Rozakis, A.M., Fernley, R., Wade, J. *et al.* (1993) The SH2 and SH3 domains of mammalian Grb2 couple the EGF receptor to the Ras activator mSos1. *Nature*, **363**, 83–5.

29 Li, W., Nishimura, R., Kashishian, A. *et al.* (1994) A new function for a phosphotyrosine phosphatase: linking GRB2-Sos to a receptor tyrosine kinase. *Mol. Cell. Biol.*, **14**, 509–17.

30 Matsuda, M., Hashimoto, Y., Muroya, K. *et al.* (1994) CRK protein binds to two guanine nucleotide-releasing proteins for the Ras family and modulates nerve growth factor-induced activation of Ras in PC12 cells. *Mol. Cell. Biol.*, **14**, 5495–500.

31 Karlovich, C.A., Bonfini, L., McCollam, L. *et al.* (1995) *In vivo* analysis of the Ras exchange factor Son of Sevenless. *Science*, **268**, 576–9.

32 Cherniack, A.D., Klarlund, J.K. and Czech, M.P. (1994) Phosphorylation of the Ras nucleotide exchange factor Son of Sevenless by mitogen-activated protein kinase. *J. Biol. Chem.*, **269**, 4717–20.

33 Aronaheim, A., Engelberg, D., Li, N. *et al.* (1994) Membrane targetting of the nucleotide exchange factor Sos is sufficient for activating the Ras signalling pathway. *Cell*, **78**, 949–61.

34 Tanaka, S., Morishita, T., Hashimoto, Y. *et al.* (1994) C3G, a guanine nucleotide releasing protein expressed ubiquitously, binds to the Src homology 3 domains of CRK and GRB2/ASH proteins. *Proc. Natl. Acad. Sci. USA*, **91**, 3443–7.

35 Sawasdikosol, S., Ravichandran, K.S., Lee, K.K. *et al.* (1995) Crk interacts with tyrosine-phosphorylated p116 upon T-cell activation. *J. Biol. Chem.*, **270**, 2893–6.

36 Shou, C., Farnsworth, C.L., Neel, B.G. and Feig, L.A. (1992) Molecular cloning of cDNAs encoding a guanine nucleotide exchange factor for p21Ras. *Nature*, **358**, 351–4.

37 Gulbins, E., Coggeshall, K.M., Langlet, C. *et al.* (1994) Activation of Ras in vitro and in intact fibroblasts by the Vav guanine nucleotide exchange protein. *Mol. Cell. Biol.*, **14**, 906–13.

38 Kikuchi, A., Kaibuchi, K., Hori, Y. *et al.* (1992) Molecular cloning of the human cDNA for a stimulatory GDP/GTP exchange protein for c-Ki-ras and smg p21. *Oncogene*, **7**, 289–93.

39 Boguski, M.S. and McCormick, F. (1993) Proteins regulating Ras and its relatives. *Nature*, **366**, 643–54.

40 Pronk, G.J. and Bos, J.L. (1994) The role of p21ras in receptor tyrosine kinase signaling. *Biochim. Biophys. Acta*, **1198**, 131–47.

41 Duchesne, M., Schweighoffer, F., Parker, F. *et al.* (1993) Identification of the SH3 domain of GAP as an essential sequence for Ras-GAP-mediated signalling. *Science*, **259**, 525–8.

42 Ellis, C., Moran, M., McCormick, F. and Pawson, T. (1990) Phosphorylation of GAP and GAP-associated proteins by transforming and mitogenic tyrosine kinases. *Nature*, **343**, 377–81.

43 Moran, M., Polakis, P., McCormick, F. *et al.* (1991) Protein tyrosine kinases regulate the phosphorylation, protein interactions, subcellular distribution and activity of p21ras GAP. *Mol. Cell. Biol.*, **11**, 1804–12.

44 Serth, J., Lautwein, A., Frech, M. *et al.* (1991) Binding of the Ras GTPase activating protein by activated EGF receptors leads to inhibition of the p21 GTPase activity *in vitro. EMBO J.*, **10**, 1325–60.

45 Serth, J., Weber, W., Frech, M. *et al.* (1992) The inhibition of the GTPase activating protein–Ras interaction by acidic lipids is due to physical association of the C-terminal domain of the GTPase activating protein with micellar structures. *Biochemistry*, **31**, 6361–5.

46 Basu, T.N., Gutmann, D.H., Fletcher, J.A. *et al.* (1992) Aberrant regulation of ras proteins in malignant tumour cells from type 1 neurofibromatosis patients. *Nature*, **356**, 713–5.

47 DeClue, J.E., Papageorge, A.G., Fletcher, J.A. *et al.* (1992) Abnormal regulation of mammalian p21ras contributes to malignant tumor growth in von Recklinghausen (type 1) neurofibromatosis. *Cell*, **69**, 265–73.

48 Bollag, G. and McCormick, F. (1991) Differential regulation of rasGAPs and neurofibromatosis gene product activities. *Nature*, **351**, 576–80.

49 Maekawa, M., Li, S., Iwamatsu, A. *et al.* (1994) A novel mammalian Ras GTPase-activating protein which has phospholipid-binding and Btk homology regions. *Mol. Cell. Biol.*, **14**, 6879–85.

50 Dickson, B., Sprenger, F., Morrison, D. and Hafen, E. (1992) Raf functions downstream of Ras1 in the Sevenless signal transduction pathway. *Nature*, **360**, 600–3.

51 Avruch, J., Zhang, X.-F. and Kyriakis, J.M. (1994) Raf meets Ras: completing the framework of a signal transduction pathway. *Trends Biochem. Sci.*, **19**, 279–83.

52 Traverse, S., Cohen, P., Paterson, H. *et al.* (1993) Specific association of activated MAP kinase kinase (Raf) with the plasma membranes of ras-transformed retinal cells. *Oncogene*, **8**, 3175–81.

53 Wartman, M. and Davis, R.J. (1994) The native structure of the activated Raf protein kinase is a membrane-bound multi-subunit complex. *J. Biol. Chem.*, **269**, 6695–701.

54 Fabian, J.R., Daar, I.O. and Morrison, D.K. (1993) Critical tyrosine residues regulate the enzymatic and biological activity of Raf-1 kinase. *Mol. Cell. Biol.*, **13**, 7170–9.

55 Kolch, W., Heidecker, G., Kochs, G. *et al.* (1993) Protein kinase Cα activates Raf-1 by direct phosphorylation. *Nature*, **364**, 249–52.

56 Morrison, D. (1994) 14–3–3: Modulators of signaling proteins? *Science*, **266**, 56–7.

57 Rodriguez, V.P., Warne, P.H., Dhand, R. *et al.* (1994) Phosphatidylinositol-3-OH kinase as a direct target of Ras. *Nature*, **370**, 527–32.

58 Hu, Q., Klippel, A., Muslin, A.J. *et al.* (1995) Ras-dependent induction of cellular

responses by constitutively active phosphatidylinositol-3 kinase. *Science*, **268**, 100–2.

59 Hawkins, P.T., Eguinoa, A., Qui, R.-G. *et al.* (1995) PDGF stimulates an increase in GTP-Rac via activation of phosphoinositide 3-kinase. *Curr. Biol.*, **5**, 393–403.

60 Kikuchi, A., Demo, S.D., Ye, Z.-H. *et al.* (1994) RalGDS family members interact with the effector loop of p21Ras. *Mol. Cell. Biol.*, **14**, 7483–91.

61 Burgering, B.M.T., Pronk, G.J., van Weeren, P.C. *et al.* (1993) cAMP antagonises p21Ras-directed activation of extracellular signal-regulated kinase 2 and phosphorylation of mSos nucleotide exchange factor. *EMBO J.*, **12**, 4211–20.

11 MAP kinases in multiple signaling pathways

Weidong Huang and
Raymond L. Erikson

The initial identification of MAP kinase (mitogen-activated protein kinase) as a protein serine/threonine kinase was based on its capacity to phosphorylate microtubule-associated protein. The fact that it is activated as a consequence of tyrosine and threonine phosphorylation stimulated widespread interest in the enzyme. In *Saccharomyces cerevisiae*, enzymes with sequence similarities have been demonstrated to be involved in pheromone response, cell wall construction and osmosensing. Multiple extracellular signals have also recently been demonstrated to activate distinct pathways utilizing related enzymes in animal cells. MAP kinase and its kin are apparently regulated by highly conserved upstream protein kinases that receive signals from effectors acting at the plasma membrane. The catalog of enzymes thus far identified indicates that the anatomy of each of the signaling pathways utilizing MAP kinases has been conserved in animal and yeast cells. This chapter will describe phosphorylation events involving Rsk (ribosomal S6 kinase), MAP kinase, Mek (MAP kinase kinase), Mek kinase and their relatives in animal cells following growth factor stimulation, ultraviolet irradiation or endotoxin stimulation. The roles of Ras and heterotrimeric G proteins in the upstream events leading to activation of these pathways are discussed in Chapters 10 and 19, respectively.

The MAP kinase pathway activated by growth factors

Insulin treatment of serum-starved 3T3-L1 adipocytes stimulates a soluble protein serine/threonine kinase that catalyzes phosphorylation of microtubule-associated protein 2 (MAP2) *in vitro*.[1] Subsequent studies revealed that this kinase activity, known as MAP kinase or Erk (extracellular signal-regulated kinase), is related to a 42 kDa protein that becomes transiently phosphorylated on tyrosine after stimulation of fibroblasts by

Signal Transduction. Edited by Carl-Henrik Heldin and Mary Purton. Published in 1996 by Chapman & Hall. ISBN 0 412 70810 8

a variety of mitogens, including epidermal growth factor (EGF), platelet-derived growth factor (PDGF), phorbol ester and insulin-like growth factor II.[2] The induction of tyrosine phosphorylation on MAP kinase by such diverse mitogenic agents suggests that it may play an important role in the pathway of signaling events responsible for the G0–G1 transition in the cell cycle. The first evidence for such a role was the identification of Rsk as a downstream target of MAP kinase.[3]

Ribosomal protein S6 is a component of the eukaryotic 40S ribosomal subunit that becomes phosphorylated on multiple serine residues in response to mitogen stimulation. Two distinct enzymes, with molecular masses of 65 kDa ($p70^{S6K}$) and 90 kDa (Rsk), appeared to contribute to this phosphorylation event.[4] Treatment of purified Rsk with serine/threonine-specific protein phosphatases (type 1 or 2A) abolishes its activity. The inactivated Rsk can then be partially reactivated *in vitro* by incubation with MAP kinase purified from insulin-stimulated cells.[3] Thus it was suggested that MAP kinase and Rsk constitute a step in the kinase pathway that transduces mitogenic signals into the nucleus.

MAP kinase is phosphorylated on threonine and tyrosine *in vivo* and can be inactivated by treatment with either protein phosphatase 2A or a tyrosine-specific protein phosphatase (CD45).[5] By a combination of two-dimensional peptide mapping and mass spectrometry, the phosphorylation sites were mapped to Thr185 and Tyr187 in the 42 kDa MAP kinase.[6] Conversion of either the threonine site to valine or the tyrosine site to phenylalanine yielded a gene product that cannot be activated *in vitro*.[7] The fact that the activation of MAP kinase requires phosphorylation of both tyrosine and threonine residues led to the speculation that the enzyme functions *in vivo* to integrate signals from a serine/threonine kinase pathway and a tyrosine kinase pathway.[5]

This hypothesis, however, was not substantiated by the subsequent identification of Mek, a 45 kDa dual-specificity kinase that activates MAP kinase by phosphorylation of both tyrosine and threonine.[8] Mek itself is phosphorylated on serine and threonine *in vivo*[9] and can be inactivated by treatment with protein phosphatase 2A but not with tyrosine-specific phosphatases.[8] The sequence of Mek is homologous to the *S. cerevisiae* STE7 protein and the *Schizosaccharomyces pombe* byr1 protein,[10] both involved in mating response pathways. Also, MAP kinase is related in sequence to the *FUS3* and *KSS1* gene products in *S. cerevisiae* and to the *spk1* gene product in *S. pombe*,[11] which are directly downstream of STE7 and byr1, respectively, in the pheromone response

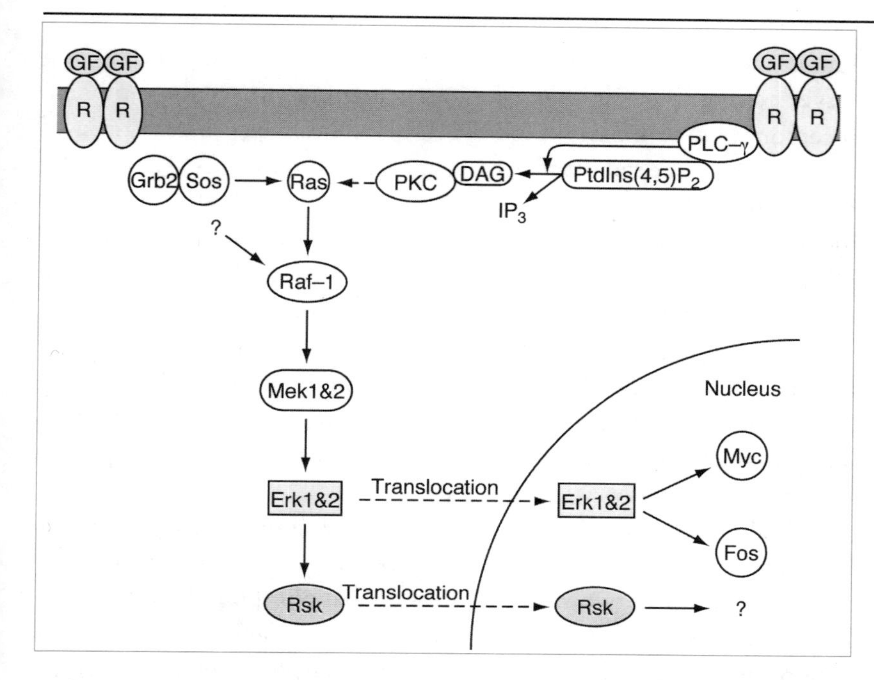

Figure 11.1 The MAP kinase pathway initially characterized in animal cells. GF, growth factor; R, growth factor receptor; PLC-γ, phospholipase C-γ; PtdIns(4, 5) P_2, phosphatidylinositol diphosphate; DAG, diacylglycerol; PKC, protein kinase C.

pathways. It was therefore apparent that the MAP kinase pathway is conserved between yeast and mammalian cells.

Since genetic analyses had placed two other protein kinases, STE11 and byr2, directly upstream of STE7 and byr1, respectively,[12,13] it was unexpected that Raf, a serine/threonine kinase unrelated to STE11 and byr2, was found to be an activator of Mek in mammalian cells.[14] Raf forms a stable complex with Mek and phosphorylates Mek on serine residues *in vitro*.[15] By peptide sequencing or mutational analyses, the sites of Mek1 phosphorylated by Raf were identified as two serine residues in subdomain VIII of the kinase catalytic domain.[16]

The activator(s) of Raf remain elusive, although several cellular components have been implicated. In its active GTP-bound state, the proto-oncogene product Ras associates with the amino-terminal region of Raf-1 and is required for the activation of the MAP kinase pathway.[17,18] Moreover, Ras is activated by the guanine nucleotide exchange factor Sos, which in turn associates with activated receptor tyrosine kinase through the SH2/SH3-containing protein Grb2.[19] Although the association with Ras *per se* does not activate Raf, the complex formation could serve to localize Raf to the plasma membrane. Experiments have shown that the recruitment of Raf-1 from cytoplasm to membrane is sufficient for its activation.[20]

Thus the MAP kinase pathway, consisting of Ras, Raf, Mek, MAP kinase and Rsk, spans from the plasma membrane to the nucleus and serves to transduce mitogenic signals downstream from membrane receptor tyrosine kinases (Figure 11.1). One notable characteristic of this pathway is the redundancy of its signaling components (Table 11.1). In murine cells, there are two Rsk genes, *rsk-1* and *rsk-2* with 84% identity,[21] two MAP kinase genes, *Erk1* (44 kDa) and *Erk2* (42 kDa) with 90% identity,[11] and two Mek genes, *Mek1* (44 kDa) and *Mek2* (45 kDa) with 80% identity.[22] The *raf* family of proto-oncogenes consists of three members, *c-raf-1*, *A-raf* and *B-raf*, which share three conserved domains (CR1, CR2 and CR3).[23–25] Homologs of each signaling component exhibit different patterns of expression in various tissues. They may also differ in their enzymatic properties and play specific roles in development.

It is also of interest that these enzymes share a similar mechanism of activation. The activation of Rsk, MAP kinase and Mek all require phosphorylation of two sites.[6,16,26] These dual phosphorylation sites all fall within subdomain VIII of the kinase catalytic domain, 5–13 residues amino-terminal to the conserved '(S/A)PE' sequence (Figure 11.2). In the crystal structures of PKA catalytic subunit, cyclin-dependent kinase 2 (Cdk2) and Erk2, this region is positioned at the apex of a loop blocking the substrate-binding site.[27–29] It has been proposed that the

Table 11.1 Summary of the major MAP kinase-related enzymes

	Subfamily		
Parameter	MAPK	SAPK	p38
Members	Erk1, Erk2	Jnk1, Jnk2	p38 (CSBP2), CSBP1[a]
Molecular mass (kDa)	44, 42	46, 55	41[b]
Phosphorylation sites in kinase subdomain VIII	**TEY**	**TPY**	**TGY**
Activating signal	NGF, PDGF, EGF, insulin, phorbol ester	UV irradiation, inflammatory cytokines, hyperosmolarity	lipopolysaccharide, hyperosmolarity
Upstream activator	Mek1, Mek2	Sek1 (MKK4 or JNKK)	MKK3
Downstream target	Rsk, TCF, c-Fos	c-Jun, ATF2	?
Regulation of gene expression	c-*myc*, c-*fos*	c-*jun*, TGF-β2, interferon β	TNF, IL-1
S. cerevisiae homolog	KSS1, FUS3	MPK1	HOG1

[a] Resulted from alternative splicing.
[b] Migrated as a 38 kDa protein in sodium dodecyl sulfate polyacrylamide gel electrophoresis (SDS–PAGE).

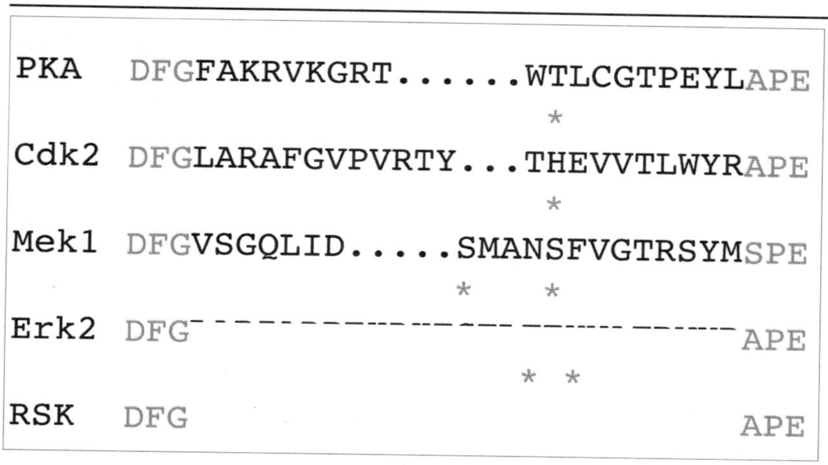

```
PKA   DFGFAKRVKGRT......WTLCGTPEYLAPE
                          *
Cdk2  DFGLARAFGVPVRTY...THEVVTLWYRAPE
                          *
Mek1  DFGVSGQLID.....SMANSFVGTRSYMSPE
                      *   *
Erk2  DFG----------------------------APE
                          *  *
RSK   DFG                          APE
```

Figure 11.2 Phosphorylation sites (labeled with asterisks) of components of the MAP kinase pathway. The residues conserved among kinases are in blue.

negatively charged phosphorylated residues in this region could interact with neighboring positively charged residues and stabilize the loop in an active conformation. Supporting this proposal, it has been shown that mutation of the Mek1 phosphorylation sites to acidic residues constitutively activates Mek1 to a degree similar to that of the fully phosphorylated wild-type protein.[30]

The functional significance of the MAP kinase pathway is highlighted by the fact that several components of the pathway are oncogenic in their constitutively active forms. Ras and Raf were originally identified as viral oncogenes before their cellular counterparts were discovered.[31,32] More recently, Mek1 mutants with constitutive activity were found to induce neuronal differentiation of PC12 cells and oncogenic transformation of fibroblast cell lines.[33] The transformed fibroblast lines exhibited increased AP-1 transcriptional activity and induced rapid tumor formation when injected into nude mice.

The observation that MAP kinases are translocated into the nucleus upon mitogen stimulation[34] suggests that they play some roles in the regulation of growth-specific gene expression. Transcriptional factors Elk-1/p62^{TCF}, c-Myc and C/EBPβ can be phosphorylated by MAP kinases *in vitro*, and these phosphorylation events have been reported to enhance their transcriptional activity.[35–37] By regulating the activities of these transcriptional factors, MAP kinases could influence the expression of a spectrum of genes including those under the control of serum-response elements (SRE). Rsk is also translocated into nucleus upon its activation by mitogen stimulation.[34] Although Rsk was originally identified as a ribosomal S6 kinase, the kinetics of S6 phosphorylation and Rsk activity suggest that S6 is unlikely to be the substrate of Rsk *in vivo*.[38] At present, the candidate cellular targets of Rsk include c-Fos[39] and the 67 kDa serum response factor (SRF) that binds SRE.[40]

The stress-activated MAP kinase pathway

The activity of the transcription factor c-Jun is unaffected by phosphorylation by MAP kinases at a carboxy-terminal site. In contrast, phosphorylation of Ser63 and Ser73 in the amino-terminal transactivation domain of c-Jun has been reported to enhance its transcriptional activity.[41] The kinase activity responsible for these phosphorylation events, termed c-Jun N-terminal kinase (Jnk) or stress-activated protein kinase (SAPK), was identified on the basis of its capacity to bind to the c-Jun transactivation domain[42] or of its ability to phosphorylate MAP2.[43] There are two well characterized proteins, Jnk1 (46 kDa) and Jnk2 (55 kDa), which are distant relatives of MAP kinases.[44–46]. As is the case for MAP kinases, the activation of Jnks requires phosphorylation of both threonine and tyrosine in subdomain VIII of the catalytic domain.[44] However, the dual phosphorylation site sequence in Jnk1 and Jnk2 (**TPY**) is distinct from that in Erk1 and Erk2 (**TEY**).

The activity of Jnks is stimulated by UV irradiation,[44] hyperosmolarity[47] and inflammatory cytokines such as tumor necrosis factor,[46] suggesting that the Jnk pathway is invoked in cellular responses to stress. It is not well understood at present how such diverse stress signals activate a common pathway.

An approach to this question entails elucidating signaling components upstream of Jnks. The fact that Jnks and Erks share significant sequence homology and similar mechanisms of activation points to the possibility that some Mek homologs may be responsible for the activation of Jnks. This hypothesis was indeed confirmed by the molecular cloning of Sek (SAPK/Erk kinase, also known as MKK4 or JNKK), a 41–43 kDa Mek-related protein that specifically activates Jnks *in vitro* and is required for the activation of Jnks *in vivo*.[48–50] Sek was originally cloned in *Xenopus* on the basis of its homology to the *Xenopus* Mek,[51] but it lacks the kinase insert domain between subdomains IX and X of its catalytic domain. The dual phosphorylation sites in Sek (serine and threonine) also differ from those in Mek (two serines).

These distinctions indicate that Sek is structurally more closely related to STE7 and byr1 than Mek. A mammalian kinase, Mek kinase, was isolated on the basis of its homology to STE11 and byr2 (upstream activators of STE7 and byr1 in *S. cerevisiae* and *S. pombe*, respectively).[52]

It was originally thought to be another activator of Mek because it activates Mek and Erk when it is overexpressed in COS cells. However, when Mek kinase is expressed in fibroblasts, it primarily activates the Jnk pathway but not the Erk pathway.[53,54] Mek kinase phosphorylates and activates Sek both *in vitro* and *in vivo*.[54] It causes Erk activation only when it is overexpressed.[53] These data imply that Mek kinase is probably a specific activator of the Sek/Jnk pathway under physiological conditions.

Upstream of STE11 in the yeast pheromone response pathway are STE20 (a kinase) and STE5 (unrelated to any known enzyme). By the use of interaction-trap assays and *in vitro* binding assays, it has been reported that STE5 serves as an anchor protein to tether STE20, STE11, STE7 and FUS3 in a multiple-kinase complex.[55] Given the similarities between Mek kinase/Sek/Jnk and STE11/STE7/FUS3, it is likely that mammalian homologs of STE20 and STE5 lie upstream of Mek kinase in the stress response pathway and activate it in a similar fashion. Recently, a 65 kDa mammalian kinase related to STE20 was cloned, which associates with and is activated by the Ras-related protein Rac1.[56] It will be of interest to determine whether this STE20 homolog activates Mek kinase or a related enzyme.

Although it has been studied in the context of stress response, the Jnk pathway (Figure 11.3) may function in other cellular processes such as thymocyte activation[57] and cell proliferation. The earliest known target of the Jnk pathway, c-Jun, is part of the AP-1 transcription factor

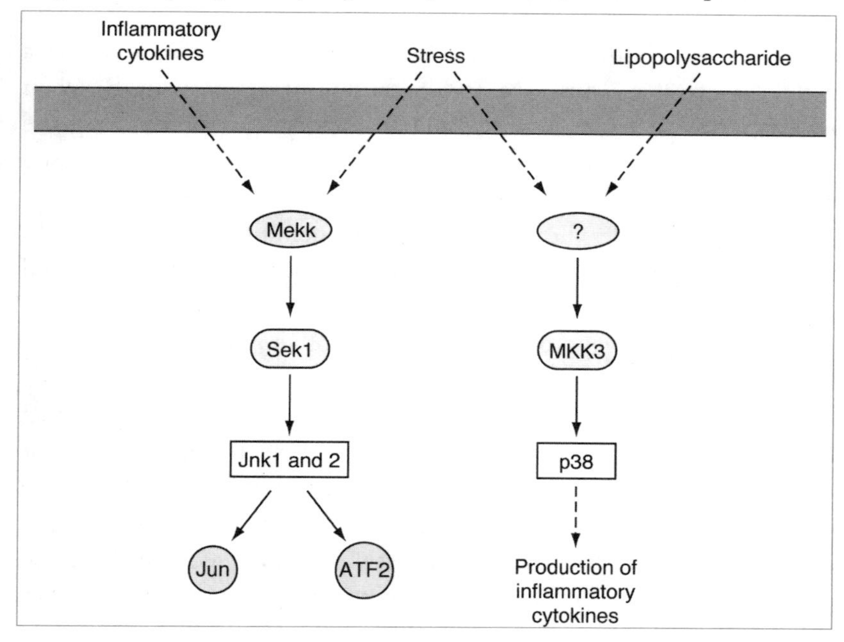

Figure 11.3 Enzymes in stress and inflammatory cytokine pathways.

that regulates the genes involved in cell proliferation[58] (see Chapter 21). ATF2, another transcription factor, is phosphorylated by Jnks on two closely spaced threonine residues in its amino-terminal activation domain.[59] The phosphorylation of these two threonine residues appears to be required for the ATF2-stimulated gene expression mediated by the retinoblastoma (Rb) tumor suppressor and the adenovirus early region 1A (E1A) oncoprotein. Together, these data point to an involvement of the Jnk pathway in cell proliferation.

This idea, however, has to be reconciled with the observation that the immediate effects of mitogenic signals such as growth factors and phorbol esters stimulate the Erk pathway but not the Jnk pathway. A clue to this puzzle comes from studies of *Drosophila* eye development. Activation of the *Drosophila* Erk pathway transforms nonneuronal cone cells into photoreceptor cell R7 in a Jun-dependent fashion.[60] Yet it has been argued that the phosphorylation of c-Jun by Erks does not activate its transcriptional activity.[41] A plausible explanation for these data is that, in some instances, the Jnk pathway functions downstream of the Erk pathway to activate c-Jun. In such cases, activation of the Jnk pathway may be a relatively slow process and involve autocrine factor(s) resulting from activation of the Erk pathway or crosstalk between pathways in as yet unidentified fashion.

The MAP kinase pathway involved in osmosensing

An *S. cerevisiae* homolog of *Mek*, *PBS2*, was originally identified as a gene that confers resistance to the antibiotic polymyxin B when overexpressed.[61] Later, it was found to be required for restoring the osmotic gradient across the plasma membrane in response to increased external osmolarity.[62] Another gene involved in this process, *HOG1*, structurally resembles the yeast MAP kinase genes *FUS3* and *KSS1*. In response to hyperosmolarity, HOG1 rapidly becomes phosphorylated on tyrosine in a PBS2-dependent fashion.[62] Thus another version of MAP kinase pathway appears to be part of the osmosensing signal transduction pathway in *S. cerevisiae*.

The mammalian counterpart of HOG1 is a 38 kDa protein that becomes tyrosine phosphorylated in response to hyperosmolarity or stimulation with endotoxic lipopolysaccharide (LPS).[63] The p38 enzyme and HOG1 share a dual phosphorylation site motif (**TGY**) distinct from those of Erks and Jnks. Also, the activation loop containing these phosphorylation sites in the kinase structure is six amino acids shorter in p38 and HOG1 than in Erks and Jnks, suggesting that the phosphorylation of HOG1 and p38 may occur through mechanisms distinct from those that control the phosphorylation of Erks and Jnks.

Two mammalian Mek homologs, Sek1 and MKK3, can phosphorylate and activate p38 both *in vitro* and when overexpressed in COS cells.[49,50] MKK3 is specific for p38. Sek1 can phosphorylate both p38 and Jnks. However, Sek1 is probably not an activator of p38 *in vivo* because co-transfection assays showed that Mek kinase (an activator of Sek1) causes activation of Jnk but not of p38.[50] The activator(s) of MKK3 and PBS2 are thus far unknown, although a receptor histidine kinase (SLN1) and its response regulator (SSK1) have been implicated upstream of PBS2/HOG1 in the yeast osmosensing pathway.[64]

The mammalian osmosensing MAP kinase pathway was recently shown to be a target of some cytokine-suppressive anti-inflammatory drugs (CSAID).[65] A group of pyridinylimidazole compounds have inhibitory effects on LPS-stimulated production of interleukin-1 (IL-1) and tumor necrosis factor α (TNFα) by human monocytes. The cellular targets of these compounds, two proteins named CSBPs (CSAID binding proteins), were identified by photoaffinity labeling. They are encoded by a single gene and one of them, CSBP2, is identical with p38. Binding of the pyridinylimidazole compounds inhibits the kinase activity of CSBPs, suggesting that CSBPs are critical for cytokine production.

If the p38 pathway is indeed involved in the regulation of inflammatory cytokine biosynthesis, then its activation is likely to lead to the activation of another MAP kinase pathway, the Jnk pathway (Figure 11.3). Such an autocrine loop may be a common way of crosstalk between different MAP kinase pathways in mammalian cells.

Summary

In this chapter we have attempted to encapsulate the information currently available on MAP kinases and their kin that are utilized in cultured animal cells. Space does not permit discussion of a vast body of evidence on their roles in development of vertebrates or invertebrates or in multiple yeast pathways. The research in this area is rapidly expanding and we anticipate that more members and other distinct pathways will soon be discovered. The question of crosstalk between the pathways is of major importance. The apparent linear nature of the pathways depicted here is likely to be an over-simplification. It should be noted that MAP kinase was originally identified as a microtubule-associated protein kinase. Enzymes related to those discussed here may play significant roles in the organization of cell structure. Other directions of keen interest include the investigation of signaling events upstream of Raf and Mek kinase and transcriptional events downstream of MAP kinase pathways. The study of this network of serine/threonine protein kinases will reveal more details about the control of cell organization, proliferation and differentiation in the near future.

References

Many laboratories have contributed to the complex of discoveries concerning the multiple MAP kinase pathways. We regret that the limits on space do not permit a complete reference list.

1 Ray, L.B. and Sturgill, T.W. (1987) Rapid stimulation by insulin of a serine/threonine kinase in 3T3-L1 adipocytes that phosphorylates microtubule-associated protein 2 *in vitro*. *Proc. Natl. Acad. Sci. USA*, **84**, 1502–6.

2 Rossomando, A.J., Payne, D.M., Webber, M.J. and Sturgill, T.W. (1989) Evidence that pp42, a major tyrosine kinase target protein, is a mitogen-activated serine/threonine protein kinase. *Proc. Natl. Acad. Sci. USA*, **86**, 6940–3.

3 Sturgill, T.W., Ray, L.B., Erikson, E. and Maller, J.L. (1988) Insulin-stimulated MAP-2 kinase phosphorylates and activates ribosomal protein S6 kinase II. *Nature*, **334**, 715–8.

4 Erikson, R.L. (1991) Structure, expression, and regulation of protein kinases involved in the phosphorylation of ribosomal protein S6. *J. Biol. Chem.*, **266**, 6007–10.

5 Anderson, N.G., Maller, J.L., Tonks, N.K. and Sturgill, T.W. (1990) Requirement for integration of signals from two distinct phosphorylation pathways for activation of MAP kinase. *Nature*, **343**, 651–3.

6 Payne, D.M., Rossomando, A.J., Martino, P. *et al.* (1991) Identification of the regulatory phosphorylation sites in pp42/mitogen-activated protein kinase (MAP kinase). *EMBO J.*, **10**, 885–92.

7 Alessandrini, A., Crews, C.M. and Erikson, R.L. (1992) Phorbol ester stimulates a protein tyrosine/threonine kinase that phosphorylates and activates the Erk-1 gene product. *Proc. Natl. Acad. Sci. USA*, **89**, 8200–4.

8 Crews, C.M. and Erikson, R.L. (1992) Purification of a murine protein tyrosine/threonine kinase that phosphorylates and activates the Erk-1 gene product: relationship to the fission yeast gene product byr1. *Proc. Natl. Acad. Sci. USA*, **89**, 8205–9.

9 Ahn, N.G., Campbell, J.S., Seger, R. *et al.* (1993) Metabolic labeling of mitogen-activated protein kinase kinase in A431 cells demonstrates phosphorylation on serine and threonine residues. *Proc. Natl. Acad. Sci. USA*, **90**, 5143–7.

10 Crews, C.M., Alessandrini, A. and Erikson, R.L. (1992) The primary structure of MEK, a protein kinase that phosphorylates the ERK gene product. *Science*, **258**, 478–80.

11 Boulton, T.G., Nye, S.H., Robbins, D.J. *et al.* (1991) ERKs: a family of protein-serine/threonine kinases that are activated and tyrosine phosphorylated in response to insulin and NGF. *Cell*, **65**, 663–75.

12 Cairns, B.R., Ramer, S.W. and Kornberg, R.D. (1992) Order of action of components in the yeast pheromone response pathway revealed with a dominant allele of the STE11 kinase and the multiple phosphorylation of the STE7 kinase. *Genes Dev.*, **6**, 1305–18.

13 Wang, Y., Xu, H.-P., Riggs, M. *et al.* (1991) byr2, a *Schizosaccharomyces pombe* gene encoding a protein kinase capable of partial suppression of the ras1 mutant phenotype. *Mol. Cell. Biol.*, **11**, 3554–63.

14 Kyriakis, J.M., App, H., Zhang, X. *et al.* (1992) Raf-1 activates MAP kinase kinase. *Nature*, **358**, 417–21.

15 Huang, W., Alessandrini, A., Crews, C.M. and Erikson, R.L. (1993) Raf-1 forms a stable complex with Mek1 and activates Mek1 by serine phosphorylation. *Proc. Natl. Acad. Sci. USA*, **90**, 10947–51.

16 Zheng, C.-F. and Guan, K.-L. (1994) Activation of MEK family kinases requires phosphorylation of two conserved Ser/Thr residues. *EMBO J.*, **13**, 1123–31.

17 Zhang, X., Settleman, J., Kyriakis, J.M. *et al.* (1993) Normal and oncogenic p21ras proteins bind to the amino-terminal regulatory domain of c-Raf-1. *Nature*, **364**, 308–13.

18 Vojtek, A.B., Hollenberg, S.M. and Cooper, J.A. (1993) Mammalian Ras interacts directly with the serine/threonine kinase Raf. *Cell*, **74**, 205–14.

19 Chardin, P., Camonis, J.H., Gale, N.W. *et al.* (1993) Human Sos1: a guanine nucleotide exchange factor for Ras that binds to Grb2. *Science*, **260**, 1338–43.

20 Stokoe, D., Macdonald, S.G., Cadwallader, K. *et al.* (1994) Activation of Raf as a result of recruitment of the plasma membrane. *Science*, **264**, 1463–7.

21 Alcorta, D.A., Crews, C.M., Sweet, L.J. *et al.* (1989) Sequence and expression of chicken and mouse rsk: homologs of *Xenopus laevis* ribosomal S6 kinase. *Mol. Cell. Biol.*, **9**, 3850–9.

22 Zheng, C.F. and Guan, K.L. (1993) Cloning and characterization of two distinct human extracellular signal-regulated kinase activator kinases, MEK1 and MEK2. *J. Biol. Chem.*, **268**, 11435–9.

23 Bonner, T.I., Oppermann, H., Seeburg, P. *et al.* (1986) The complete coding sequence of the human raf oncogene and the corresponding structure of the c-raf-1 gene. *Nucleic Acids Res.*, **14**, 1009–15.

24 Beck, T.W., Huleihel, M., Gunnell, M.A. *et al.* (1987) The complete coding sequence of the human A-raf-1 oncogene and transforming activity of a human A-raf carrying retrovirus. *Nucleic Acids Res.*, **15**, 595–609.

25 Ikawa, S., Fukui, M., Ueyama, Y. *et al.* (1988) B-raf, a new member of the raf family, is activated by DNA rearrangement. *Mol. Cell. Biol.*, **8**, 2651–4.

26 Sutherland, C., Campbell, D.G. and Cohen, P. (1993) Identification of insulin-stimulated protein kinase-1 as the rabbit equivalent of rsk-2. *Eur. J. Biochem.*, **212**, 581–8.

27 Knighton, D.R., Zheng, J., Eyck, L.F.T. *et al.* (1991) Crystal structure of the catalytic subunit of cyclic adenosine monophosphate-dependent protein kinase. *Science*, **253**, 407–20.

28 De Bondt, H.L., Rosenblatt, J., Jancarik, J. *et al.* (1993) Crystal structure of cyclin-dependent kinase 2. *Nature*, **363**, 595–602.

29 Zhang, F., Strand, A., Robbins, D. *et al.* (1994) Atomic structure of the MAP kinase ERK2 at 2.3 A resolution. *Nature*, **367**, 704–11.

30 Huang, W., Kessler, D.S. and Erikson, R.L. (1995) Biochemical and biological analysis of Mek1 phosphorylation site mutants. *Mol. Biol. Cell.*, **6**, 237–45.

31 Chang, E.H., Furth, M.E., Scolnick, E.M. and Lowy, D.R. (1982) Tumorigenic transformation of mammalian cells induced by a normal human gene homologous to the oncogene of Harvey murine sarcoma virus. *Nature*, **297**, 479–83.

32 Bonner, T.I., Kerby, S.B., Sutrave, P. *et al.* (1985) Structure and biological activity of human homologs of the raf/mil oncogene. *Mol. Biol. Cell.*, **5**, 1400–7.

33 Cowley, S., Paterson, H., Kemp, P. and Marshall, C.J. (1994) Activation of MAP kinase kinase is necessary and sufficient for PC12 differentiation and for transformation of NIH 3T3 cells. *Cell*, **77**, 841–52.

34 Chen, R., Sarnecki, C. and Blenis, J. (1992) Nuclear localization and regulation of erk- and rsk-encoded protein kinases. *Mol. Cell. Biol.*, **12**, 915–27.

35 Gille, H., Sharrocks, A.D. and Shaw, P.E. (1992) Phosphorylation of transcription factor p62TCF by MAP kinase stimulates ternary complex formation at c-fos promoter. *Nature*, **358**, 414–7.

36 Seth, A., Gonzalez, F.A., Gupta, S. *et al.* (1992) Signal transduction within the nucleus by mitogen-activated protein kinase. *J. Biol. Chem.*, **267**, 24796–804.

37 Nakajima, T., Kinoshita, S., Sasagawa, T. *et al.* (1993) Phosphorylation at threonine-235 by a ras-dependent mitogen-activated protein kinase cascade is essential for transcription factor NF-IL6. *Proc. Natl. Acad. Sci. USA*, **90**, 2207–11.

38 Blenis, J., Chung, J., Erikson, E. *et al.* (1991) Distinct mechanisms for the activation of the RSK kinases/MAP2 kinase/pp90rsk and pp70–S6 kinase signaling systems are indicated by inhibition of protein synthesis. *Cell Growth Differ.*, **2**, 279–85.

39 Chen, R.H., Abate, C. and Blenis, J. (1993) Phosphorylation of the c-Fos transrepression domain by mitogen-activated protein kinase and 90-kDa ribosomal S6 kinase. *Proc. Natl. Acad. Sci. USA*, **90**, 10952–6.

40 Rivera, V.M., Miranti, C.K., Misra, R.P. *et al.* (1993). A growth factor-induced kinase phosphorylates the serum response factor at a site that regulates its DNA-binding activity. *Mol. Cell. Biol.*, **13**, 6260–73.

41 Smeal, T., Binetruy, B., Mercola, D. *et al.* (1992) Oncoprotein-mediated signaling cascade stimulates c-Jun activity by phosphorylation of serines 63 and 73. *Mol. Cell. Biol.*, **12**, 3507–13.

42 Hibi, M., Lin, A., Smeal, T. *et al.* (1993) Identification of an oncoprotein- and

UV-responsive protein kinase that binds and potentiates the c-Jun activation domain. *Genes Dev.*, **7**, 2135–48.

43 Kyriakis, J.M. and Avruch, J. (1990) pp54 microtubule-associated protein 2 kinase. A novel service/threonine protein kinase regulated by phosphorylation and stimulated by poly-L-lysine. *J. Biol. Chem.*, **265**, 17355–63.

44 Derijard, B., Hibi, M., Wu, I. *et al.* (1994) JNK1: a protein kinase stimulated by UV light and Ha-Ras that binds and phosphorylates the c-Jun activation domain. *Cell*, **76**, 1025–37.

45 Kyriakis, J.M., Banerjee, P., Nikolakaki, E. *et al.* (1994) The stress-activated protein kinase subfamily of c-Jun kinases. *Nature*, **369**, 156–60.

46 Sluss, H.K., Barrett, T., Derijard, B. and Davis, R.J. (1994) Signal transduction by tumor necrosis factor mediated by JNK protein kinases. *Mol. Cell. Biol.*, **14**, 8376–84.

47 Galcheva-Gargova, Z., Derijard, B., Wu, I. and Davis, R.J. (1994) An osmosensing signal transduction pathway in mammalian cells. *Science*, **265**, 806–8.

48 Sanchez, I., Hughes, R.T., Mayer, B.J. *et al.* (1994) Role of SAPK/ERK kinase-1 in the stress-activated pathway regulating transcription factor c-Jun. *Nature*, **372**, 794–8.

49 Derijard, B., Raingeaud, J., Barrett, T. *et al.* (1995) Independent human MAP kinase signal transduction pathways defined by MEK and MKK isoforms. *Science*, **267**, 682–5.

50 Lin, A., Minden, A., Martinetto, H. *et al.* (1995) Identification of a dual specificity kinase that activates the Jun kinases and p38-Mpk2. *Science*, **268**, 286–91.

51 Yashar, B.M., Kelley, C., Yee, K. *et al.* (1993) Novel members of the mitogen-activated protein kinase activator family in *Xenopus laevis*. *Mol. Cell. Biol.*, **13**, 5738–48.

52 Lange-Carter, C.A., Pleiman, C.M., Gardner, A.M. *et al.* (1993) A divergence in the MAP kinase regulatory network defined by MEK kinase and Raf. *Science*, **260**, 315–9.

53 Minden, A., Lin, A., McMahon, M. *et al.* (1994) Differential activation of ERK and JNK mitogen-activated protein kinases by Raf-1 and MEKK. *Science*, **266**, 1719–22.

54 Yan, M., Dai, T., Deak, J.C. *et al.* (1994) Activation of stress-activated protein kinase by MEKK1 phosphorylation of its activator SEK1. *Nature*, **372**, 798–800.

55 Choi, K., Satterberg, B., Lyons, D.M. and Elion, E.A. (1994) Ste5 tethers multiple protein kinases in the MAP kinase cascade required for mating in S. cerevisiae. *Cell*, **78**, 499–512.

56 Manser, E., Leung, T., Salihuddin, H. *et al.* (1994) A brain serine/threonine protein kinase activated by Cdc42 and Rac1. *Nature*, **367**, 40–6.

57 Su, B., Jacinto, E., Hibi, M. *et al.* (1994) JNK is involved in signal integration during costimulation of T lymphocytes. *Cell*, **77**, 727–36.

58 Bohmann, D., Bos, T.J., Admon, A. *et al.* (1987) Human proto-oncogene c-jun encodes a DNA binding protein with structural and functional properties of transcription factor AP-1. *Science*, **238**, 1386–92.

59 Gupta, S., Campbell, D., Derijard, B. and Davis, R.J. (1995) Transcription factor ATF2 regulation by the JNK signal transduction pathway. *Science*, **267**, 389–93.

60 Bohmann, D., Ellis, M.C., Staszewski, L.M. and Mlodzik, M. (1994) *Drosophila* Jun mediates Ras-dependent photoreceptor determination. *Cell*, **78**, 973–86.

61 Boguslawski, G. and Polazzi, J.O. (1987) Complete nucleotide sequence of a gene conferring polymyxin B resistance on yeast: similarity of the predicted polypeptide to protein kinases. *Proc. Natl. Acad. Sci. USA*, **84**, 5848–52.

62 Brewster, J.L., Valoir, T.D., Dwyer, N.D. *et al.* (1993) An osmosensing signal transduction pathway in yeast. *Science*, **259**, 1760–3.

63 Han, J., Lee, J.-D., Bibbs, L. and Ulevitch, R.J. (1994) A MAP kinase targeted by endotoxin and heperosmolarity in mammalian cells. *Science*, **265**, 808–11.

64 Maeda, T., Wurgler-Murphy, S.M. and Saito, H. (1994) A two-component system that regulates an osmosensing MAP kinase cascade in yeast. *Nature*, **369**, 242–5.

65 Lee, J.C., Laydon, J.T., McDonnell, P.C. *et al.* (1994) A protein kinase involved in the regulation of inflammatory cytokine biosynthesis. *Nature*, **372**, 739–46.

12 Function of phospholipases in signal transduction

Sue Goo Rhee and
Edward A. Dennis

The phospholipases play important roles in cellular metabolism, including the biosynthesis and degradation of membrane lipids. However, certain phospholipases also play a key part in generating the lipid second messengers implicated in signal transduction processes[1] (Figure 12.1). This chapter will focus on phospholipase A_2 (PLA_2), phospholipase C (PLC) and phospholipase D (PLD) because they have been the most studied both as enzymes and as signal transducers.

These enzymes are defined and named by the position they attack on the phospholipid backbone. PLA_1 catalyzes the hydrolysis of the fatty acid group on the first (sn-1) position and PLA_2 on the middle (sn-2) position, while PLC catalyzes the hydrolysis of the phosphodiester bond on the third (sn-3) position of the glycerol and PLD catalyzes the

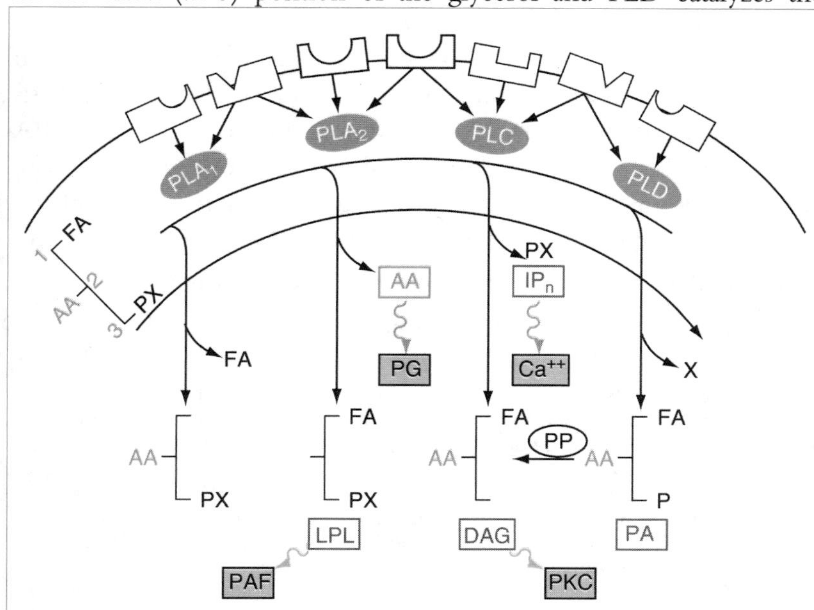

Figure 12.1 Possible roles of phospholipases in the production of lipid and lipid-derived second messengers. Abbreviations: PLA_1, phospholipase A_1; PLA_2, phospholipase A_2; PLC, phospholipase C; PLD, phospholipase D; PP, phosphatidate phosphatase; AA, arachidonic acid; FA, fatty acid; LPL, lysophospholipids; PAF, platelet activating factor; IP_n inositol phosphates; DAG, diacylglycerol; PA, phosphatidic acid. [Adapted with permission from Dennis, E.A., Rhee, S.G., Billah, M.M. and Hunnun, Y.A. (1991) FASEB J., **5**, 2068–77.]

Signal Transduction. Edited by Carl-Henrik Heldin and Mary Purton. Published in 1996 by Chapman & Hall. ISBN 0 412 70810 8

hydrolysis of the phosphodiester bond between the phosphate and the polar group (which can be choline, ethanolamine, serine, glycerol or inositol).

Phospholipase A_2

A wide variety of extracellular signals have been shown to lead to cell activation and enhanced PLA_2 activity. The signal transduction process can be direct or mediated by second messengers and often involves G proteins and protein tyrosine kinases. The products of PLA_2 action, arachidonic acid and lysophospholipids, have been implicated as downstream effectors in signal transduction, but they also serve as precursors of other lipid mediators including prostaglandins, leukotrienes and platelet activating factor (PAF).[2] The PLA_2s were the first of the phospholipases to be identified, purified and studied mechanistically. As such, PLA_2 has served as the model or prototype for understanding how the phospholipases interact with their phospholipid substrate.[3] However, in recent years, many new and novel forms of PLA_2 have been identified (Table 12.1).

Secreted PLA_2s

The secreted PLA_2s ($sPLA_2$s) are small, water-soluble proteins of 13–18 kDa, consisting of a single polypeptide chain containing five to eight disulfide bonds, which makes them extremely stable. These extracellular enzymes are found in snake and bee venoms and the mammalian pancreas and have been classified into groups (Table 12.1) on the basis of their sequences and disulfide bonding patterns.[4] The group IIA enzyme has been implicated in signal transduction processes. The pancreatic group I enzyme was originally thought to play a role in digestion but the recent discovery of cell surface receptors for this

Table 12.1 Characteristics of the major groups of phospholipase A_2[a]

Group	Source	Location	Size (kDa)	Ca^{2+} requirement
IA	Cobras, kraits	Secreted	13–15	<mM
IB	Porcine/human pancreas	Secreted	13–15	<mM
IIA	Rattlesnakes, vipers, human synovial platelets	Secreted	13–15	<mM
IIB	Gaboon viper	Secreted	13–15	<mM
III	Bees, lizards	Secreted	16–18	<mM
IV	Raw 264.7/rat kidney, human U937/platelets	Cytosolic	85	<μM
?	Canine/human myocardium P338D$_1$ macrophages	Cytosolic	40 80	None

[a] Adapted with permission from Dennis, E.A. (1994) *J. Biol. Chem.*, **269**, 13057–60.

enzyme[5] suggests that it may also play a role in signal transduction.

X-ray crystallographic structures have been solved for many of the secreted PLA_2s (see Figure 12.2a). All of these enzymes have a catalytic site containing a histidine and aspartic acid which, with a water molecule, catalyze the hydrolysis of the fatty acyl chain. Catalysis is

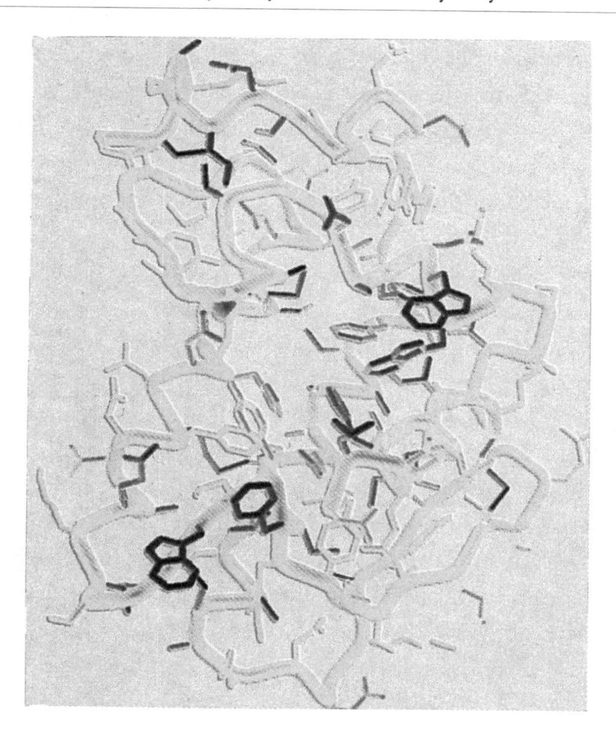

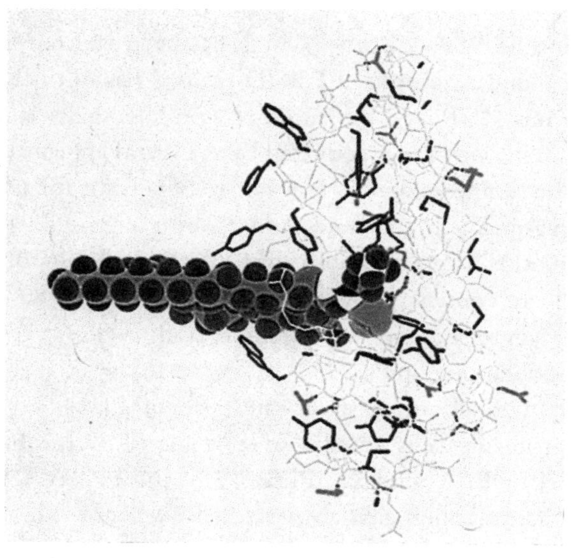

Figure 12.2 X-ray crystallographic structure of cobra venom phospholipase A_2. (a) Structure of the refined Indian cobra PLA_2. [Reproduced with permission from Fremont, D.H., Anderson, D., Wilson, I.A. *et al.* (1993) *Proc. Natl. Acad. Sci. USA*, **90**, 342–6.] (b) Cobra venom PLA_2 with bound Ca^{2+} showing a space-filling model of dimyristoylphosphatidylethanol amine bound in the catalytic site. The ends of the fatty acid chains stick out of the enzyme and are presumably associated with the micelle or membrane. Normal phospholipid fatty acid chains would be 4–6 carbons longer. [Reproduced with permission from Dennis, E.A. (1994) *J. Biol. Chem.*, **269**, 13057–60.] Reproduced here in black and white. See original publications for colour versions.

assisted by an essential Ca^{2+} which is bound in the active site of the enzyme.[6] The water-soluble PLA_2 associates with phospholipid polar groups in membranes and micelles and then binds a phospholipid substrate in its active site. The bulk of the hydrophobic fatty acid chains are sequestered in the micelle or membrane; hence these enzymes show little specificity for the fatty acid chain hydrolyzed (Figure 12.2a). In signal transduction processes, this enzyme is believed to be secreted by cells; it then binds to the peptidoglycan exterior of cells to carry out its hydrolysis. This requires that the free arachidonic acid liberated by this process act on cells in an autocrine or paracrine fashion or be taken up to be converted into prostaglandins and other eicosanoids.[7]

Group IV cytosolic PLA_2

The cytosolic group IV PLA_2 ($cPLA_2$)[8,9] shows a preference for arachidonic acid in the sn-2 position of phospholipids and so is a prime candidate for signal transduction processes. This 85 kDa enzyme contains no disulfide bonds. Ca^{2+} is not required for catalysis, but the enzyme contains a CalB domain which appears to aid in the translocation to membranes in the presence of submicromolar concentrations of Ca^{2+}. This enzyme carries out both lysophospholipase and transacylase reactions.[10] Its mechanism of action, which appears to involve Ser228, is different from that of the $sPLA_2$s.[11] The group IV PLA_2 is regulated by phosphorylation at Ser505 by a MAP kinase cascade.[12] Its presence and activation in a variety of cell types have been demonstrated.

Ca^{2+} Independent PLA_2

Another cytosolic PLA_2 ($iPLA_2$), which has been isolated from myocardial tissue[13] and macrophage $P388D_1$ cells,[14] has distinct properties from the group IV PLA_2, the most obvious being its lack of Ca^{2+} dependence. This enzyme is activated by ATP and appears to act as an aggregate. The canine enzyme is reported to be specific for arachidonyl-containing plasmalogen substrates and to be composed of an active subunit of 40 kDa associated with four 80 kDa subunits of phosphofructokinase.[13] In contrast, the macrophage $iPLA_2$ has an 80 kDa molecular mass, aggregates as a tetramer and lacks specificity for arachidonyl-containing lipids.[14]

Most cells contain more than one type of PLA_2. For example, $P388D_1$ macrophage cells contain at least an $sPLA_2$ (probably group II), a group IV $cPLA_2$ and an $iPLA_2$.[15] All three enzymes have the potential for action when cells are activated. Recent evidence suggests that when these cells are activated with lipopolysaccharide (LPS)/PAF, the $sPLA_2$ is secreted and releases arachidonate on the outside of cells and another PLA_2, probably $cPLA_2$, does the same on the inside of cells.[16] Investigators are now trying to distinguish the specific PLA_2 targets of different activation protocols and the specific role of each of the PLA_2s in these events.

Phospholipase C

The hydrolysis of a minor membrane phospholipid, phosphatidylinositol 4,5-bisphosphate [PtdIns(4,5)P_2], by a specific PLC is one of the earliest key events by which more than 100 extracellular signaling molecules are known to regulate functions of their target cells. The hydrolysis produces two intracellular messengers, inositol trisphosphate [Ins(1,4,5)P_3] and diacylglycerol (DAG). Ins(1,4,5)P_3 induces the release of Ca^{2+} from internal stores and DAG activates protein kinase C (PKC).[17,18]

PLC isozymes

PLC, like PLA_2, exists in multiple isoforms. The ten mammalian PLC enzymes identified to date are all single polypeptides and can be divided into three types [β (4), γ (2) and δ (4)] on the basis of size and amino acid sequence.[18] There are only two regions of high similarity between the various isotypes: X (~170 amino acid residues) is 60% identical and Y (260 residues) is 40% identical (Figure 12.3). These regions appear to constitute the catalytic domain. The sequence between the X and Y domains is fairly short (50–70 residues) in the PLC-β and PLC-δ isozymes. In PLC-γ isozymes this region is much longer (~400 residues)

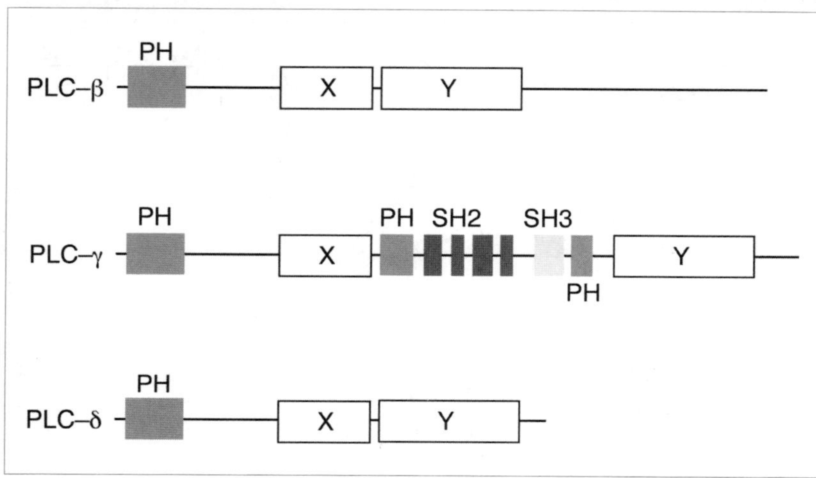

Figure 12.3 Domains structure of the PLC family. All members contain highly similar X and Y regions (boxed), which constitute the catalytic domain. Src homology domains, SH2 and SH3, and pleckstrin homology (PH) domain are indicated.

and contains both SH2 and SH3 domains.[19] All mammalian PLC isozymes have an amino-terminal region of ~300 residues preceding the X region which contains a pleckstrin homology (PH) domain. The PLC-γ isozymes contain another PH domain that is split by the SH2 and SH3 domains (Figure 12.3). (See Chapter 9 for a fuller description of SH2, SH3 and PH domains.)

PLC isozymes containing the X and Y catalytic domains have also been cloned from other organisms such as *Dictyostelium* (PLC-δ type) and *Drosophila* (PLC-β type).[18] All PLC isozymes containing X and Y domains hydrolyze only the phospholipids phosphatidylinositol 4-phosphate [PtdIns(4)P], PtdIns(4,5)P_2 and phosphatidylinositol (PtdIns) in a calcium-dependent manner. However, at low calcium concentrations PtdIns(4,5)P_2 is the preferred substrate for these isozymes. Some bacteria secrete a smaller (298 amino acid residues) soluble PLC that does not contain regions equivalent to the X and Y domains. These PLCs hydrolyze PtdIns and glycerophosphatidylinositol (GroIns) but not PtdIns(4,5)P_2.[20]

Various signaling molecules activate PLC isozymes through binding to a receptor. There appear to be at least two distinct mechanisms to link receptor occupancy to PLC activation (Figure 12.4): PLC-γ

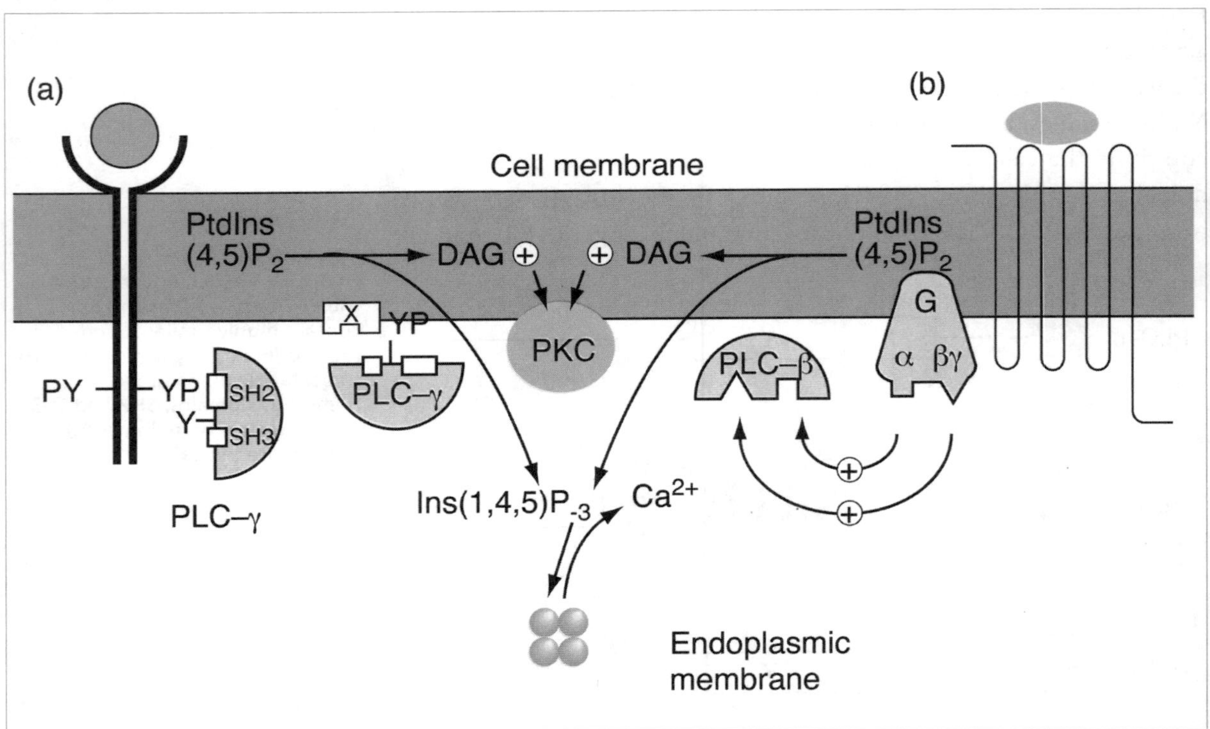

Figure 12.4 Mechanisms of activation of (a) PLC-γ by protein tyrosine kinase and (b) PLC-β isozymes by heterotrimeric G proteins. Abbreviations: YP, phosphotyrosine; X, a putative membrane-associated protein with proline-rich sequence that might serve as the site of interaction with the SH3 domain of PLC-γ; DAG, diacylglycerol; PKC, protein kinase C.

isozymes are activated by tyrosine phosphorylation and PLC-β isozymes by heterotrimeric G proteins.[18]

Activation of PLC-γ isozymes

Polypeptide growth factors, such as platelet-derived growth factor (PDGF), epidermal growth factor (EGF), fibroblast growth factor (FGF) and nerve growth factor (NGF), are known to stimulate turnover of PtdIns(4,5)P_2 by activating PLC-γ1 in a wide variety of cells.[18] When a receptor binds its cognate growth factor, it dimerizes and its intrinsic tyrosine kinase activity is activated (see Chapter 1). This leads to tyrosine phosphorylation of numerous proteins including the receptor itself and PLC-γ1. The receptor autophosphorylation creates high-affinity binding sites for several SH2-containing proteins including PLC-γ1. For example, PLC-γ1 binds to only one (Tyr1021) of eight autophosphorylation sites on the PDGF receptor β-chain.[21] In the FGF and NGF receptors, PLC-γ1 binds specifically to *P*Tyr766[22] and *P*Tyr785 respectively. However, in the EGF receptor individual autophosphorylation sites are not specific for the recognition and association of PLC-γ1.[24]

PDGF, EGF, FGF and NGF all phosphorylate PLC-γ1 at Tyr771, Tyr783 and Tyr1254.[25] Substitution of phenylalanine for Tyr783 completely blocked the activation of PLC by PDGF but did not prevent association of the enzyme with the PDGF receptor.[25] The binding of PLC-γ1 to receptors was disrupted when Tyr1021 (PDGF receptor) or Tyr766 (FGF receptor) were replaced with phenylalanine[22,26] and there was no growth-factor dependent production of Ins(1,4,5)P_3. However, these mutants mediated readily detectable amounts of growth-factor dependent PLC-γ1 tyrosine phosphorylation. Thus, for activation by growth factors, PLC-γ1 requires not only tyrosine phosphorylation but also association with the growth factor receptor. Tyrosine phosphorylation of PLC-γ1 appears to promote its association with actin components of cytoskeleton.[27,28] The SH3 domain of PLC-γ1 has been shown to be responsible for targeting it to actin microfilament networks.[19] Whether this cytoskeletal association serves to bring the enzyme to its substrate or whether it promotes interaction with another protein component essential for its activation is unknown.

Non-receptor protein tyrosine kinases also phosphorylate and activate PLC-γ isozymes in response to the ligation of certain cell surface receptors[18] (see Chapters 2, 3 and 4).

PLC-γ2, which is mainly expressed in hematopoietic cells, is also phosphorylated at two residues, Tyr753 and Tyr759, by the growth factor receptors in addition to non-receptor protein tyrosine kinases.

Activation of PLC-β isozymes

PLC-β isozymes are activated by the GTP-bound α subunits of the G_q class of heterotrimeric G proteins.[29–31] The receptors that are known to utilize this $G_q\alpha$/PLC-β pathway include those for thromboxane A_2, bradykinin, bombesin, angiotensin II, histamine, vasopressin, muscarinic acetylcholine (m1, m2 and m3), α_1-adrenergic agonist and thyroid stimulating hormone. Some PLC-β isozymes are also activated by the βγ subunits of G proteins, for example the m2 and m4 subtypes of the muscarinic acetylcholine receptor family and the receptor for interleukin-8 are known to utilize this pathway.[29]

Different regions of PLC-β isozymes interact with $G_q\alpha$ subunit and Gβγ dimers.[32] Whereas the carboxy-terminal region following the Y region is important for activation by $G_q\alpha$ subunits, the site of interaction with Gβγ dimers appears to be localized to the amino-terminal region preceding the X region which also contains the PH domain. The β-adrenergic receptor kinase has been shown to bind to Gβγ dimers through regions including their PH domains.[33] The presence of separate binding sites indicates that PLC-β isozymes can be activated by $G_q\alpha$ and Gβγ subunits in an additive manner.

The PLC-β isozymes differ in their sensitivities to $G_q\alpha$ or Gβγ subunits. Whereas $G_q\alpha$ subunits activate PLC-β isozymes according to the hierarchy PLC-β1≥ PLC-β3 > PLC-β4 > PLC-β2, the sensitivity to Gβγ subunits decreases in the order PLC-β3 > PLC-β2 > PLC-β1.[29] The Gβγ subunits do not activate PLC-β4.[31] These results suggest that cell-specific expression of PLC-β isozymes and G-protein subunits contribute to diversity in the type and magnitude of enzyme responses observed.

PLC-β isozymes are not the only type of PLC that can respond to G-protein-coupled receptors. The angiotensin II receptor, which has been shown to stimulate PLC-β isozymes through $G_q\alpha$ subunits in rat liver, also activates PLC-γ1 through tyrosine phosphorylation in rat aortic vascular muscle cells.[34] The receptor for PAF, another G-protein-coupled receptor, promotes the tyrosine phosphorylation of PLC-γ1.[35]

Maximal activation of PLC-β isozymes in permeabilized HL-60 cells was found to require PtdIns transfer protein (PI-TP) in addition to the receptor and the G protein.[36] As the supply of substrate PtdIns(4,5)P_2, which is synthesized from PtdIns, is likely to be limited in the vicinity of activated PLC, PI-TP is thought to ensure a steady substrate supply for PLC-β and other types of PLC.

Phospholipase D

Phospholipase D (PLD) catalyzes the hydrolysis of phospholipids at their terminal phosphodiester bond, thus producing phosphatidic acid (PA), and releasing the free polar head group. Stimulation of PLD occurs in a wide variety of cell types treated with hormones and growth factors.[37] PA has been implicated as a biologically active molecule and can be further metabolized by PA phosphohydrolase to form DAG, a protein kinase C activator.[37] PLD attacks the phospholipid substrate to form a transient phosphatidyl–PLD intermediate. With water as an acceptor, PA is the product; however, in the presence of a primary alcohol, the phosphatidyl moiety is preferentially transferred to the alcohol, generating phosphatidyl alcohol.[38] The production of phosphatidyl alcohol has proven to be extremely useful assay for the detection of PLD in intact cells and in crude homogenates, because PA can also be generated by *de novo* synthesis and also from DAG by DAG kinase. In addition, the metabolism of phosphatidyl alcohols is much slower than that of PA and thus they accumulate in stimulated cells.

PLD isozymes

PLD was first detected and purified from plants, and subsequently purified and cloned from several prokaryotes.[37] There is compelling evidence for the existence of multiple types of cellular signal-activated PLDs in mammalian cells. Most mammalian PLD activity is found associated with membranes and appears to be specific to phosphatidylcholine (PC), whereas cytosolic PLC hydrolyzes phosphatidylethanolamine (PE), PI and PC.[37] Furthermore, PC-specific PLD activity from rat brain membranes could be separated into two distinct forms, one completely dependent on sodium oleate for activity and the other activated by the addition of ADP-ribosylation factor (ARF) and GTP-γS in the presence of PtdIns(4,5)P_2.[39] Additional support for the existence of multiple forms of PLD was derived from the fact that mammalian PLD is activated via multiple pathways involving G proteins, Ca^{2+}, unsaturated fatty acids, PKC or protein tyrosine kinases. However, there is little information on the biochemical and molecular properties of mammalian PLDs. To date, only the ARF-dependent PLD has been cloned[40]. Nevertheless, there is growing evidence for two

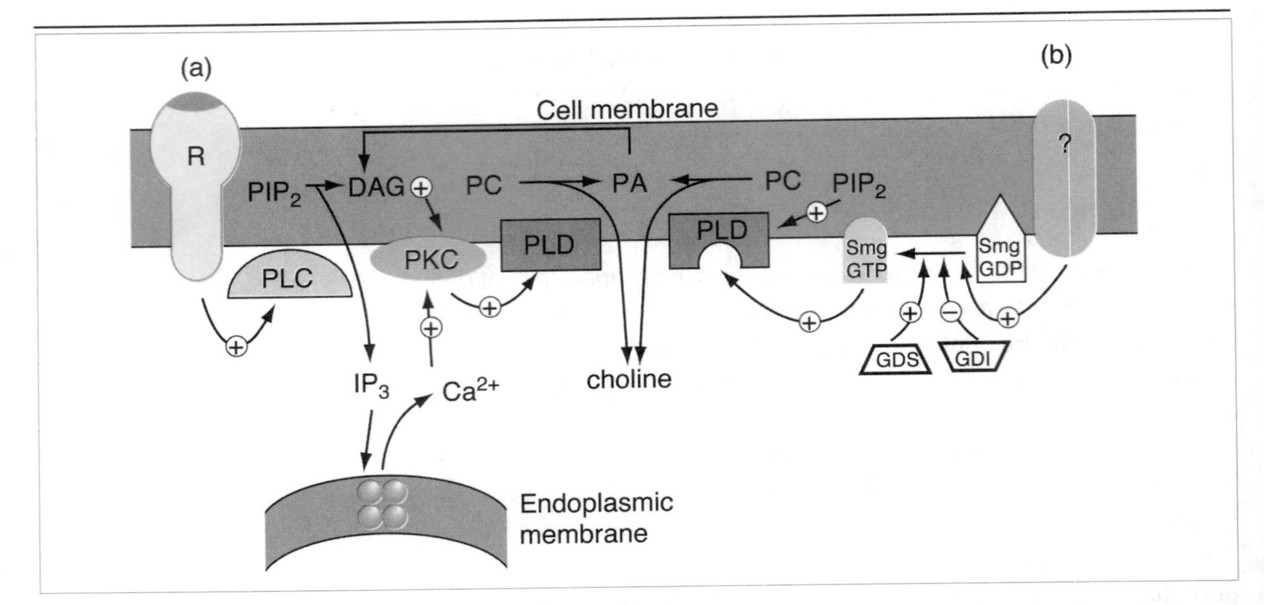

Figure 12.5 Two distinct activation mechanisms of PLD: (a) dependent on PKC and (b) dependent on Smg. Abbreviations: R, a receptor; DAG, diacylglycerol; PKC, a protein kinase C; PC, phosphatidylcholine; PA, phosphatidic acid; Smg GTP (or GDP), a GTP (or GDP)-bound low-molecular-mass G protein; GDS, a GDP/GTP dissociation stimulator; and GDI, a GDP/GTP dissociation inhibitor PIP_2, PtdIns (4,5)P_2.

different mechanisms for the activation of PLD, one by protein kinase C and the other by low-molecular-mass G proteins (Smg), in activated cells (Figure 12.5).

Activation by protein kinase C

In many cells, activation of PKC by phorbol esters (non-physiological substitutes for DAG) is sufficient for PLD activation. In addition, down-regulation of PKC by long-term exposure to phorbol ester and PKC inhibitors frequently, but not always, blocks activation of PLD by agonists and phorbol esters, suggesting the involvement of both PKC-dependent and PKC-independent pathways for PLD activation. It has long been believed that activation of PLC (and by implication PKC) occurs before activation of PLD, based on the observation that the activation of PLD is seen in response to a variety of calcium-mobilizing agonists that are known to activate PLC and Ins(1,4,5)P_3 production.[37] While in many cases the sequence of events may well be PLC → PKC → PLD, PLD activation also occurs in the absence of any PtdIns(4,5)P_2 hydrolysis, suggesting the existence of other mechanisms for PLD activation.

The interrelation between PLC and PLD has been studied in cells activated by PDGF.[41,42] Both PLC and PLD were stimulated by PDGF in cells expressing wild-type PDGF receptors but not in cells expressing kinase-deficient receptors, indicating that tyrosine phosphorylation is required for activation of both PLC and PLD. Substitution of Phe for Tyr1021 in the PDGF receptor caused loss of PDGF stimulation of both PLC and PLD. This loss of stimulation was reversed by a mutant

PDGF receptor that was able to bind PLC-γ1 but not other signaling proteins. Furthermore, receptors in which association with (PI 3-kinase) or protein phosphatase 1D was individually restored were unable to mediate PDGF stimulation of PLC or PLD. These data indicate that the binding of signal transduction proteins other than PLC-γ1 is not required for the activation of PLD by PDGF. Treatment of the cells with a PKC inhibitor or depletion of cellular PKC by pretreatment with phorbol esters resulted in a loss of PLD activation by PDGF. In another experiment, ten-fold overexpression of PLC-γ1 resulted in the proportional increase in PLD activity in response to PDGF stimulation. All of these results are consistent with the sequential activation of PLC → PKC → PLD in PDGF-treated cells.

The PKC-dependent activation was demonstrated directly when incubation of purified PKC isozymes with PLD present in CCL39 lung fibroblast membranes stimulated PLD activity.[43] The activation required either phorbol ester or Ca^{2+}. Surprisingly, however, the activation occurred in the absence of both exogenous ATP and autophosphorylation of PKC. This phosphorylation-independent mechanism raised the possibility that PKC may bind directly to PLD and activate it by an allosteric mechanism. There are also reports suggesting the phosphorylation-dependent activation of PLD by PKC.[37] One example of phosphorylation-dependent activation of PLD by PKC is the enzyme present in nuclei.[44] In this system, PLD activity could only be measured in the presence of ATP, which acted as a phosphoryl donor in a PKC-mediated reaction. Only after PLD has been isolated can the mechanisms of activation be tested directly. Furthermore, if different isoforms exist, their isolation should allow examination of the possibility that they are differentially regulated by PKC isoforms.

Activation by low-molecular-mass G proteins

Evidence for the involvement of G proteins in the regulation of PLD activity comes mainly from the observation that, in both permeabilized cells and cell-free systems, PLD can be stimulated by GTP-γS.[37] Independence of this GTP-γS-induced PLD activation from PLC was demonstrated under conditions where PLC activation was nullified. The failure of aluminum fluoride to activate PLD in most cell types suggested that PLD-activating G protein might be a low-molecular-mass G protein (Smg) rather than a heterotrimeric one. Furthermore, the activation of PLD by GTP-γS was enhanced by an Smg-GDS that promotes GDP–GTP exchange on Smg, and was blocked by a Rho-GDI that blocks such exchange.[45] The Rho-GDI and Smg-GDS are not totally specific and appear to be capable of regulating multiple Smg proteins from different families. Recent studies identified members of the Rho and ARF subfamilies as the PLD-regulating Smg protein.[45–49]

Brain cytosol reconstituted PLD activity in permeabilized, cytosol-depleted HL-60 cells[46] and in isolated HL-60 plasma membranes and membrane extracts[47] that had lost their PLD activity. The activating factor was purified and identified as ARF in both cases. ARF was initially identified as an activator of cholera toxin ADP-ribosyltransferase activity and is now known to play a critical role in vesicular membrane trafficking in all eukaryotic cells. All three subclasses of the ARF family were shown to activate PLD.[39] Studies with recombinant ARF proteins revealed that myristoylation of ARF, although not essential, enhances the PLD-activating activity. The amino-terminal 73 amino acid residues of ARF are also necessary for activation of PLD, whereas this region is not critical for activation of cholera toxin.[48]

The inhibition by Rho-GDI of GTP-γS-dependent activation of PLD on HL60 membranes first suggested that members of the Rho family were PLD-activating Smg.[45] Rho-GDI both blocks GDP–GTP exchange on Rho proteins and causes their dissociation from membranes. Dose-dependent inhibition of GTP-γS-stimulated PLD by Rho-GDI was also observed in liver membranes, and full reconstitution of GTP-γS-stimulated PLD in Rho-GDI-washed liver membranes was achieved with purified RhoA.[49] There was partial reconstitution with Racl, another Smg, but no enhancement with Cdc42 or, surprisingly, ARF. These findings support a role for members of the Rho family in the activation of membrane-associated PLD and suggest that major form of PLD in liver membranes is different from the ARF-regulated enzyme. Indeed, the Rho-regulated enzyme appears to be present only in membranes, whereas the ARF-responsive enzyme is found in both cytosol and membranes.[50] For the detection of both ARF- and Rho-dependent PLD activities, the presence of PtdIns(4,5)P_2 in substrate vesicles was essential.[39,47,51] This observation defines a novel function of PtdIns(4,5)P_2 as a cofactor for PLD and suggests that PtdIns(4,5)P_2 synthesis and hydrolysis could be important determinants in regulating PLD action in signal transduction.

At present there is no evidence that links ARF or Rho with receptor-mediated activation of PLD in intact cells. Such evidence will be required before those Smg proteins can be ascribed a physiological role in signal-dependent activation of PLD.

The identification of Smg proteins as PLD-activating G proteins and the PLC-γ/PKC sequence as a PLD-activating pathway in PDGF-treated cells casts some doubt on the earlier suggestion of direct involvement of heterotrimeric G proteins and protein tyrosine kinases in receptor–PLD coupling. However, by analogy to the PLC family, the evidence for multiple PLD isozymes suggests that there may be multiple modes of PLD regulation. Thus, roles of heterotrimeric G proteins and protein tyrosine kinases as direct mediators of receptor–PLD coupling cannot be excluded until pure PLDs can be tested in reconstituted systems.

Lipid second messengers

In summary, the phospholipases generate a large variety of lipid-derived second messengers. Signal transduction processes that are initiated by a wide variety of agonists, often leading to activation of G proteins and protein tyrosine kinases, activate PLA_2, PLC or PLD and their products, arachidonate, lysophospholipids, Ins(1,4,5)P_3, DAG and PA, which play important roles in activating other processes such as Ca^{2+} mobilization and PKC activation. In other cases the lipid-derived products give rise to further metabolites such as PAF and various prostaglandins that can be agonists in their own right. The important role of each of these lipid-derived messengers in cell activation and the detailed mechanisms of their regulation are areas of very active current investigation.

References

1 Dennis, E.A., Rhee, S.G., Billah, M.M. and Hunnun, Y.A. (1991) Role of phospolipases in generating lipid second messengers in signal transduction. *FASEB J.*, **5**, 2068–77.

2 Dennis, E.A. (1987) The regulation of eicosanoid production: role of phospholipases and inhibitors. *Bio/Technology*, **5**, 1294–1300.

3 Dennis, E.A. (1994) Diversity of group types, regulation, and function of phospholipase A_2. *J. Biol. Chem.*, **269**, 13057–60.

4 Davidson, F.F. and Dennis, E.A. (1990) Evolutionary relationships and implications for the regulation of phospholipase A_2: from snake venom to human secreted forms. *J. Mol. Evol.*, **31**, 228–38.

5 Hanasaki, K. and Arita, H. (1992) Characterization of a high affinity binding site for pancreatic-type phospholipase A_2 in the rat. Its cellular and tissue distribution. *J. Biol. Chem.*, **267**, 6414–20.

6 Fremont, D.H., Anderson, D., Wilson, I.A. *et al.* (1993) The crystal structure of phospholipase A_2 from Indian cobra reveals a novel trimeric association. *Proc. Natl. Acad. Sci. USA*, **90**, 342–6.

7 Asmis, R. and Dennis, E.A. (1994) PAF stimulates cAMP formation in $P388D_1$ macrophage-like cells via the formation and secretion of prostaglandin E_2 in an autocrine fashion. *Biochim. Biophys. Acta*, **1224**, 295–301.

8 Clark, J.D., Lin, L.-L., Kriz, R.W. *et al.* (1991) A novel arachidonic acid-selective cytosolic PLA2 contains a Ca^{2+}-dependent translocation domain with homology to PKC and GAP. *Cell*, **65**, 1043–51.

9 Sharp, J.D., White, D.L., Chiou, X.G. *et al.* (1991) Molecular cloning and expression of human Ca^{2+}-sensitive cytosolic phospholipase A_2. *J. Biol. Chem.*, **266**, 14850–3.

10 Reynolds, L., Hughes, L., Louis, A.I. *et al.* (1993) Metal ion and salt effects on the phospholipase A_2, lysophospholipase, and transacylase activity of human cytosolic phospholipase A_2. *Biochim. Biophys. Acta*, **1167**, 272–80.

11 Sharp, J.D., Pickard, T., Chiou, X.G. *et al.* (1994) Serine 228 is essential for catalytic activities of 85-kDa cytosolic phospholipase A_2. *J. Biol. Chem.*, **269**, 23250–4.

12 Lin, L.-L., Wartman, M., Lin, A.F. *et al.* (1993) Cytoplasmic PLA_2 is phosphorylated and activated by MAP kinase. *Cell*, **72**, 269–78.

13 Hazen, S.L., Stuppy, R.J. and Gross, R.W. (1990) Purification and characterization of canine myocardial cytosolic phospholipase A_2. A calcium-independent phospholipase with absolute f1–2 regiospecificity for diradyl glycerophospholipids. *J. Biol. Chem.*, **265**, 10622–30.

14 Ackermann, E.J., Kempner, E.S. and Dennis, E.A. (1994) Ca^{2+}-independent cytosolic phospholipase A_2 from the macrophage-like $P388D_1$ cells: isolation and characterization. *J. Biol. Chem.*, **269**, 9227–33.

15 Barbour, S. and Dennis, E.A. (1993) Antisense inhibition of group II phospholipase A_2 expression blocks the production of prostaglandin E_2 by $P388D_1$ cells. *J. Biol. Chem.*, **268**, 21875–82.

16 Balsinde, J., Barbour, S.E., Bianco, I.D. and Dennis, E.A. (1994) Arachidonic acid mobilization in $P388D_1$ macrophages is controlled by two distinct Ca^{2+}-dependent phospholipase A_2s. *Proc. Natl. Acad. Sci. USA*, **91**, 11060–4.

17 Berridge, M.J. (1993) Inositol trisphosphate and calcium signaling. *Nature*, **361**, 315–325.

18 Rhee, S.G. and Choi, K.D. (1992) Regulation of inositol phospholipid-specific phospholipase C isozymes. *J. Biol. Chem.*, **267**, 12393–6.

19 Cohen, G.B., Ren, R. and Baltimore, D. (1995) Molecular binding domains in signal transduction. *Cell*, **80**, 237–48.

20 Kuppe, A., Evans, L.M., McMillen, D.A. and Griffith, O.H. (1989) Phosphatidylinositol-specific phospholipases C of *Bacillus cereus*: cloning, sequencing, and relationship to other phospholipases. *J. Bacteriol.*, **171**, 6077–83.

21 Claesson-Welsh, L. (1994) Platelet-derived growth factor receptor signals. *J. Biol. Chem.*, **51**, 32023–6.

22 Mohammadi, M., Dionne, C.A., Li, W. *et al.* (1992) Point mutation in FGF receptor eliminates phosphatidylinositol hydrolysis without affecting mitogenesis. *Nature*, **358**, 681–4.

23 Obermeier, A., Halfter, H., Wiesmüller, K.-H. *et al.* (1993) Tyrosine 785 is a major determinant of Trk–substrate interaction. *EMBO J.* **12**, 933–41.

24 Soler, C., Beguinot, L. and Carpenter, G. (1994) Individual epidermal growth factor receptor autophosphorylation sites do not stringently define association motifs for several SH2-containing proteins. *J. Biol. Chem.*, **269**, 12320–4.

25 Kim, K.H., Kim, J.W., Zilberstein, A. *et al.* (1991) PDGF stimulation of inositol phospholipid hydrolysis requires PLC-γ1 phosphorylation on tyrosine residues 783 and 1254. *Cell*, **65**, 435–41.

26 Valius, M. and Kazlauskas, A. (1993) Phospholipase C-γ1 and phosphatidylinositol 3 kinase are the downstream mediators of the PDGF receptor's mitogenic signal. *Cell*, **73**, 321–34.

27 McBride, K., Rhee, S.G. and Jaken, S. (1991) Immunocytochemical localization of phospholipase C-γ in rat embryo fibroblasts. *Proc. Natl. Acad. Sci. USA*, **88**, 7111–5.

28 Yang, L.J., Rhee, S.G. and Williamson, J.R. (1994) Epidermal growth factor-induced activation and translocation of phospholipase C-γl to the cytoskeleton in rat hepatocytes. *J. Biol. Chem.*, **269**, 7156–62.

29 Rhee, S.G., (1994) Regulation of phosphoinositide-specific phospholipase C by G protein, in *Signal-Activated Phospholipases* (ed. Liscovitch, M.) R.G. Landes, Austin, TX, pp. 1–12.

30 Jhon, D.-Y., Lee, H.-H., Park, D. *et al.* (1993) Cloning, sequencing, purification, and Gq-dependent activation of phospholipase C-β3. *J. Biol. Chem.*, **268**, 6654–61.

31 Lee, C.-W., Lee, K.-H., Lee, S.B. *et al.* (1994) Regulation of phospholipase C-β4 by ribonucleotides and the α subunit of Gq. *J. Biol. Chem.*, **268**, 25335–8.

32 Lee, S.B., Shin, S.H., Hepler, J.R. *et al.* (1993) Activation of phospholipase C-β2 mutants by G protein αq and βγ subunits. *J. Biol. Chem.*, **268**, 25952–7.

33 Touhara, K., Inglese, J., Pitcher, J.A. *et al.* (1994) Binding of G protein βγ-subunits to pleckstrin homology domains. *J. Biol. Chem.*, **269**, 10217–20.

34 Marrero, M.B., Paxton, W.G., Duff, J.L. *et al.* (1994) Angiotensin II stimulates tyrosine phosphorylation of phospholipase C-γ1 in vascular smooth muscle cells. *J. Biol. Chem.*, **269**, 10935–9.

35 Dhar, A. and Shukla, S.D. (1994) Electrotransection of $pp60^{v\text{-}Src}$ monoclonal antibody inhibits activation of phospholipase C in platelets, *J. Biol. Chem.*, **269**, 9123–7.

36 Thomas, G.M.H., Cunningham, E., Fensome, A. *et al.* (1993) An essential role for phosphatidylinositol transfer protein in phospholipase C-mediated inositol lipid signaling. *Cell*, **74**, 919–28.

37 Liscovitch, M. and Chalifa, V. (1994) Signal-activated phospholipase D, in *Signal-Activated Phospholipases* (Ed. Liscovitch, M.), R.G. Landes, Austin, TX, pp. 31–63.

38 Pai, J.-K., Siegel, M.I., Egan, R.W. and Billah, M.M. (1988) Phospholipase D

catalyzes phospholipid metabolism in chemotactic peptide-stimulated HL-60 granulocytes. *J. Biol. Chem.*, **263**, 12472–7.

39 Massenburg, D., Han, J.-S., Liyanage, M. *et al.* (1994) Activation of rat brain phospholipase D by ADP-ribosylation factors 1, 5, and 6: separation of ADP-ribosylation factor-dependent and oleate-dependent enzymes. *Proc. Natl. Acad. Sci. USA*, **91**, 11718–22.

40 Hammond, S.M., Altshuller, Y.M., Sung, T.-C. *et al.* (1995) Human ADP-ribosylation factor-activated phoshatidylcholine-specific phospholipase D defines a new and highly conserved gene family. *J. Biol. Chem.*, **270**, 29640–3.

41 Lee, Y.H., Kim, H.S., Pai, J.-K. *et al.* (1994) Activation of phospholipase D induced by platelet-derived growth factor is dependent upon the level of phospholipase C-γ1. *J. Biol. Chem.*, **269**, 26842–7.

42 Yeo, E.-J., Kazlauskas, A. and Exton, J.H. (1994) Activation of phospholipase C-γ is necessary for stimulation of phospholipase D by platelet-derived growth factor. *J. Biol. Chem.*, **269**, 27823–6.

43 Conricode, K.M., Brewer, K.A. and Exton, J.H. (1992) Activation of phospholipase D by protein kinase C. *J. Biol. Chem.*, **267**, 7199–202.

44 Balboa, M.A., Balsinde, J., Dennis, E.A. and Insel, P.A. (1995) A phospholipase D-mediated pathway for generating diacylglycerol in nuclei from Madin–Darby canine kidney cells. *J. Biol. Chem.*, **270**, 11738–40.

45 Bowman, E.P., Uhlinger, D.J. and Lambeth, J.D. (1993) Neutrophil phospholipase D is activated by a membrane-associated Rho family small molecular weight GTP-binding protein. *J. Biol. Chem.*, **268**, 21509–12.

46 Cockcroft, S., Thomas, G.M., Fensome, A. *et al.* (1994) Phospholipase D is a downstream effector of ARF in granulocytes. *Science*, **263**, 523–6.

47 Brown, H.A., Butowski, S., Moomaw, C.R. *et al.* (1993) ADP-ribosylation factor, a small GTP-dependent regulatory protein stimulates phospholipase D activity. *Cell*, **75**, 1137–44.

48 Zhang, G.-F., Patton, W.A., Lee, F.-J.S. *et al.* (1995) Different ARF domains are required for the activation of cholera toxin and phospholipase D. *J. Biol. Chem.*, **270**, 21–4.

49 Malcolm, K.C., Ross, A.H., Qui, R.-G. *et al.* (1994) Activation of rat liver phospholipase D by the small GTP-binding protein RhoA. *J. Biol. Chem.*, **269**, 25951–4.

50 Siddiqi, A.R., Smith, J.L., Ross, A.H. *et al.* (1995) Regulation of phospholipase D in HL60 cells. *J. Biol. Chem.*, **270**, 8466–73.

51 Liscovitch, M., Chalifa, V., Pertile, P. *et al.* (1994) Novel function of phosphatidylinositol 4,5-bisphosphate as a cofactor for brain membrane phospholipase D. *J. Biol. Chem.*, **269**, 21403–6.

13 Multiple roles for phosphoinositide 3-kinases in signal transduction

George Panayotou

The elucidation of the diverse biochemical pathways that are propagated within cells is paramount to our understanding of the mode of action of a vast array of extracellular mediators that are important in cell signaling. An enzyme that has been consistently implicated in signaling cascades within the cell is the phosphoinositide 3-kinase (PI3K).[1–3] This enzymatic activity can phosphorylate the 3′-hydroxyl group on the inositol ring of phosphatidylinositol (PtdIns) or the same group of PtdIns 4-phosphate or PtdIns 4,5-bisphosphate

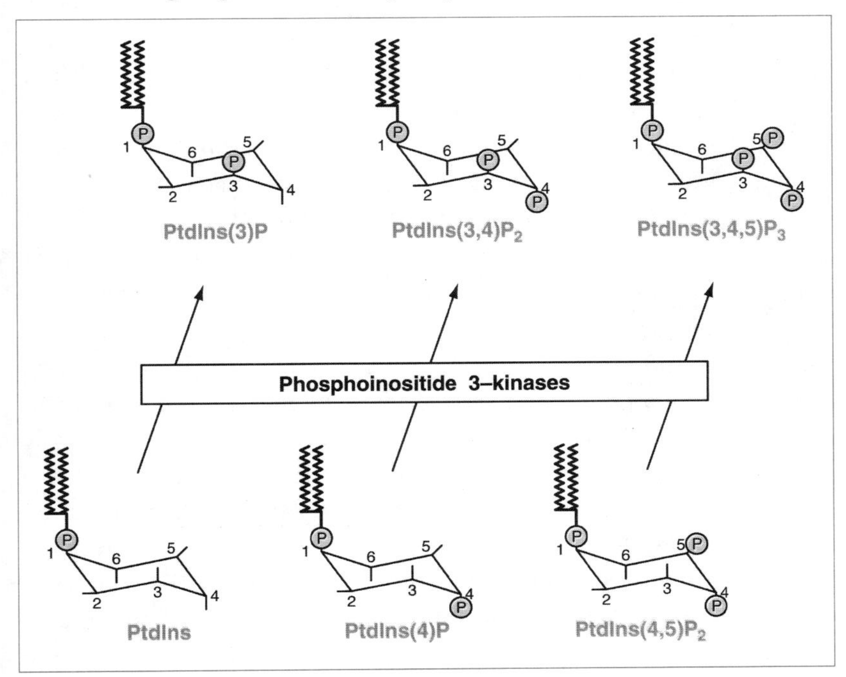

Figure 13.1 The phospholipids generated by the action of phosphoinositide 3-kinases.

Signal Transduction. Edited by Carl-Henrik Heldin and Mary Purton. Published in 1996 by Chapman & Hall. ISBN 0 412 70810 8

(Figure 13.1). PtdIns(4,5)P_2 is also a substrate of phospholipase C (PLC), an enzyme which is central in signal transduction pathways, being responsible for the generation of the second messengers inositol trisphosphate and diacylglycerol.

The appearance of 3′-phosphorylated phosphoinositides in response to extracellular signals or oncogenic transformation was first demonstrated by Cantley's group.[4,5] Subsequently, these phospholipids have been identified in many cell types and it has been suggested that PtdIns(3,4,5)P_3 is the most important second messenger species.[6] Discovering how these phospholipids initiate intracellular signaling cascades remains a major challenge in the field of signal transduction. No specific phospholipase has been described to date that would generate smaller signaling entities in analogy with the breakdown of PtdIns(4,5)P_2 by PLC. It is possible that the presence of a highly charged head group on membrane-attached phospholipids may have substantial effects on the local structure and stability of the lipid bilayer, thus initiating cell surface processes that are required in signaling. However, the most widely accepted hypothesis is that the products of PI3K can specifically interact with proteins and thus modulate their activity or localization.

The heterodimeric p85α/p110α PI3K

The first full characterization of PI3K came through the purification and cDNA cloning of the enzyme from bovine brain,[7,8] which showed that PI3K consists of two subunits, p85α and p110α (Figure 13.2). Analysis of these proteins showed that the catalytic activity resides in the 110 kDa subunit, while the primary structure of the p85α subunit suggested a regulatory role. p85α is a modular protein, consisting of several independently folding domains. These include two SH2 domains, which are conserved in many molecules involved in signal transduction and mediate protein–protein interactions by recognizing with high affinity phosphotyrosine residues within a defined amino acid sequence[9,10] (see Chapter 9). Growth factor receptors with intrinsic protein–tyrosine kinase activity are known, upon activation to, autophosphorylate on several tyrosine residues, which can then form distinct binding sites for proteins containing SH2 domains (see Chapter 1). While the phosphotyrosine binding site is common to all SH2 domains, specificity is provided by additional interactions of several amino acid residues downstream of the phosphotyrosine.[10] In the case of the p85α–p110α PI3K complex, several studies have defined the interaction motif with different receptors or other signaling molecules, using

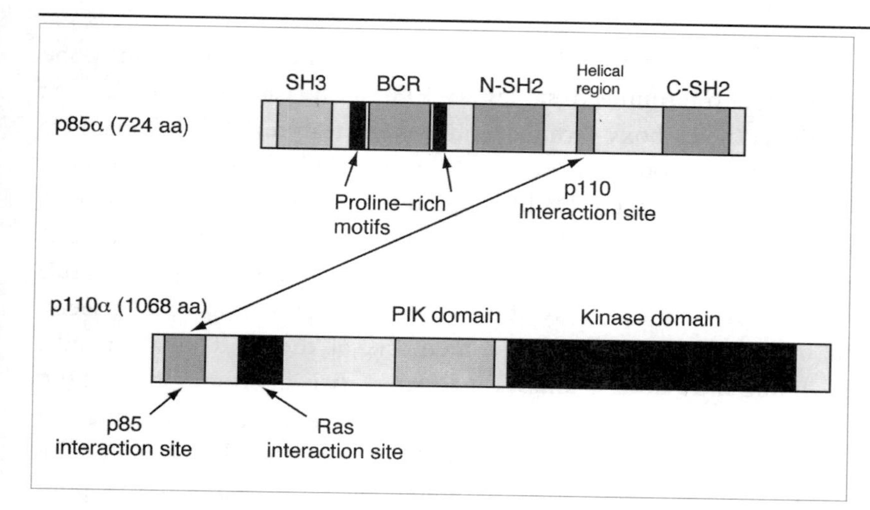

Figure 13.2 Schematic structure of the regulatory p85α subunit and the catalytic p110α subunit of the heterodimeric PI 3-kinase, indicating domains important for their function. aa, Amino acid residues.

site-directed mutagenesis of tyrosine-phosphorylation sites. A universal feature of all high affinity sites is the presence of a methionine residue downstream of the phosphotyrosine (Tyr–X–X–Met). It is clear therefore that the role of the SH2 domains in the p85α subunit is to interact with activated receptors or other tyrosine-phosphorylated molecules and thus recruit the associated catalytic subunit.

p85α also contains an SH3 domain, which is again conserved in many signaling proteins as well as molecules interacting with the cytoskeleton.[9] This domain also mediates protein–protein interactions through the specific recognition of proline-rich sequences[11] (see Chapter 9). Two proline-rich sequences have also been found on p85α itself, suggesting that they could either recognize other molecules containing SH3 domains or could serve as a binding partner of the p85α SH3, in an intramolecular and possibly regulated fashion.

Another modular region of p85α is flanked by the two proline-rich motifs. This is called the Bcr region by virtue of its similarity (albeit limited) with part of the break point cluster region (bcr) gene. This motif is conserved in other proteins, such as Rho GAP, that display GTPase-activating activity towards some of the small G proteins.[12] However, no such activity has yet been reported for the p85α Bcr region.

The sequence between the two SH2 domains of p85α is predicted to adopt a helical conformation,[13] and could play a role in positioning the two SH2 domains in an optimal alignment for their interactions with target proteins. It also contains the binding site for the p110α subunit (see Figure 13.2), as mapped by sequential truncations of p85α down to a small region of 35 amino acids that retains the ability to bind recombinant p110α.

The catalytic subunit has a corresponding interaction site for p85α at the amino terminus (residues 20–108).[13] The catalytic domain itself resides at the carboxy-terminal end. Several features of this domain are conserved in protein kinases [such as the DFG (Asp–Phe–Gly) motif that forms part of the ATP-binding site]; however, there are also many differences which may define the unique identity of lipid kinases. A sequence outside the kinase domain [termed the phosphoinositide kinase (PIK) domain] may also be relevant in this respect as it seems to be conserved among other members of the lipid kinase family, including the PtdIns 4-kinase[14] (see later). It is interesting that p110α has been shown to be a dual-specificity kinase, phosphorylating not only inositol lipids but also its associated regulatory subunit p85α. This phosphorylation (at Ser608) is absolutely dependent on the prior formation of the p85α–p110α complex and may be important in regulating the lipid kinase activity.[15]

The heterodimeric structure of the p85α–p110α PI3K suggests possible models for its activation in the cell. Binding to membrane-associated receptors via the SH2 domains of p85α brings the catalytic subunit close to its substrates and that may be the most crucial parameter for activation *in vivo*. Phosphorylation of p85α on tyrosine residues by associated tyrosine kinases does not appear to contribute to activation but the tight binding of the SH2 domains has been shown to result in a modest enhancement of the p110α catalytic activity.[16] However, structural studies have revealed that the conformational changes induced by the binding events are rather limited and do not seem to be compatible with an allosteric activation mechanism.[17,18]

A family of PI 3-kinases

The multitude of biological responses in which PI3K is implicated (see Figure 13.5) makes it difficult to imagine that a single enzymatic entity could be involved and suggests that at least some distinct regulatory subunits should exist in order to allow differential coupling of PI3K activity to various stimuli. Indeed, the cloning of the p85α subunit of the heterodimeric PI3K revealed the existence of a related protein, termed p85β, with similar overall structure but distinct amino acid sequence.[7] Although no differential function of this homolog has been reported so far, its discovery supported the notion of other family members. The subsequent cloning of the p110α subunit and the search for similar sequences among known proteins showed a clear relationship between p110α and the product of the *VPS34* gene of *Saccharomyces cerevisiae*, which was subsequently demonstrated to possess an intrinsic 3-kinase activity.[19] This protein is important for vesicular trafficking and targeting of proteins to the yeast vacuole. It requires for its function a tightly coupled protein called VPS15, which possesses an intrinsic protein kinase activity and may serve an adaptor or regulatory function for the PI3K activity of VPS34.

The similarity of p110α to VPS34 suggested that in mammalian cells a PI3K activity may also be involved in membrane trafficking events, such as receptor-mediated endocytosis.[20] Such an involvement has been shown in the case of the platelet-derived growth factor (PDGF) receptor whose internalization after PDGF activation is dependent on PI3K.[21] It is not clear, however, if this extends to other receptors or indeed other trafficking events. The picture becomes complicated since it has recently been shown that a human homolog of VPS34 does exist and is also found in a complex with a protein similar to VPS15.[22] This demonstrates that the control of vesicular trafficking may be conserved at the molecular level between yeast and humans. It is interesting in this respect that both VPS34 and its human homolog have a distinct substrate specificity, as compared with the p85α–p110α PI3K: both phosphorylate exclusively phosphatidylinositol but not any of the 4-phosphorylated phosphoinositides which are substrates of p110α. The human homolog has been termed PtdIns 3-kinase to emphasize this fact (see Figures 13.1 and 13.3). This is a clear demonstration of how different 3-phosphorylated lipids may be involved in distinct pathways within the cell. Apart from the difference in substrate specificity, the VPS34 family also shows a distinct requirement for divalent cations for

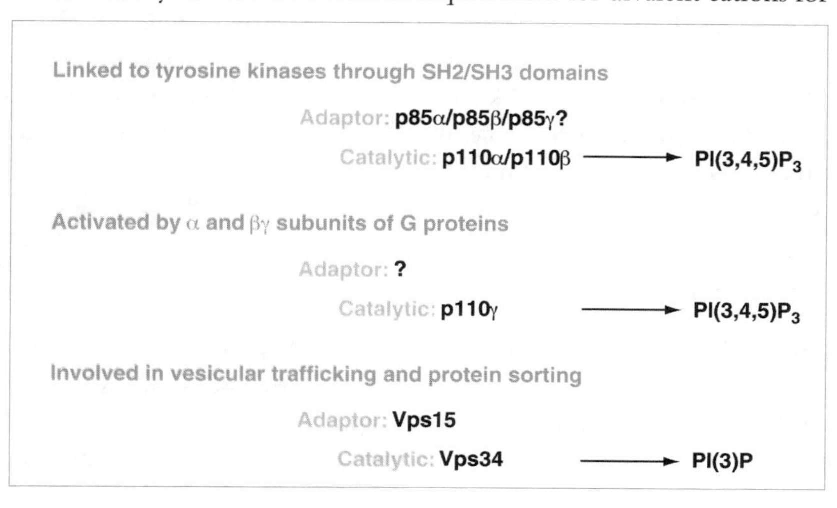

Figure 13.3 A family of PI 3-kinases: three distinct classes are listed, together with the characterized adaptor and catalytic subunits and the major phospholipid products thought to be responsible for their biological effects.

its activation and moreover there are differences regarding the ability of PI3K inhibitors to affect its activity.

The discovery of novel members of the PI3K family is facilitated by the availability of sequence information that can be used to screen cDNA libraries. This approach has led to the cloning of a p110α-related protein, termed p110β.[23] It is interesting that no specific difference has been reported so far between this protein and p110α in terms of its coupling to p85α, substrate specificity, sensitivity to inhibitors and tissue distribution. It remains possible that some redundancy exists within the PI3K family, involving interchangeable binding between p85α or p85β and p110α or p110β (Figure 13.3).

Recent work using the same approach has extended the PI3K family to include a distinct catalytic entity that appears to be involved in signaling by G-protein-coupled receptors (Figure 13.3). This novel protein, called p110γ, does not associate with any p85 species and may not even use an alternative adaptor. A clue for its function comes from the finding that it is activated *in vitro* by both α and βγ subunits of heterotrimeric G proteins.[24]

There is 36% identity between p110γ and p110α, the similarity being highest at the catalytic domain. The two proteins exhibit a similar pattern of substrate phosphorylation utilizing PtdIns, PtdIns4*P* and PtdIns(4,5)P_2. The properties of p110γ suggest that it may be directly stimulated in response to activation of heterotrimeric G-protein-linked receptors. An alternative approach to the identification of such a type of PI3K has been its direct purification from tissues using an assay based on βγ subunit activation.[25] As more members of the PI3K family emerge, the evolutionary relationships between them can be measured in more detail. Figure 13.4 shows the similarity between the catalytic domains of the PI3K family, other lipid kinases, such as the PtdIns 4-kinase, and other related proteins whose lipid kinase activity has not been demonstrated yet, including the Tor/FRAP family.[26] The latter are thought to be direct targets of the action of the immunosuppressant rapamycin, which causes cell-cycle arrest in both yeast cells and mammalian T cells.

Biological role

The study of signal transduction pathways has made apparent the inherent complexity and diversity of these processes. The issues of specificity and redundancy, cell-type dependent variations, regulation by competitive interactions and the transient nature of most signaling

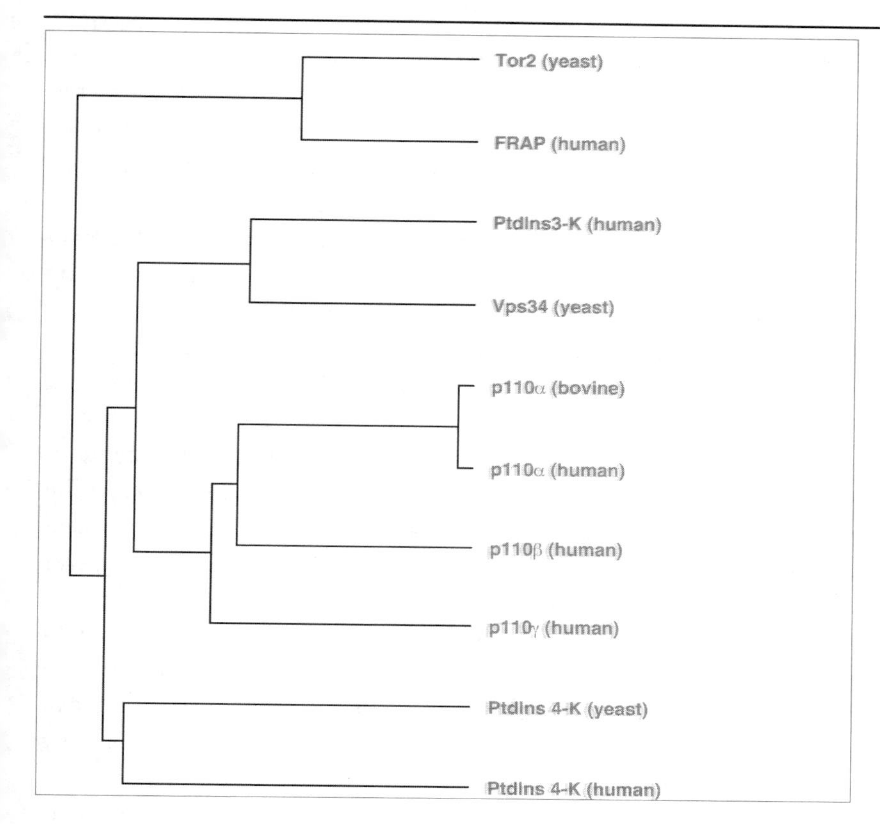

Figure 13.4 A dendrogram comparing the sequence similarity within the catalytic domains of members of the lipid kinase family. Note that a lipid kinase activity has not been reported to date for the Tor/FRAP family. Yeast denotes *Saccharomyces cerevisiae*.

events exemplify the need for extreme care when attempting to investigate the necessity (or sufficiency) of any distinct molecule in a particular signaling pathway. This is particularly true for PI3K, since this activity is found in most cell types and tissues and has been implicated in many and diverse biological responses.

The currently accepted methodology for showing at least an involvement of PI3K in a particular cellular process encompasses several complementary approaches. First, the process investigated should result in the appearance of 3′-phosphorylated lipids and this can be demonstrated by *in vivo* labeling of cells with radioactive precursors followed by HPLC analysis of cellular extracts and detection of labeled deacylated lipids by comparison with standards. The importance of PI3K activation can then be evaluated by preventing its recruitment by specific cell-surface molecules, such as growth factor receptors. For example, by mutating specific residues that form part of a docking site for PI3K and expressing the mutant receptors in a suitable cell system, their ability to initiate a specific biological response in the absence of PI3K activation can be examined. This approach depends, of course, on the precise understanding of the mode of interaction between a receptor or other molecule and the PI3K. One shortcoming is the often encountered redundancy in the interactions between receptors and signaling molecules, as a single binding/recruitment site can serve not only PI3K but also other proteins.[27]

An alternative approach is to prevent the endogenous PI3K from becoming recruited and activated. This can be achieved either by microinjecting neutralizing antibodies or by expressing a mutated regulatory p85α subunit of the PI3K which lacks the site of interaction with the catalytic subunit and can therefore act in a dominant negative fashion. The use of inhibitors that show specificity for PI3K activity has also proved a popular tool in implicating this enzyme in biological responses. In particular, wortmannin, a fungal metabolite that covalently associates with the catalytic subunit of PI3K, has been shown to inhibit the enzyme at nanomolar concentrations, far lower than the amounts required to inhibit other enzymes.[28,29] Its capacity to cross the plasma membrane and inhibit endogenous PI3K activity in intact cells has resulted in its widespread use in cell culture systems. Clearly, however, a larger selection of inhibitors needs to be discovered to overcome the problems of specificity for different PI3K family members. Finally, several techniques exist for the discovery of direct binding interactions between PI3K and other intracellular enzymes, including the yeast two-hybrid system or the use of immobilized recombinant enzyme or domains to probe cell lysates for binding partners which can then be identified by sequencing analysis. If these partners have already been assigned specific roles in a particular biological response, then it can be implied that the PI3K may also be involved in the same pathway.

Using the above approaches, many different cellular processes have been shown to engage PI3K (Figure 13.5). They include DNA synthesis,[30,31] cell survival,[32] membrane ruffling,[33,34] cell chemotaxis,[35,36] the oxidative burst in neutrophils,[6,37] receptor internalization,[21] oocyte maturation,[38] insulin-activated glucose transport[39,40] and vesicular trafficking.[19] The stimuli that couple to the activation of PI3K in the cell range from small, soluble, extracellular molecules, such as formyl–Met–Leu–Pro peptide (FMLP), thrombin and many growth factors, to plasma membrane-associated proteins such as the heterotrimeric G proteins, the small GTP-binding proteins and cytoplasmic, non-receptor tyrosine kinases which are often involved in oncogenic transformation.[3] Of the many diverse responses that these molecules regulate, DNA synthesis in response to growth factors has attracted much attention. The results have often been contradictory, depending on the cell type examined and the experimental approaches used, reflecting the complexity of the responses involved in mitogenesis and highlighting the importance of using complementary approaches, as discussed above.[31,41] While studies using site-directed mutagenesis of growth factor receptor residues responsible for recruiting PI3K have been given diverse interpretations, a recent study using microinjection of neutralizing anti-PI3K antibodies has shown a correlation of this activity and the mitogenic activity of some growth factors such as PDGF and EGF but not others, e.g. CSF-1 and bombesin.[42]

It should be noted that most of the biological responses listed in Figure 13.5 involve some form of cytoskeletal rearrangement or movement of membrane components. It is possible, therefore, that the primary target of PI3K products may be components involved in these rearrangements, which, in turn, are necessary (although most probably not sufficient) for the diverse physiological responses.

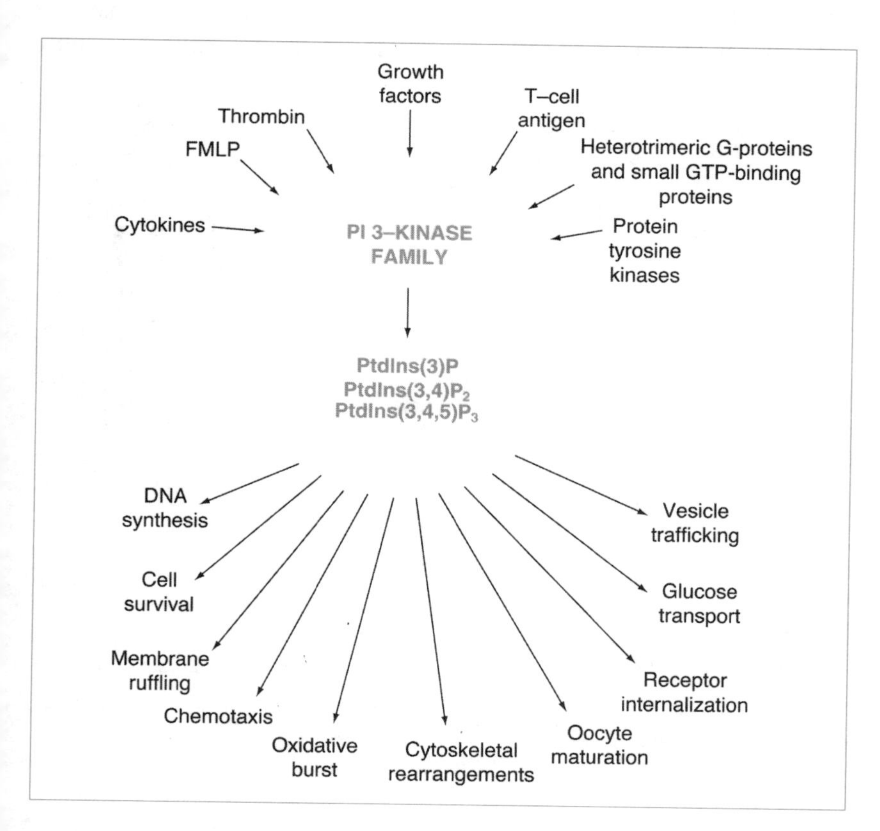

Figure 13.5 Summary of the biological responses in which activation of PI 3-kinases has been implicated. Some of the stimuli that elicit these responses are shown at the top.

Upstream and downstream regulators

Signal transduction pathways are characterized by a series of steps where enzymes or adaptor proteins recruit and activate downstream molecules. Unraveling the precise steps that regulate PI3K and its effectors remains a very important task which is only beginning to be elucidated. Figure 13.6 summarizes the current state of affairs regarding the molecules that have been shown to activate PI3K and those that are candidate downstream targets. It has to be stressed that a definitive connection between PI3K and downstream effectors has not yet been established in an unambiguous way, such as a genetically defined pathway, and the physiological relevance of the interactions presented here is not certain.

As mentioned previously, the role of growth factor receptors with intrinsic tyrosine kinase activity in recruiting and activating PI3K is well established. Several other molecules have been shown to interact directly with the PI3K or lie upstream in some biological responses. The cytoplasmic tyrosine kinases of the Src family, which are themselves activated by transmembrane receptors, utilize their SH3 domain to recognize the first proline-rich region of p85α, thus forming a complex

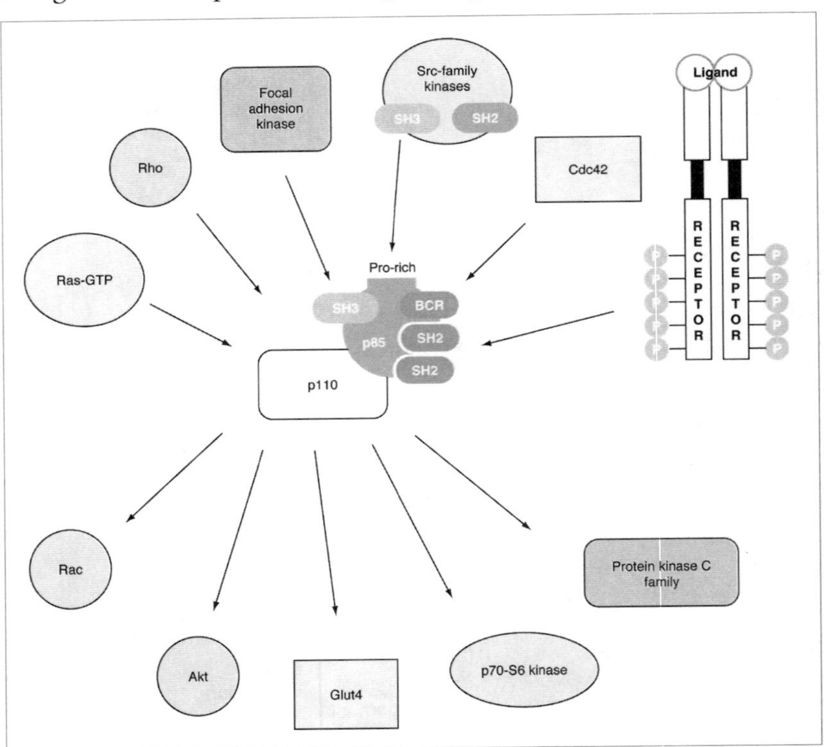

Figure 13.6 Proteins that have been implicated as upstream activators or downstream targets of the PI 3-kinase.

with and activating PI3K.[43,44] Indeed the activation of PI3K by antigen receptor was blocked in permeabilized cells that were treated with a peptide containing the proline-rich region. A different mechanism of activation was reported recently and involves the focal adhesion kinase (FAK), which itself contains a proline-rich motif. In thrombin-activated platelets, this proline-rich sequence may interact with the SH3 domain of p85α, resulting in activation.[45] An association between FAK and PI3K has also been seen in PDGF-stimulated fibroblasts.[46]

Further examples of activation by direct binding have been provided by small GTP-binding proteins: the Ras-related Cdc42, which regulates cytoskeletal reorganization during the yeast cell cycle, binds to p85α in an interaction that probably involves the Bcr-homology domain of p85α and the effector domain of Cdc42.[47] Perhaps more importantly, given its central role in mammalian cell growth, Ras in its active, GTP-bound state can form a complex directly with the p110α catalytic subunit.[48,49] The site of interaction has been mapped using deletion mutants and is located close to the p85α binding site at the amino terminus of p110α (J. Downward, personal communication; see Figure 13.2). Finally, the small GTP-binding protein Rho has been implicated as an upstream regulator of PI3K activity in thrombin-activated platelets, although a direct interaction in not evident.[50,51] Given the importance of the small GTP-binding proteins in many signal transduction pathways, these interactions may prove to be particularly interesting. It should be noted that another member of this family, Rac, is probably downstream of PI3K, being involved in the regulation of membrane ruffling.[33,34,52]

The involvement of PI3K in insulin-mediated responses is well established and a possible molecular link is provided by the activation (in a PI3K-dependent manner) of glucose uptake by the translocation of GLUT1 and GLUT4 glucose transporters to the plasma membrane.[39] This has been shown both by the use of wortmannin and the expression of a dominant-negative p85α. Transfected PDGF receptors can also translocate GLUT4 and mutation of the PI3K interaction sites on this receptor abrogates this effect.[53] Another response that is mediated both by insulin and PDGF receptors is the activation of p70-S6 kinase, an event which is thought to be important for progression through the cell cycle. The use of inhibitors and receptor mutants has again shown a requirement for PI3K in this event.[54] The same pathway is also sensitive to rapamycin, an inhibitor of the Tor proteins that share homologous regions with lipid kinases. However, the points at which PI3K inhibitors and rapamycin exert their effects are distinct.[55]

Although several molecules are candidate downstream effectors of PI3K, direct targets of the 3-phosphorylated inositides have been much more difficult to identify. Using different phosphoinositides and recombinant protein kinase C isoforms, one group has reported that calcium-independent forms of PKC can be activated by PtdIns(3,4)P_2 and PtdIns(3,4,5)P_3 specifically.[56] However, another study has shown that this activation is not restricted to the calcium-insensitive members of the family and may be a feature of many other PKC isoforms,[57] emphasizing the need to confirm that indeed such interactions occur *in vivo*. In any event, given the central role of PKCs in cell regulation, these observations may have a substantial significance for the function of PI3K.

Another molecule that may be a direct target of 3-phosphoinositides is the serine/threonine kinase Akt. This kinase is activated by PDGF in a process that requires the PI3K binding sites on the PDGF receptor and is sensitive to wortmannin.[58] PtdIns 3-phosphate directly stimulates the kinase activity of Akt and activation depends on a structurally intact pleckstrin homology (PH) domain (see Chapter 9), suggesting that the PH domain of this kinase may be a direct target of PI3K.

It is obvious that much more needs to be done before a clear understanding of the mechanisms of action of PI3K will emerge. In this respect, a very useful approach will undoubtedly prove to be the use of genetically manipulatable organisms, such as *Drosophila*, in order to identify the important components of the pathways involving PI3K.

Acknowledgements

The author thanks Michael J. Fry for critical reading of the manuscript and many useful suggestions. Many thanks are also due to the members of Mike Waterfield's laboratory for their help and discussions, Ritu Dhand for inspiration in preparing some of the figures and P. Parker, J. Downward and P. Tsichlis for communicating results prior to publication.

Summary

In recent years, PI3K activity has been implicated in diverse biological processes. A family of related enzymes is beginning to emerge, providing an explanation for this diversity. The main tasks ahead include the characterization of new members of the PI3K family, the understanding of how differences in structure reflect diversity in function, the delineation of the precise pathways in which these enzymes fit in through the identification of upstream and downstream regulatory proteins and the development of specific and potent inhibitors that should not only help us understand the function of these enzymes but, given the role of PI3K in cell growth and many metabolic processes, may also have therapeutic value. A novel member of the PI3K family has recently been identified in *Drosophilia*. PI3K_68D has the unique feature of a carboxy-terminal C2 domain which follows the catalytic domain. The function and regulation of this enzyme is at present unknown.[59]

References

Owing to space limitations, many important papers could not be referenced, and the author apologizes to the researchers involved.

1 Stephens, L.R., Jackson, T.R. and Hawkins, P.T. (1993) Agonist-stimulated synthesis of phosphatidylinositol (3,4,5)-trisphosphate – a new intracellular signaling system. *Biochim. Biophys. Acta*, **1179**, 27–75.

2 Downes, C.P. and Carter, A.N. (1991) Phosphoinositide 3-kinase: a new effector in signal transduction? *Cell. Signaling*, **3**, 501–13.

3 Fry, M.J. (1994) Structure, regulation and function of phosphoinositide 3-kinases. *Biochim. Biophys. Acta*, **1226**, 237–68.

4 Whitman, M., Kaplan, D.R., Schaffhausen, B. *et al.* (1985) Association of phosphatidylinositol kinase activity with polyoma middle-T component for transformation. *Nature*, **315**, 239–42.

5 Auger, K.R., Serunian, L.A., Soltoff, S.P. *et al.* (1989) PDGF-dependent tyrosine phosphorylation stimulates production of novel polyphosphoinositides in intact cells. *Cell*, **57**, 167–75.

6 Stephens, L.R., Hughes, K.T. and Irvine, R.F. (1991) Pathway of phosphatidylinositol (3,4,5)-trisphosphate synthesis in activated neutrophils. *Nature*, **351**, 33–9.

7 Otsu, M., Hiles, I., Goot, I. *et al.* (1991) Characterization of two 85 kd proteins that associate with receptor tyrosine kinases, middle-T/pp60c-Src complexes, and PI3-kinase. *Cell*, **65**, 91–104.

8 Hiles, I.D., Otsu, M., Volinia, S. *et al.* (1992) Phosphatidylinositol 3-kinase – structure and expression of the 110kd catalytic subunit. *Cell*, **70**, 419–29.

9 Pawson, T. and Schlessinger, J. (1993) SH2 and SH3 domains. *Curr. Biol.*, **3**, 434–42.

10 Songyang, Z., Shoelson, S.E., Chaudhuri, M. *et al.* (1993) SH2 domains recognize specific phosphopeptide sequences. *Cell*, **72**, 767–78.

11 Yu, H.T., Chen, J.K., Feng, S.B. *et al.* (1994) Structural basis for the binding of proline-rich peptides to SH3 domains. *Cell*, **76**, 933–45.

12 Fry, M.J. (1992) Defining a new GAP family. *Curr. Biol.*, **2**, 78–80.

13 Dhand, R., Hara, K., Hiles, I. *et al.* (1994) PI 3-kinase: structural and functional analysis of intersubunit interactions. *EMBO J.*, **13**, 511–21.

14 Flanagan, C.A., Schnieders, E.A., Emerick, A.W. *et al.* (1993) Phosphatidylinositol 4-kinase – gene structure and requirement for yeast cell viability. *Science*, **262**, 1444–8.

15 Dhand, R., Hiles, I., Panayotou, G. *et al.* (1994) PI 3-kinase is a dual specificity enzyme: autoregulation by an intrinsic protein-serine kinase activity. *EMBO J.*, **13**, 522–33.

16 Rordorf-Nikolic, T., Van Horn, D.J., Chen, D.X. *et al.* (1995) Regulation of phosphatidylinositol 3′-kinase by tyrosyl phosphoproteins – full activation requires occupancy of both SH2 domains in the 85-kDa regulatory subunit. *J. Biol. Chem.*, **270**, 3662–66.

17 Hensmann, M., Booker, G.W., Panayotou, G. *et al.* (1994) Phosphopeptide binding to the N-terminal SH2 domain of the p85 alpha subunit of PI 3′-kinase: a heteronuclear NMR study. *Protein Sci.*, **3**, 1020–30.

18 Kuriyan, J. and Cowburn, D. (1993) Structures of SH2 and SH3 domains. *Curr. Opinion Struct. Biol.* **3**, 828–37.

19 Schu, P.V., Takegawa, K., Fry, M.J. *et al.* (1993) Phosphatidylinositol 3-kinase encoded by yeast VPS34 gene essential for protein sorting. *Science*, **260**, 88–91.

20 Panayotou, G. and Waterfield, M.D. (1992) Phosphatidylinositol 3-kinase: a key enzyme in diverse signaling processes. *Trends Cell Biol.*, **2**, 358–60.

21 Joly, M., Kazlauskas, A., Fay, F.S. and Corvera, S. (1994) Disruption of PDGF receptor trafficking by mutation of its PI-3 kinase binding sites. *Science*, **263**, 684–7.

22 Volinia, S., Dhand, R., Vanhaesebroeck, B. *et al.* (1995) A human phosphatidylinositol 3-kinase complex related to the yeast Vps34p/Vps15p protein sorting system. *EMBO J.*, **14**, 3339–48.

23 Hu, P., Mondino, A., Skolnik, E.Y. and Schlessinger, J. (1993) Cloning of a novel, ubiquitously expressed human phosphatidylinositol 3-kinase and identification of its binding site on p85. *Mol. Cell. Biol.*, **13**, 7677–88.

24 Stoyanov, B., Volinia, S., Hanck, T. *et al.* (1995) Cloning and characterisation of a G protein-activated human phosphatidylinositol 3-kinase. *Science*, **269**, 690–3.

25 Stephens, L., Smrcka, A., Cooke, F.T. *et al.* (1994) A novel phosphoinositide 3-kinase activity in myeloid-derived cells is activated by G protein beta gamma subunits. *Cell*, **77**, 83–93.

26 Brown, E.J., Albers, M.W., Shin, T.B. *et al.* (1994) A mammalian protein targeted by G1-arresting rapamycin-receptor complex. *Nature*, **369**, 756–8.

27 Nishimura, R., Li, W., Kashishian, A. *et al.* (1993) Two signaling molecules share a phosphotyrosine-containing binding site in the platelet-derived growth factor receptor. *Mol. Cell. Biol.*, **13**, 6889–96.

28 Arcaro, A. and Wymann, M.P. (1993) Wortmannin is a potent phosphatidylinositol 3-kinase inhibitor – the role of phosphatidylinositol 3,4,5-trisphosphate in neutrophil responses. *Biochem. J.*, **296**, 297–301.

29 Woscholski, R., Kodaki, T., Mckinnon, M. *et al.* (1994) A comparison of demethoxyviridin and wortmannin as inhibitors of phosphatidylinositol 3-kinase. *FEBS Lett.*, **342**, 109–14.

30 Valius, M. and Kazlauskas, A. (1993) Phospholipase-C-gamma-1 and phosphatidylinositol-3 kinase are the downstream mediators of the PDGF receptor's mitogenic signal. *Cell*, **73**, 321–34.

31 Fantl, W.J., Escobedo, J.A., Martin, G.A. *et al.* (1992) Distinct phosphotyrosines on a growth factor receptor bind to specific molecules that mediate different signaling pathways. *Cell*, **69**, 413–23.

32 Yao, R.J. and Cooper, G.M. (1995) Requirement for phosphatidylinositol-3-kinase in the prevention of apoptosis by nerve growth factor. *Science*, **267**, 2003–6.

33 Wennstrom, S., Hawkins, P., Cooke, F. *et al.* (1994) Activation of phosphoinositide 3-kinase is required for PDGF-stimulated membrane ruffling. *Curr. Biol.*, **4**, 385–93.

34 Kotani, K., Yonezawa, K., Hara, K. *et al.* (1994) Involvement of phosphoinositide 3-kinase in insulin-or IGF-1-induced membrane ruffling. *EMBO J.*, **13**, 2313–21.

35 Kundra, V., Escobedo, J.A., Kazlauskas, A. *et al.* (1994) Regulation of chemotaxis by the platelet-derived growth factor receptor-beta. *Nature*, **367**, 474–6.

36 Wennstrom, S., Siegbahn, A., Yokote, K. *et al.* (1994) Membrane ruffling and chemotaxis transduced by the PDGF beta-receptor require the binding site for phosphatidylinositol 3′-kinase. *Oncogene*, **9**, 651–60.

37 Vlahos, C.J., Matter, W.F., Brown, R.F. *et al.* (1995) Investigation of neutrophil signal transduction using a specific inhibitor of phosphatidylinositol 3-kinase. *J. Immunol.*, **154**, 2413–22.

38 Muslin, A.J., Klippel, A. and Williams, L.T. (1993) Phosphatidylinositol 3-kinase activity is important for progesterone-induced xenopus oocyte maturation. *Mol. Cell. Biol.*, **13**, 6661–6.

39 Hara, K., Yonezawa, K., Sakaue, H. *et al.* (1994) 1-Phosphatidylinositol 3-kinase activity is required for insulin-stimulated glucose transport but not for RAS activation in CHO cells. *Proc. Natl. Acad. Sci. USA*, **91**, 7415–9.

40 Cheatham, B., Vlahos, C.J., Cheatham, L. *et al.* (1994) Phosphatidylinositol 3-kinase activation is required for insulin stimulation of pp70 S6 kinase, DNA synthesis, and glucose transporter translocation. *Mol. Cell. Biol.*, **14**, 4902–11.

41 Heidaran, M.A., Pierce, J.H., Lombardi, D. *et al.* (1991) Deletion or substitution within the alpha platelet-derived growth factor receptor kinase insert domain: effects on functional coupling with intracellular signaling pathways. *Mol. Cell. Biol.*, **11**, 134–42.

42 Roche, S., Koegl, M. and Courtneidge, S.A. (1994) The phosphatidylinositol 3-kinase alpha is required for DNA synthesis induced by some, but not all, growth factors. *Proc. Natl. Acad. Sci. USA*, **91**, 9185–9.

43 Pleiman, C.M., Hertz, W.M. and Cambier, J.C. (1994) Activation of phosphatidylinositol-3′-kinase by SRC-family kinase SH3 binding to the p85 subunit. *Science*, **263**, 1609–12.

44 Liu X.Q., Marengere, L.E.M., Koch, C.A. and Pawson, T. (1993) The v-Src SH3-domain binds phosphatidylinositol 3′-kinase. *Mol. Cell. Biol.*, **13**, 5225–32.

45 Guinebault, C., Payrastre, B., Racaud-Sultan, C. *et al.* (1995) Integrin-dependent translocation of phosphoinositide 3-kinase to the cytoskeleton of thrombin-activated platelets involves specific interactions of p85α with actin filaments and focal adhesion kinase. *J. Cell. Biol.*, **129**, 831–42.

46 Chen, H.C. and Guan, J.L. (1994) Stimulation of phosphatidylinositol 3′-kinase association with focal adhesion kinase by platelet-derived growth factor. *J. Biol. Chem.*, **269**, 31229–33.

47 Zheng, Y., Bagrodia, S. and Cerione, R.A. (1994) Activation of phosphoinositide 3-kinase activity by Cdc42Hs binding to p85. *J. Biol. Chem.*, **269**, 18727–30.

48 Rodriguez-Viciana, P., Warne, P.H., Dhand, R. *et al.* (1994) Phosphatidylinositol-3-OH kinase as a direct target of ras. *Nature*, **370**, 527–32.

49 Kodaki, T., Woscholski, R., Hallberg, B. *et al.* (1994) The activation of phosphatidylinsolitol 3-kinase by ras. *Curr. Biol.*, **4**, 798–806.

50 Zhang, J., King, W.G., Dillon, S. *et al.* (1993) Activation of platelet phosphatidylinositide 3-kinase requires the small GTP-binding protein Rho. *J. Biol. Chem.*, **268**, 22251–4.

51 Kumagai, N., Morii, N., Fujisawa, K. *et al.* (1993) ADP-ribosylation of rho p21 inhibits lysophosphatidic acid-induced protein tyrosine phosphorylation and phosphatidylinositol 3-kinase activation in cultured Swiss 3T3 cells. *J. Biol. Chem.*, **268**, 24535–8.

52 Ridley, A.J., Paterson, H.F., Johnston, C.L. *et al.* (1992) The small GTP-binding protein rac regulates growth factor induced membrane ruffling. *Cell*, **70**, 401–10.

53 Kamohara, S., Hayashi, H., Todaka, M. *et al.* (1995) Platelet-derived growth factor triggers translocation of the insulin-regulatable glucose transporter (type 4) predominantly through phosphatidylinositol 3-kinase binding sites on the receptor. *Proc. Natl. Acad. Sci. USA*, **92**, 1077–81.

54 Chung, J.K., Grammer, T.C., Lemon, K.P. *et al.* (1994) PDGF- and insulin-dependent pp70(S6k) activation mediated by phosphatidylinositol-3-OH kinase. *Nature*, **370**, 71–5.

55 Monfar, M., Lemon, K.P., Grammer, T.C. *et al.* (1995) Activation of pp70/85 s6 kinases in interleukin-2-responsive lymphoid cells is mediated by phosphatidylinositol 3-kinase and inhibited by cyclic AMP. *Mol. Cell. Biol.*, **15**, 326–337.

56 Toker, A., Meyer, M., Reddy, K.K. *et al.* (1994) Activation of protein kinase C family members by the novel polyphosphoinositides PtdIns-3,4-P-2 and PtdIns-3,4,5-P-3. *J. Biol. Chem.*, **269**, 32358–67.

57 Palmer, R.H., Dekker, L.V., Woscholski, R. *et al.* (1995) Activation of PRK1 by phosphatidylinositol 4,5-bisphosphate and phsophatidylinositol 3,4,5-trisphosphate. A comparison with protein kinase C isotypes. *J. Biol. Chem.*, **270**, 22412–16.

58 Franke, T.F., Yang, S.I., Chan, T.O. *et al.* (1995) The protein kinase encoded by the Akt proto-oncogene is a target of the platelet-derived growth factor (PDGF)-activated phosphatidylinositol 3-kinase (PI 3-kinase). *Cell*, in press.

59 MacDougall, L.K., Domin, J. and Waterfield, M.D. (1995) A family of phosphinositide 3-kinases in *Drosophila* identifies a new mediator of signal transduction. *Curr. Biol.*, **5**, 1404–15.

14 Calcium signaling

Michael J. Berridge
and Martin D.
Bootman

The divalent cation calcium (Ca^{2+}) is used by cells to regulate many of their activities. Right at the beginning of life, a rapid surge of Ca^{2+} at fertilization initiates the development of a new organism. Once cells take on their specialized functions,[1] calcium is used as an intracellular signal to control many diverse processes including muscle contraction, secretion, metabolism, neuronal excitability and proliferation.

The basic mechanism of Ca^{2+} action is simple: when the Ca^{2+} concentration is kept low (10–100 nM range) cells remain quiescent, but when the Ca^{2+} concentration is raised (500–1000 nM range) cells are activated to perform their particular functions. Cells contain sensors such as calmodulin (CAM) and troponin C (TnC), that are responsible for detecting the Ca^{2+} rise and transducing the information into specific cellular responses.

Whether or not a cell is activated is therefore determined by a balance between the calcium-ON and calcium-OFF mechanisms. The former are responsible for pouring Ca^{2+} into the cytoplasm, from both internal and external sources. The latter remove cytoplasmic Ca^{2+} by pumping it either out of the cell or back into the internal stores, thus maintaining the *status quo*.

In this chapter we shall consider these ON and OFF mechanisms and also the spatial and temporal patterns of stimulated intracellular Ca^{2+} signals and the sensory systems responsible for detecting and responding to Ca^{2+} signals. Finally, we shall briefly describe the pivotal role of Ca^{2+} in controlling a diverse array of physiological processes (Figure 14.1).

Calcium ON mechanisms

Cells have access to two major sources of Ca^{2+}: extracellular and organellar. Since the Ca^{2+} concentration outside of the cell is 10 000-fold greater than inside, and the inside of the cell is negatively charged with respect to the outside, there is a large electrochemical gradient favouring Ca^{2+} entry. This gradient is exploited by some cell types to drive a rapid

Signal Transduction. Edited by Carl-Henrik Heldin and Mary Purton. Published in 1996 by Chapman & Hall. ISBN 0 412 70810 8

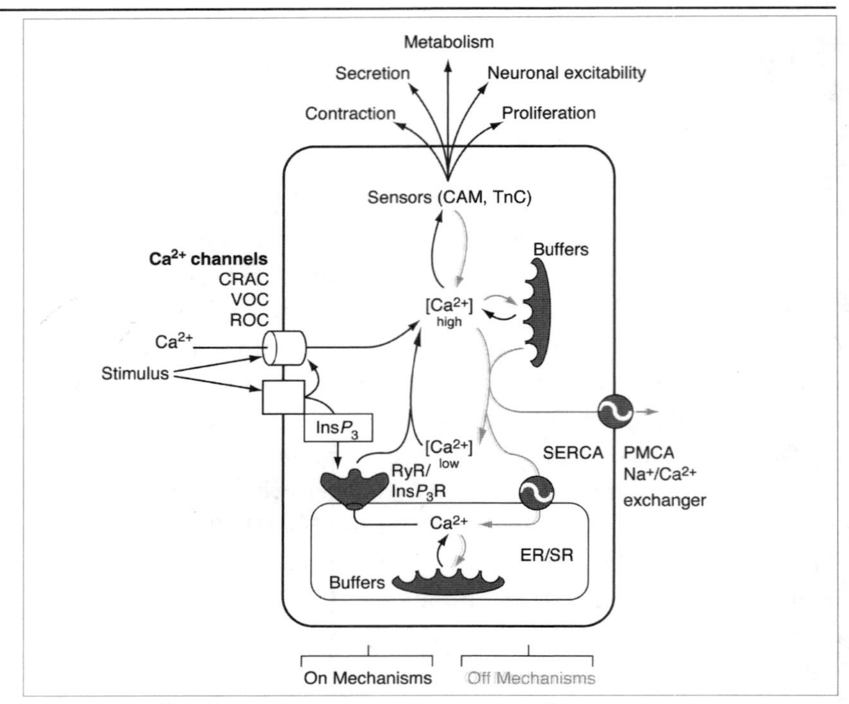

Figure 14.1 Summary of the major ON and OFF mechanisms responsible for regulating the concentration of intracellular Ca^{2+}. Stimuli raise the level of Ca^{2+} by activating the ON mechanisms, which promote either the entry of external Ca^{2+}, or the release of Ca^{2+} from intracellular stores (ER/SR). Changes in Ca^{2+} are damped by buffers located both in the cytoplasm and in the ER/SR compartments. The OFF mechanisms restore the low resting level of Ca^{2+} by either pumping it out of the cell or back into the stores. The effects of an elevated Ca^{2+} concentration are mediated by sensors such as calmodulin (CAM) or troponin C (TnC), to regulate a wide range of cellular activities (see Figure 14.4 for details).

influx of Ca^{2+}. In addition, many cells use Ca^{2+} sequestered within the endoplasmic reticulum (ER) or sarcoplasmic reticulum (SR) in muscle cells. These two sources of Ca^{2+} can be used either separately or in combination.

The ON mechanisms of Ca^{2+} signaling depend upon the activation of Ca^{2+} channels that control either the entry of extracellular Ca^{2+} or the release of Ca^{2+} from internal stores (Figure 14.1).

Plasma membrane calcium channels

The plasma membrane contains a variety of channels[1,2] capable of regulating the influx of external Ca^{2+}, which have been named according to the way in which their channel activities are controlled (Table 14.1a).

Voltage-operated channels (VOCs) are found primarily in excitable cells (nerve, muscle and certain endocrine cells), where they open in response to membrane depolarization and allow extracellular Ca^{2+} to enter the cell.[2,3] These channels can be regulated both by receptor-mediated and intracellular messenger (e.g. cyclic AMP) systems.[4]

Receptor-operated channels (ROCs) are activated by the binding of extracellular agonists, which are usually neurotransmitters, e.g.

glutamate and ATP. A classical example of the ROCs are the *N*-methyl-D-aspartic acid (NMDA) receptors which open in response to glutamate to gate Ca^{2+} entry. The Ca^{2+} entering neurons through NMDA receptors plays an important role in modifying synaptic plasticity.

The plasma membrane also contains channels that are controlled indirectly by the state of fullness of the ER Ca^{2+} store, through a mechanism known as capacitative Ca^{2+} entry.[1,5,6] In this scheme, the ER Ca^{2+} stores behave like a capacitor in that they inhibit Ca^{2+} entry when they are fully loaded, but as they discharge they begin to open **calcium-release activated Ca^{2+} (CRAC) channels**. The mechanism by which Ca^{2+} stores modulate CRACs in the plasma membrane is unclear. One suggestion is that the empty Ca^{2+} stores release a messenger, **calcium influx factor (CIF)**, which diffuses to the membrane and opens the CRACs.[7] An alternative model, known as conformational coupling, suggests that $InsP_3$ receptors can 'sense' the Ca^{2+} content of the ER stores,

Table 14.1 The major calcium channels responsible for gating Ca^{2+} flux into the cytoplasm[a]

Channels	Type	Activators	Inhibitors	Single-channel conductance (pS)	Channel kinetics	Tissue distribution
(a) Plasma membrane calcium channels						
Voltage-operated channels (VOC)	T-type	Low voltage		8	Transient	Muscle, nerve pacemakers
	P-type		FTX	10–12	Sustained	Neurons
	N-type	High voltage	ωCgTx	12–20	Transient	Neurons (transmitter release)
	L-type	High voltage BAYK 8644	Nifedipine	25	Sustained	Muscle, nerve
Receptor-operated channels (ROC)	ATP-sensitive	Adenine nucleotides				Neurons, smooth muscle
	Glutamate-sensitive	NMDA				Neurons
Ca^{2+} release-activated channels (CRAC)		Ca^{2+} store depletion	La^{3+}	<1	Sustained	Widespread
(b) Intracellular calcium channels						
	$InsP_3R$ (types 1–4)	$InsP_3$, Ca^{2+}	Heparin, caffeine	20–80	Sustained	Widespread
	RyR1	Ca^{2+}, ryanodine, L-type VOC, caffeine	Ruthenium red	145	Sustained	Skeletal muscle, some neurons
	RyR2	Ca^{2+}, ryanodine, cADPR, caffeine		145	Sustained	Cardiac muscle, neurons
	RyR3	Caffeine, cADPR		?	Sustained	Widespread

[a] Abbreviations: FTX, funnel-web spider toxin; ωCgTx, ω-conotoxin; cADPR, cyclic adenosine diphosphate ribose.

and can transmit this information via a direct protein–protein interaction to the channels in the plasma membrane.[8,9]

Intracellular calcium channels

There are two families of intracellular channels responsible for releasing Ca^{2+} from the intracellular stores: **inositol 1,4,5-trisphosphate receptors (Ins*P*₃Rs)** and **ryanodine receptors (RyRs)**[1,10] (Table 14.1b). These two receptors probably evolved from a common protein ancestor, as they display considerable similarity in primary and secondary structure and share several physiological properties (see below).

Two alternative mechanisms have evolved to control the opening of these channels: information is transferred from the cell surface to the channels either through a direct protein–protein interaction, such as occurs in skeletal muscle,[11] or through diffusible Ca^{2+}-mobilizing intracellular messengers such as Ca^{2+} itself, $InsP_3$ and cyclic adenosine diphosphate ribose (cADPR)[10,12,13] (Box 14.1).

$InsP_3$Rs consist of three domains: an $InsP_3$-binding amino-terminal domain, a regulatory domain containing ATP-binding and phosphorylation sites and a carboxy-terminal domain containing six transmembrane regions.[14] The binding of $InsP_3$ to the amino-terminal domain induces a conformational change in the protein, leading to Ca^{2+} channel opening. The transmembrane regions are responsible for the aggregation of four subunits into the functional tetrameric receptor protein, and also serve to form the Ca^{2+} channel. RyRs have a similar three-domain structure. Electron microscopy of both $InsP_3$Rs and RyRs supports the concept of a tetrameric protein, since the receptors were found to resemble pinwheel-like structures with four radial arms protruding from a central hub.[15]

Currently, four different $InsP_3$R genes have been described, with further diversity being generated by two alternative splice sites in the gene products. The expression and splicing of the various $InsP_3$R isoforms is controlled in a tissue-specific manner, suggesting that the behavior of these Ca^{2+} channels can be tissue-dependent.[16] $InsP_3$Rs are largely located on the surface of the ER, but may also be distributed on other intracellular membranes, e.g. the nuclear membrane. There are indications that $InsP_3$Rs can also be located on the plasma membrane, but such instances are rare and are mainly restricted to olfactory neurons, where they play a critical role in sensory transduction.[17]

RyRs were first identified in muscle cells owing to their strong affinity for the plant alkaloid ryanodine. They are now known to

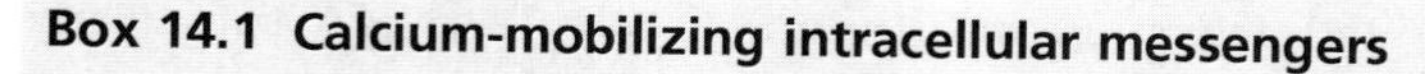

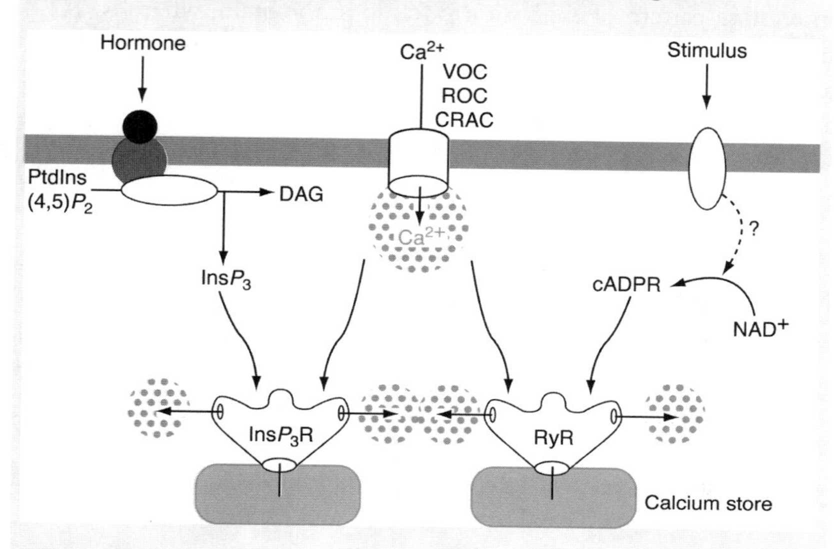

Inositol 1,4,5-trisphosphate (InsP_3). The precursor of this messenger is the minor membrane phospholipid phosphatidylinositol 4,5-bisphosphate [PtdIns(4,5)P_2], which is cleaved in response to hormonal stimulation to give diacylglycerol (DAG) and InsP_3. The latter diffuses into the cytoplasm, binds to its receptor and releases stored Ca^{2+}.[10]

Cyclic adenosine diphosphate ribose (cADPR). This messenger is formed from the cellular metabolite NAD^+, and potently activates Ca^{2+} release from ryanodine receptors (RyR).[12,13] There is uncertainty concerning whether extracellular stimuli act to increase the intracellular level of cADPR, or how this process might be activated.

Calcium. The most important diffusible messenger for activating InsP_3R and RyR is Ca^{2+} itself. The process of Ca^{2+}-induced Ca^{2+} release (CICR) is of fundamental importance in regulating the way in which cells mobilize Ca^{2+} from their internal stores. Trigger Ca^{2+} can either enter from the outside through plasma membrane channels (Figures 14.2b and c), or it can be relayed from one intracellular channel to the next, thus generating regenerative Ca^{2+} waves (Figures 14.2d and e). It should be stressed that although Ca^{2+} can act as the sole trigger for activating RyRs, InsP_3Rs usually require the simultaneous presence of Ca^{2+} and InsP_3.

Co-agonist concept. The two intracellular Ca^{2+} channels depicted above are under dual regulation by separate messengers, which act in a co-ordinated way to induce channel opening. An important action of InsP_3 and cADPR is to increase the sensitivity of their respective receptors to the stimulatory action of Ca^{2+}. Such dual regulation means that InsP_3R and RyR can integrate information from two separate sources. This might be particularly important in the nervous system, where these channels might function as coincidence detectors, to integrate separate neural inputs during the acquisition of memory.

function in Ca^{2+} signaling in a wide variety of cell types.[11,18,19] Molecular studies have identified a family of RyRs which have tissue-specific expression and activation mechanisms (Table 14.1b). A particularly interesting pattern of expression is found in the brain, where RyR1 is mainly restricted to the cerebellar Purkinje neurons, whereas RyR2 and RyR3 are more widely, but differentially, distributed in the central and peripheral neurons.

Calcium OFF mechanisms

Once the ON mechanisms have generated a Ca^{2+} signal by introducing Ca^{2+} into the cytoplasm, the OFF mechanisms begin the process of recovery by returning Ca^{2+} either to the stores or back to the external medium (Figure 14.1). These recovery pathways have to be extremely active because they need to remove not only the free cytosolic Ca^{2+} ions, but also the 100-fold larger amount that is bound to the Ca^{2+} buffers and Ca^{2+} sensors (see below) (Figure 14.1). The removal of cytoplasmic Ca^{2+} is carried out by pumps and exchangers.

Calcium pumps

The **Na^+/Ca^{2+} exchanger** is found in the plasma membrane and utilizes the energy stored in the electrochemical Na^+ gradient to extrude Ca^{2+} from the cell.[20] This extrusion mechanism is particularly well developed in excitable cells such as neurons, cardiac muscle and smooth muscle. The Na^+/Ca^{2+} exchanger can sometimes operate in the reverse direction and cause Ca^{2+} influx, as occurs during a cardiac action potential; a transient increase in Na^+ concentration beneath the plasma membrane reverses the direction of the exchanger, resulting in a rapid influx of Ca^{2+}.[21] Once the Na^+ gradient dissipates, the exchanger reverts back to its role of Ca^{2+} extrusion.

In most cells, Ca^{2+} extrusion depends upon **plasma membrane Ca^{2+} ATPases (PMCAs),** which utilize the energy of ATP to transport Ca^{2+} against the enormous electrochemical gradient that exists across the plasma membrane.[22] The PMCA enzyme is a member of the P-type ATPases (i.e. it forms a covalent phosphorylated intermediate during the reaction cycle).

Several PMCA isoforms have been described. There are four separate genes encoding PMCAs, with further diversity caused by alternative splicing of the gene products, such that there are approximately 20 different transcripts of the four isogenes.[22] PMCA1 and PMCA2 are present in most cells and may be 'housekeeping' isoforms, whereas PMCA3 and PMCA4 are found in more specialized cells.

A characteristic of PMCA is its regulation by a variety of factors including calmodulin (CAM), acidic phospholipids and protein kinases such as cyclic AMP-dependent protein kinase (PKA) and protein kinase C (PKC). Since CAM is a Ca^{2+}-binding protein and increases pump activity, this molecule conveys positive feedback on PMCAs, i.e. Ca^{2+}-stimulated Ca^{2+} extrusion. When activated by the Ca^{2+}–CAM complex, both the sensitivity and capacity of Ca^{2+} pumping are greatly increased, owing to a reduction in K_m (from 10–20 μM to 0.5 μM) and an increase in V_{max}.

The **sarco/endoplasmic reticulum Ca^{2+} ATPase (SERCA)** pumps are located in the membranes of intracellular stores, where they function to sequester the Ca^{2+} introduced into the cytoplasm as described above (Figure 14.1). Although the SERCAs are encoded by a separate gene family, they have many structural and functional similarities to the PMCA family and are also P-type ATPases. There are three different SERCA genes, with further diversity generated by alternative splicing.[23]

A major difference between the two families of Ca^{2+} pump is that the SERCAs lack the regulatory CAM-binding domain. Instead, regulation is mediated by an accessory membrane-spanning protein phospholamban, which inhibits SERCA1 and SERCA2 (but not SERCA3) activity by blocking the active site. This inhibition is relaxed when phospholamban is phosphorylated by PKA or Ca^{2+}- and CAM-dependent protein kinase (CAM kinase II).

The activity of SERCA pumps can be inhibited by agents such as thapsigargin, cyclopiazonic acid and 2,5-di(*tert*-butyl)-1,4-hydroquinone. These inhibitors have provided valuable tools for dissecting the various mechanisms that control intracellular Ca^{2+} signals.

Ca^{2+}-binding proteins

Cells possess a great variety of Ca^{2+}-binding proteins that contribute to Ca^{2+}-mediated signaling, either by buffering free Ca^{2+} ions and thus shaping the cellular response, or by acting as sensors that mediate the messenger role of Ca^{2+} (Figure 14.1).[24] Proteins capable of buffering Ca^{2+} are found in both the cytoplasm and within the ER/SR.

Only a small proportion of the Ca^{2+}-binding molecules in the cytoplasm act as sensors, the remainder seem to function as buffers. In chromaffin cells for example, approximately 98–99% of the Ca^{2+} entering the cell is rapidly bound by the cytosolic buffers.[25] The stimulus-evoked Ca^{2+} elevation therefore only represents a small proportion of the total Ca^{2+} entering the cytoplasm. The Ca^{2+} bound to these buffers is in equilibrium with free Ca^{2+} in the cytoplasm so, as the extrusion mechanisms remove cytoplasmic Ca^{2+}, the bound Ca^{2+} dissociates from the buffers and can also be removed. Some of the molecules that might perform such a buffering role are the proteins parvalbumin, calretinin and calbindin.

The Ca^{2+} that is pumped into the ER/SR is rapidly buffered by specific storage proteins such as calsequestrin (CSQ) and calreticulin (CR), located in the lumen of these Ca^{2+} stores (Figure 14.1). CSQ is found mainly in muscle and cerebellar Purkinje neurons, whereas CR is found in most eukaryotic cells. In keeping with their storage role, they have a low affinity for Ca^{2+} (K_d in the mM range) but a high capacity (25–50 Ca^{2+} ions per molecule).[23] The existence of these Ca^{2+}-binding proteins serves to buffer the Ca^{2+} near a set value, which may be important not only to ensure consistency in the amount of Ca^{2+} released from these stores, but also to ensure proper operation of other functions in the ER/SR which also require Ca^{2+}. In addition to CSQ and CR, the ER lumen contains a host of other Ca^{2+}-binding proteins such as endoplasmin and BiP, which play a role in the folding and maturation of secretory proteins.

Spatiotemporal aspects of calcium signaling

The ON mechanisms depend upon the opening of channels in the plasma membrane and ER/SR, to allow Ca^{2+} to flow into the cytoplasm (Figure 14.1). Using sensitive imaging techniques, the opening of either individual or small groups of channels has been visualized as highly localized 'spheres' of Ca^{2+}, which have been variously termed quantum emission domains (QEDs) (Figure 14.2a) in synaptic terminals,[26,27] sparks in cardiac myocytes[28] (Figure 14.2c), quantum bumps in photoreceptors[29] or puffs in *Xenopus* oocytes.[30] These localized unitary events all have characteristic kinetics: the cytoplasmic Ca^{2+} concentration rapidly increases as Ca^{2+} flows through the open channel, and then decreases exponentially as the Ca^{2+} diffuses slowly away, once the channel has closed (Figures 14.3a and b). A large number of these unitary events have to be synchronized in order to generate global Ca^{2+} signals.

Cells employ two major mechanisms for entraining the activity of their Ca^{2+} channels. First, action potentials serve to co-ordinate the opening of the VOCs in the plasma membrane of excitable cells. For example, at nerve terminals, the QEDs (Figure 14.2a) are activated

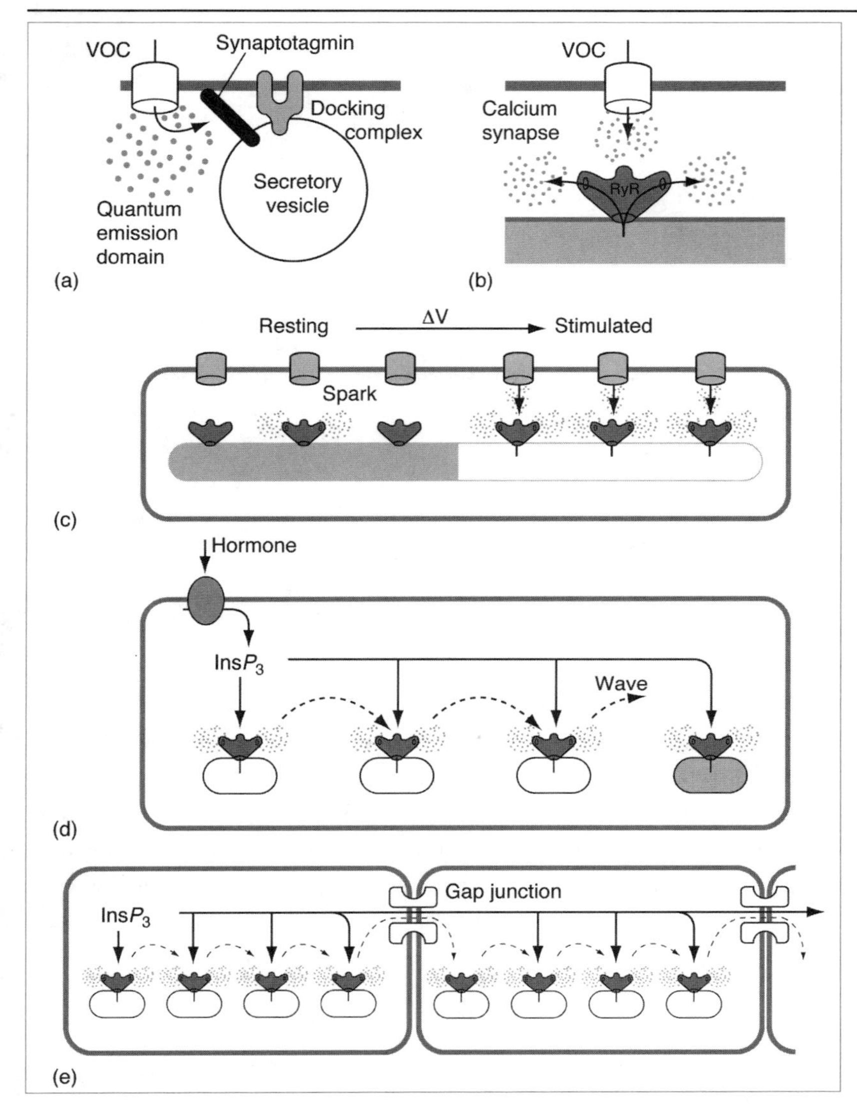

Figure 14.2 Spatial aspects of Ca^{2+} signaling. (a) Quantum emission domain (QED); (b) Ca^{2+} synapse, as proposed for cardiac muscle; (c) excitation–contraction coupling in cardiac muscle; (d) intracellular Ca^{2+} wave; (e) intercellular Ca^{2+} wave.

synchronously to produce the uniform sub-plasmalemmal Ca^{2+} elevation responsible for triggering the exocytotic response (Figure 14.4c).[31] In heart cells, the RyRs, which can display random sparks in the absence of stimulation,[28] are synchronized and activated through local interactions with VOCs (calcium synapses) (Figures 14.2b and c)[32] to give the global elevation of Ca^{2+} (Figure 14.3c) necessary to trigger contraction. Second, the co-ordinated release of Ca^{2+} from intracellular stores is achieved through the regenerative process of CICR (see Box 14.1) (Figure 14.2d). The hormone-stimulated elevation of InsP_3 activates the intracellular Ca^{2+} stores to release Ca^{2+}. Initially this starts in a

Figure 14.3 Temporal aspects of Ca^{2+} signaling. Traces of $[Ca^{2+}]_i$ increases recorded from intact cells. For (a)–(c), the $[Ca^{2+}]_i$ changes are shown using an arbitrary scale, although (b) and (c) reflect the actual relative sizes of the recorded $[Ca^{2+}]_i$ signals. (a) Quantum emission domain (QED) from squid giant synapse [reproduced from Silver, R.B., Sugimori, M., Lang, E.J. and Llinas, R. (1994) *Biol. Bull.*, 187, 293–9]; (b) Ca^{2+} spark recorded in a cardiac myocyte [reproduced from Cheng, H., Lederer, W.J. and Cannell, M.B. (1993) *Science*, 262, 740–4]; (c) an electrically evoked global elevation of $[Ca^{2+}]_i$ in cardiac muscle [reproduced from Cheng, H., Lederer, W.J. and Cannell, M.B. (1993) *Science*, 262, 740–4]; (d) hormone-evoked Ca^{2+} spikes in a single HeLa cell (M. Bootman, unpublished observation).

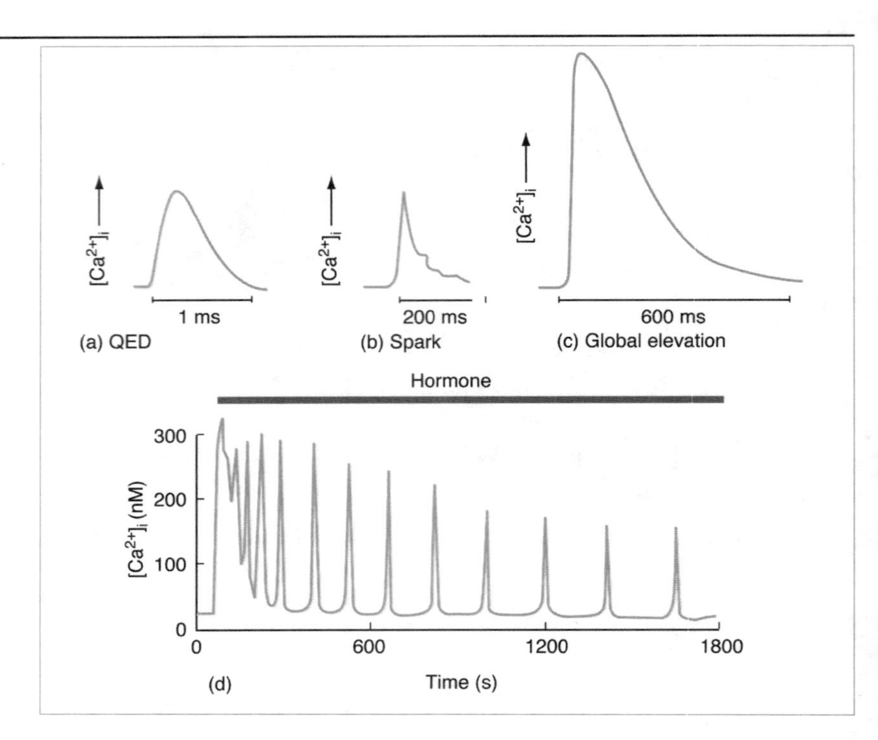

localized region, but then spreads across the cell as a wave.[1,33,34] In many cell types (e.g. HeLa cells), this process is cyclical, giving rise to trains of repetitive Ca^{2+} spikes (Figure 14.3d).[35,36] In those cases where cells are connected together via gap junctions, such Ca^{2+} signals can be communicated from one cell to the next through an intercellular wave (Figure 14.2e), to co-ordinate the activity of local cell populations.[37]

Calcium sensors

There are two groups of proteins concerned with detecting Ca^{2+} signals and transducing the information into changes in specific cellular processes, the annexin protein family and the EF-hand proteins. Between them, these two groups of proteins mediate most of the intracellular effects of Ca^{2+}.

Annexins

The annexins are a heterogeneous group of proteins that share a common property of interacting with membranes in a Ca^{2+}-dependent manner.[38,39] During the formation of the phospholipid–Ca^{2+}–annexin ternary complex, each annexin can bind 2–10 Ca^{2+} ions. These annexins have a much lower affinity for Ca^{2+} than the EF-hand proteins so their action might be restricted to domains near membranes bearing Ca^{2+} release/entry channels where the Ca^{2+} concentration is relatively high.

At least 12 annexins (I–XII) have been identified. These proteins share some structural homology, in that they have a canonical motif comprising about 70 amino acids repeated four times (eight times in the case of annexin VI).

Some of the proposed functions of the annexins include the inhibition of phospholipase A_2 and attachment of cytoskeletal proteins to membranes. Annexins I and VII may accelerate vesicle aggregation during exocytosis and annexin V may function to form Ca^{2+}-selective channels.

EF-hand proteins and their targets

The EF-hand proteins derive their name from the helix–loop–helix motif of the first Ca^{2+}-binding domain to be identified.[24,42] The loop between the E and F α-helices of parvalbumin consists of 12 amino acids which provide the ligands (carbonyl oxygens) organized in a precise orientation, so as to bind Ca^{2+} with a high specificity. This group of proteins is very large, with more than 30 subfamilies, and comprises

Box 14.2 Calmodulin action

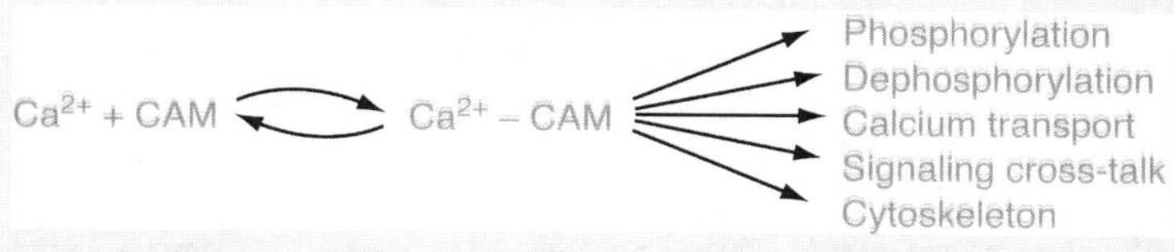

Phosphorylation. One of the major actions of calmodulin (CAM) is to stimulate a variety of protein kinases. Some of these have a broad substrate specificity (e.g. CAM kinase II),[40] whereas others act on specific targets [e.g. phosphorylase kinase, myosin light chain kinase (MLCK) and elongation factor kinase].

Dephosphorylation. Calcineurin, a CAM-dependent protein phosphatase, plays a critical role in the activation of lymphocytes (Figure 14.4e), and is the target of the immunosuppressant drugs cyclosporin and FK-506.[41]

Ca^{2+} transport. Ca^{2+} uses CAM to regulate its own extrusion, by activating PMCA pumps (see text for details).

Signaling cross-talk. A characteristic feature of signaling pathways is that they frequently interact with each other at many different levels. In the case of Ca^{2+}, the interaction often requires CAM:

- **Cyclic AMP.** There are five isozymes of adenylate cyclase, which catalyze the production of the intracellular messenger cyclic AMP from ATP. Ca^{2+}–CAM can activate both type I and III isoforms. High levels of the type I isoform have been found in the hippocampus and neocortex, where it has been implicated in the mechanisms of memory. A similar function has been described in *Aplysia*, where a Ca^{2+}-sensitive adenylate cyclase enzyme functions as a coincidence detector.
- **Ins*P*3.** The intracellular InsP_3 concentration is modulated by Ca^{2+}–CAM, which stimulates an InsP_3 kinase to give inositol 1,3,4,5-tetrakisphosphate.
- **Nitric oxide (NO).** Synthesis of this local hormone by NO synthase is activated by Ca^{2+}–CAM.
- **Cytoskeleton.** A variety of cytoskeletal proteins, such as Tau, fodrin and neuromodulin, are regulated by Ca^{2+} through CAM.

some of the buffers mentioned earlier (e.g. parvalbumin and calbindin), and the two major Ca^{2+} sensors troponin C (TnC) and CAM. These two sensors are symmetrical dumbbell-shaped molecules containing two EF hands at each end, separated by an α-helical structure. As these proteins bind to Ca^{2+}, they undergo the conformational changes responsible for activating their downstream targets (Box 14.2). The function of TnC is restricted to skeletal and cardiac muscle, where it

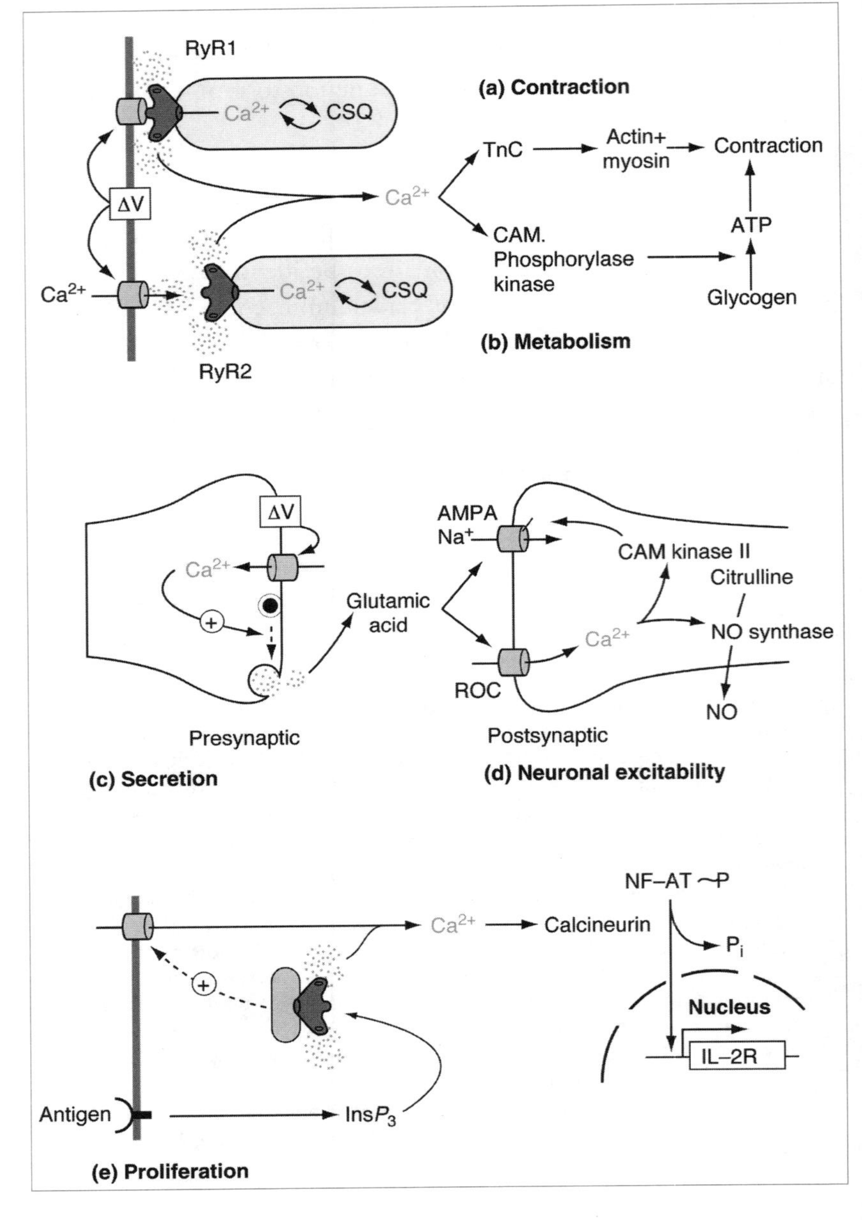

Figure 14.4 The role of Ca^{2+} in regulating a variety of cellular processes. See text for details.

mediates excitation–contraction coupling by controlling the interaction between actin and myosin (Figure 14.4a). In contrast, CAM is a multifunctional Ca^{2+} sensor responsible for activating a great variety of processes (Box 14.2).[43,44] It is one of the most abundant intracellular proteins, and is found in all eukaryotic cells.

Calcium signaling pathways

To understand the ways in which cells use Ca^{2+} as a messenger, it is helpful to consider how the flow of information summarized in Figure 14.1 is used to control some specific cellular processes (Figure 14.4).

Contraction

Excitation–contraction coupling in skeletal muscle begins with depolarization (ΔV) of the plasma membrane. This alters the conformation of the voltage-sensing dihydropyridine receptors which, in turn, activate RyR1 channels by a direct protein–protein interaction, thereby releasing Ca^{2+} from the SR (Figure 14.4a).[11] In the case of cardiac cells, depolarization of the plasma membrane opens L-type VOCs which provide the pulse of trigger Ca^{2+} responsible for activating the RyR2 through Ca^{2+} synapses (Figures 14.2b and 14.4a). The Ca^{2+} released by RyR activation can bind to troponin C to initiate contraction through the interaction between actin and myosin.

Metabolism

In addition to triggering contraction, the Ca^{2+} released during excitation–contraction coupling is also used to activate metabolism by stimulating phosphorylase kinase (Figure 14.4b).

Secretion

A classical example of stimulus-secretion coupling is the release of neurotransmitters (e.g. glutamic acid) from synaptic terminals (Figure 14.4c). As the action potential invades the terminal, the depolarization (ΔV) opens VOCs, organized as QEDs (Figure 14.2a), to trigger the exocytosis of synaptic vesicles.

Neuronal excitability

In neurons, postsynaptic ROCs such as the ionotropic NMDA receptors mediate the influx of Ca^{2+} (Figure 14.4d), which can have long-term effects on synaptic plasticity by activating CAM kinase II to modify other ionotropic receptors, e.g. AMPA (D,L-α-amino-3-hydroxy-5-methyl-4-isoxalonepropionic acid) receptors, and by generating nitric oxide (NO) (see Box 14.2).

Proliferation

Activation of lymphocytes by antigens begins with the generation of $InsP_3$ which releases Ca^{2+} from intracellular stores and promotes capacitative Ca^{2+} entry, to provide the Ca^{2+} signal required to initiate gene transcription (Figure 14.4e). Ca^{2+} activates the phosphatase calcineurin, leading to the dephosphorylation of the nuclear factor of activated T cells (NF-AT).[41] When dephosphorylated, this transcription factor enters the nucleus and promotes the transcription of interleukin-2 receptor (IL-2R) gene.

Summary

Many cellular processes are regulated by the second messenger Ca^{2+} which is derived from two separate sources. It can enter from the outside through channels in the plasma membrane or it can be released from internal stores. These ON mechanisms are counteracted by OFF mechanisms that remove Ca^{2+} from the cytoplasm. The ON and OFF mechanisms are organized so as to generate brief pulses of calcium which often recur to give trains of repetitive spikes. In addition, each spike can be organized as a wave in that the Ca^{2+} signal initiates in one region of the cell and then spreads in a regenerative manner through a process of calcium-induced calcium release. The full significance of these complex spatiotemporal patterns remains to be determined. One intriguing possibility is that the generation of spikes may indicate that calcium signaling is based on a digital system with information being encoded through frequency rather than amplitude. Some support for a frequency-modulated (fm) system has come from numerous observations that the frequency of Ca^{2+} spiking varies with hormone concentration. A major challenge for the future is to understand first the encoding mechanism responsible for generating Ca^{2+} spikes and then to find out how the cell decodes these digital signals into changes in the activity of specific processes.

Finally, it is important to remember that elevated levels of Ca^{2+}, especially if maintained for long periods, can be cytotoxic, as occurs during ischemia in the heart and brain. It seems that cells resolve the paradox between the signaling and cytotoxic effects of Ca^{2+} by evolving a signaling system based on frequency rather than amplitude modulation.

References

1 Clapham, D.E. (1995) Calcium signaling. *Cell*, **80**, 259–68.

2 McCleskey, E.W. (1994) Calcium channels: cellular roles and molecular mechanisms. *Curr. Opinion Neurobiol.*, **4**, 304–12.

3 Hofmann, F., Biel, M. and Flockerzi, V. (1994) Molecular basis for Ca^{2+} channel diversity. *Annu. Rev. Neurosci.*, **17**, 399–418.

4 McDonald, T.F., Pelzer, S., Trautwein, W. and Pelzer, D.J. (1994) Regulation and modulation of calcium channels in cardiac, skeletal and smooth muscle cells. *Physiol. Rev.*, **74**, 365–507.

5 Penner, R., Fasolato, C. and Hoth, M. (1993) Calcium influx and its control by calcium release. *Curr. Opinion Neurobiol.*, **3**, 368–74.

6 Putney, J.W., Jr, and Bird, G. St. J. (1993) The signal for capacitative calcium entry. *Cell*, **75**, 199–201.

7 Randriamampita, C. and Tsien, R.Y. (1993) Emptying of intracellular Ca^{2+} stores releases a novel small messenger that stimulates Ca^{2+} influx. *Nature*, **364**, 809–14.

8 Irvine, R.F. (1990) 'Quantal' Ca^{2+} release and the control of Ca^{2+} entry by inositol phosphates – a possible mechanism. *FEBS Lett.*, **263**, 5–9.

9 Berridge, M.J. (1990) Calcium oscillations. *J. Biol. Chem.*, **265**, 9583–6.

10 Berridge, M.J. (1993) Inositol trisphosphate and calcium signalling. *Nature*, **361**, 315–25.

11 Meissner, G. (1994) Ryanodine receptor/Ca^{2+} release channels and their regulation by endogenous factors. *Annu. Rev. Physiol.*, **56**, 485–508.

12 Galione, A. and White, A. (1994) Ca^{2+} release induced by cyclic ADP-ribose. *Trends Cell Biol.*, **4**, 431–436.

13 Lee, H.C. (1994) Cyclic ADP-ribose: a calcium mobilizing metabolite of NAD^+. *Mol. Cell. Biochem.*, **138**, 229–35.

14 Taylor, C.W. and Richardson, A. (1991) Structure and function of inositol 1,4,5-trisphosphate receptors. *Pharmacol. Ther.*, **51**, 97–137.

15 Chadwick, C.C., Saito, A. and Fleischer, S. (1990) Isolation and characterisation of the inositol trisphosphate receptor from smooth muscle. *Proc. Natl. Acad. Sci. USA*, **87**, 2132–6.

16 De Smedt, H., Missiaen, L., Parys, J.B. *et al.* (1994) Determination of the relative amounts of inositol trisphosphate receptor mRNA isoforms by polymerase chain reaction. *J. Biol. Chem.*, **269**, 21691–8.

17 Zufall, Z., Firestein, S. and Shepherd, G.M. (1994) Cyclic nucleotide-gated ion channels and sensory transduction in olfactory receptor neurons. *Annu. Rev. Biophys. Biomol. Struct.*, **23**, 577–607.

18 McPherson, P.S. and Campbell, K.P. (1993) The ryanodine receptor/Ca^{2+} release channel. *J. Biol. Chem.*, **268**, 13765–8.

19 Coronado, R., Mortissette, J., Sukhareva, M. and Vaughan, D.M. (1994) Structure and function of ryanodine receptors. *Am. J. Physiol.*, **266**, C1485–C1504.

20 Philipson, K.D. and Nicoll, D.A. (1992) Sodium–calcium exchange. *Curr. Opinion Cell Biol.*, **4**, 678–83.

21 Lipp, P. and Niggli, E. (1994) Sodium current-induced calcium signals in isolated guinea-pig ventricular myocytes. *J. Physiol.*, **474**, 439–46.

22 Carafoli, E. (1994) Biogenesis: plasma membrane calcium ATPase: 15 years of work on a purified enzyme. *FASEB J.*, **8**, 993–1002.

23 Pozzan, T., Rizzuto, R., Volpe, P. and Meldolesi, J. (1994) Molecular and cellular physiology of intracellular calcium stores. *Physiol. Rev.*, **74**, 595–636.

24 Baimbridge, K.G., Celio, M.R. and Rogers, J.H. (1992) Calcium-binding proteins in the nervous system. *Trends Neurosci.*, **15**, 303–8.

25 Neher, E. and Augustine, G.J. (1992) Calcium gradients and buffers in bovine chromaffin cells. *J. Physiol.*, **450**, 273–301.

26 Llinás, R., Sugimori, M. and Silver, R.B. (1992) Microdomains of high calcium concentration in a presynaptic terminal. *Science*, **256**, 677–9.

27 Silver, R.B., Sugimori, M., Lang, E.J. and Llinas, R. (1994) Time-resolved imaging of Ca^{2+}-dependent aequorin luminescence of microdomains and QEDs in synaptic preterminals. *Biol. Bull.*, **187**, 293–9.

28 Cheng, H., Lederer, W.J. and Cannell, M.B. (1993) Calcium sparks: elementary events underlying excitation–contraction coupling in heart muscle. *Science*, **262**, 740–4.

29 Hardie, R.C. and Minke, B. (1993) Novel Ca^{2+} channels underlying transduction on Drosophila photoreceptors: implications for phosphoinositide-mediated Ca^{2+} mobilization. *Trends Neurosci*, **16**, 371–6.

30 Yao, Y., Choi, J. and Parker, I. (1995) Quantal puffs of intracellular Ca^{2+} evoked by inositol trisphosphate in *Xenopus* oocytes. *J. Physiol.*, **482**, 533–53.

31 Smith, S.J. and Augustine, G.J. (1988) Calcium ions, active zones and synaptic transmitter release. *Trends Neurosci.*, **11**, 458–64.

32 Stern, M.D. (1992) Theory of excitation–contraction coupling in cardiac muscle. *Biophys. J.*, **63**, 497–517.

33 Amundson, J. and Clapham, D. (1993) Calcium waves. *Curr. Opinion Neurobiol.*, **3**, 375–82.

34 Cheek, T.R. (1991) Calcium regulation and homeostasis. *Curr. Opinion Cell Biol.*, **3**, 199–205.

35 Bootman, M.D., Cheek, T.R., Moreton, R.B. *et al.* (1994) Smoothly graded Ca^{2+} release from inositol 1,4,5-trisphosphate-sensitive Ca^{2+} stores. *J. Biol. Chem.*, **269**, 24783–91.

36 Fewtrell, C. (1993) Ca^{2+} oscillations in non-excitable cells. *Annu. Rev. Physiol.*, **55**, 427–54.

37 Sanderson, M.J., Charles, A.C., Boitano, S. and Dirksen, E.R. (1994) Mechanisms and function of intercellular calcium signaling. *Mol. Cell. Endocrinol.*, **98**, 173–87.

38 Heizmann, C.W. and Hunziker, W. (1991) Intracellular calcium-binding proteins: more sites than insights. *Trends Biochem. Sci.*, **16**, 98–103.

39 Swairjo, M.A. and Seaton, B.A. (1994) Annexin structure and membrane interactions: a molecular perspective. *Annu. Rev. Biophys. Biomol. Struct.*, **23**, 193–213.

40 Schulman, H. (1993) The multifunctional Ca^{2+}/calmodulin-dependent protein kinases. *Curr. Opinion Cell Biol.*, **5**, 247–53.

41 Crabtree, G.R. and Clipstone, N.A. (1994) Signal transmission between the plasma membrane and nucleus of T lymphocytes. *Annu. Rev. Biochem.*, **63**, 1045–83.

42 Nakayama, S. and Kretsinger, R.H. (1994) Evolution of the EF-hand family of proteins. *Annu. Rev. Biophys. Biomol. Struct.*, **23**, 473–507.

43 Gnegy, M.E. (1993) Calmodulin in neurotransmitter and hormone action. *Annu. Rev. Pharmacol. Toxicol.*, **32**, 45–70.

44 James, P., Vorherr, T. and Carafoli, E. (1995) Calmodulin-binding domains: just two faced or multi-faceted. *Trends Biochem. Sci.*, **20**, 38–42.

15 Cyclic AMP and cyclic GMP in cell signaling

Sharron H. Francis
and Jackie D. Corbin

Cyclic AMP (cAMP) and cyclic GMP (cGMP) comprise the cyclic nucleotide family of second messengers (Box 15.1). The discovery of cAMP by Sutherland and studies to elucidate its mechanism of action provided the basis for the concept of 'second messengers'.[1] Cellular cAMP and cGMP levels are determined by the activities of cyclases, which catalyze their synthesis, and phosphodiesterases (PDE), which hydrolyze them.[1,2] Cyclases and PDEs are highly regulated and are targets for many drugs that are used to treat cardiovascular diseases, asthma and other maladies. In most eukaryotes, intracellular receptors for cyclic nucleotides include cAMP- and cGMP-dependent protein kinases (PKA and PKG), cyclic nucleotide-gated cation channels and cGMP-binding PDEs (Figure 15.1).[1–9]

Adenylyl cyclases and guanylyl cyclases

A family of membrane-bound adenylyl cyclases converts ATP into cAMP, and hormones (such as epinephrine, glucagon and vasopressin) activate these enzymes (Figure 15.1).[10] Hormone-activated adenylyl cyclases typically include three components: a receptor that is specific for a given hormone, heterotrimeric G proteins (either inhibitory or stimulatory) and a cyclase catalytic component. Hormone binding to the receptor stimulates exchange of GDP for GTP on the α-subunit of inactive heterotrimeric G proteins.[11] The α-GTP dissociates from the βγ components and activates the cyclase. An inherent GTPase activity of the α-subunit subsequently cleaves the GTP to GDP, and inactive α-GDP reassociates with the βγ complex. Certain cyclases are activated by Ca^{2+}/calmodulin and are implicated in learning.[12,13] The complex regulation of adenylyl cyclases integrates many physiological signals that modulate cAMP levels.

Signal Transduction. Edited by Carl-Henrik Heldin and Mary Purton. Published in 1996 by Chapman & Hall. ISBN 0 412 70810 8

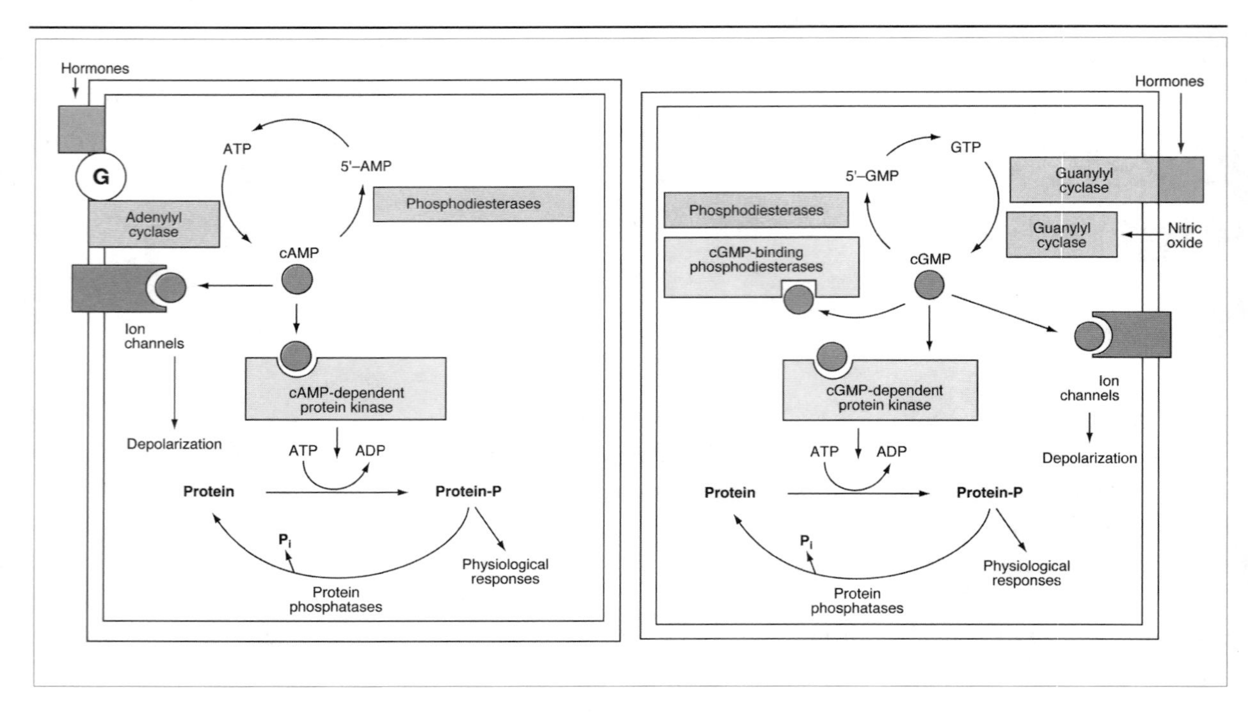

Figure 15.1 Schematic summary of intracellular receptors for cAMP and cGMP.

Guanylyl cyclases convert GTP to cGMP and occur as both membrane-bound and cytosolic forms[14,15] (Figure 15.1). Membrane-bound guanylyl cyclases are homodimeric proteins which contain receptor elements in the extracellular amino terminus, a transmembrane domain and a carboxy-terminal cyclase domain. Membrane-bound guanylyl cyclases are activated by atrial natriuretic peptide, brain and heart natriuretic peptide, type C natriuretic peptide and guanylins, each of which binds to a specific receptor portion of a guanylyl cyclase. Cytosolic guanylyl cyclases are heme-containing heterodimers that are distinct in structure and regulation from membrane-bound cyclases. Activation of cytosolic cyclases by nitric oxide and related vasodilators (such as nitroglycerin used for relief of angina) is mediated by the heme moiety by an unknown mechanism.

Cyclic nucleotide phosphodiesterases

A large family of both membrane-bound and cytosolic PDEs converts cAMP and cGMP to 5′-nucleoside phosphates.[18,19] In *Drosophila* mutants with learning deficits, the defective gene (*dunce*) encodes a PDE in normal flies.[20] PDEs have structurally similar carboxy-terminal catalytic domains but have diverse amino termini. PDE catalytic functions vary in specificities and affinities for cAMP and cGMP and are regulated by myriad mechanisms, including Ca^{2+}/calmodulin, phosphorylation and allosteric cGMP binding (see below).[18,19] PDEs are targets for many agents that inhibit catalysis, including methylxanthines such as caffeine in coffee and tea.

cAMP signaling via cAMP-dependent protein kinases

In mammals, PKAs mediate most of the effects of cAMP, but cAMP-gated channels are important in some tissues (Figure 15.1). Cyclic AMP activates PKA, which phosphorylates serines/threonines in proteins containing the consensus sequence Arg–Arg–X–Ser/Thr–X. The phosphorylation site sequence typically suffices for high affinity interaction with the kinase.[21] The phosphoprotein product retains a large portion ($\Delta G = -6.5$ kcal/mol) of the free energy of hydrolysis of ATP ($\Delta G = -8.4$ kcal/mol), compared with free phosphoserine ($\Delta G = -2.9$ kcal/mol).

PKA phosphorylation of proteins frequently causes a functional effect, and by this process hormones can modulate metabolic pathways. However, some phosphorylations are functionally 'silent.' The hundreds of known physiological substrates for PKA include phosphorylase kinase, glycogen synthase, hormone-sensitive lipase, pyruvate kinase, phospholamban, PDEs, pyruvate kinase, tyrosine hydroxylase, Ca^{2+} channels, cAMP response element binding protein (CREB), phosphatase inhibitor 1, cholesterol ester hydrolase, acetyl CoA carboxylase, ATP-citrate lyase, 6-phosphofructo-2-kinase and myriad hormone receptors and ion channels.[22,23]

Phosphorylations catalyzed by PKA and PKG are reversed by protein phosphatases[24,25] (Figure 15.1) that have complex regulatory schemes and exhibit diverse and selective phosphoprotein substrate specificities (see Chapter 18). The balance between phosphorylation of substrates by the kinases and their dephosphorylation by protein phosphatases determines the physiological impact of cyclic nucleotide signals.[22]

Regulation of cAMP-dependent protein kinases

In mammalian cells, PKA concentrations range from 0.2 to 2 μM, which is only slightly higher than the cAMP concentration. There is likely to be little free cAMP in the cell. In the absence of cAMP, mammalian PKA is composed of a dimer of regulatory subunits (R) and two monomeric catalytic subunits (C). R and C interact with an affinity of

Box 15.1 Structural models of cAMP and cGMP.

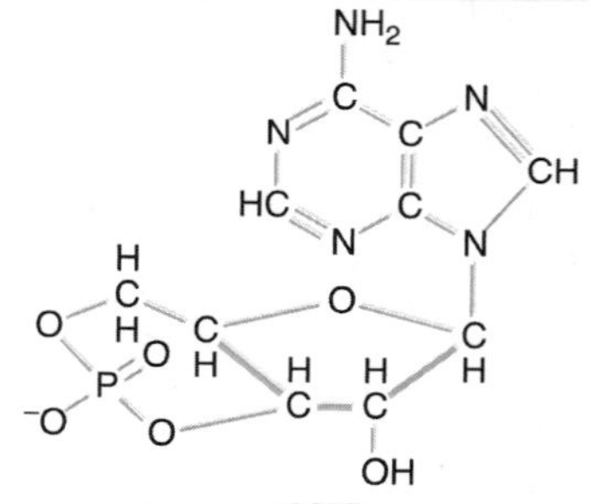

cAMP

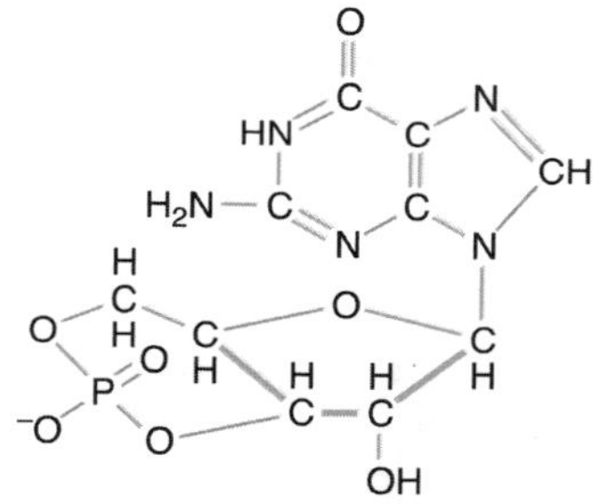

cGMP

The cyclic phosphate moiety of cAMP and cGMP is critical for interaction with cyclic nucleotide PDEs, for activation of PKA/PKG, and for interaction with cyclic nucleotide-gated channels. The cyclic phosphate rings are high-energy bonds that approximate the energy in the γ-phosphate of ATP or GTP.[16,17] They are cleaved by cyclic nucleotide PDEs but are resistant to other phosphoesterases such as intestinal phosphatases and ribonucleases.[2] In solution, cAMP and cGMP are in equilibrium between two conformations, *syn* and *anti*, based on the angle of orientation between the purine and the ribose,[17] and specific confomers of these nucleotides are preferred for interaction with sites on the kinases and PDEs.[6]

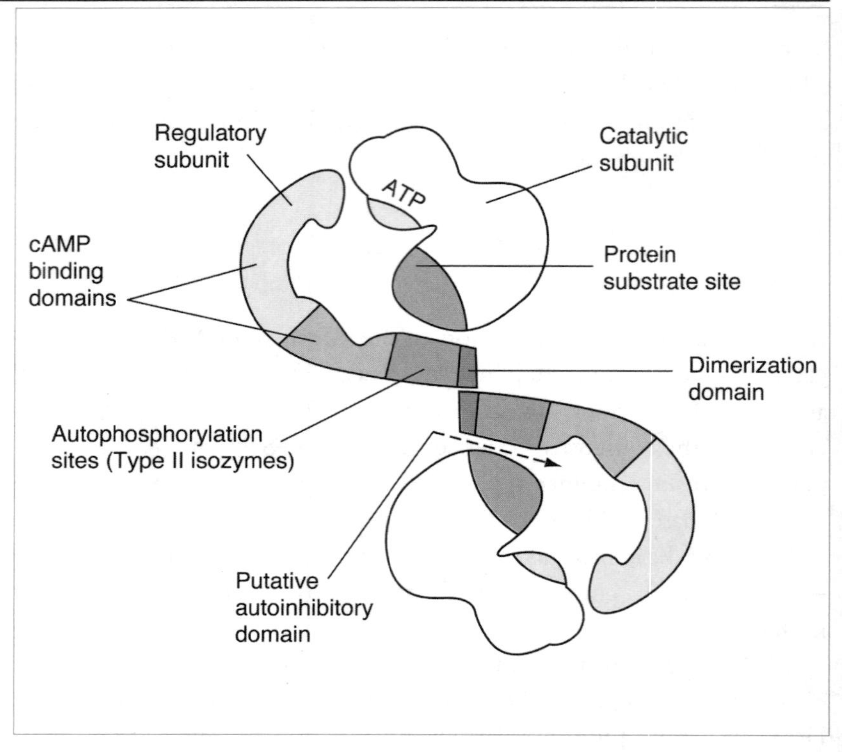

Figure 15.2 Working model of PKA. In the absence of cAMP, PKA exists as a heterotetramer composed of a dimeric regulatory subunit (R) and two catalytic subunits (C). The R subunits most commonly form homodimers through a sequence near the amino terminus. The autoinhibitory/autophosphorylation domain is situated ~50–100 residues from the amino terminus and precedes the two cyclic nucleotide binding sites in the linear sequence. The two binding sites are ~100 amino acids in length and exhibit positive cooperativity.

<0.2 nM;[26,27] the interaction involves a consensus phosphorylation sequence (or pseudosubstrate sequence) in R that acts as a competitive inhibitory substrate (or substrate-like site) for C and is known as the 'autoinhibitory domain'.[28] This mechanism of autoinhibition, involving a substrate-like sequence, is utilized by many protein kinases, although other elements may also contribute. Autoinhibitory domains in type I PKAs and PKGs involve pseudosubstrate sites, whereas those of type II PKAs have a phosphorylatable serine.

Each subunit of R (and PKGs) contains two structurally and kinetically distinct cyclic nucleotide-binding sites (Figure 15.2) that apparently arose by gene duplication prior to the divergence of PKA and PKG. These sites are homologous in sequence and structure with the cAMP-binding site in the bacterial cAMP-regulated catabolite gene activator protein (CAP)[6]. The X-ray crystallographic structures of CAP[6] and a truncated R[50] have been determined. Binding of two cAMP molecules per R subunit decreases the affinity of R for C by 10 000–100 000-fold,[27] and PKA dissociates into dimeric R and two active monomers of C.

The X-ray crystallographic structure of C reveals a bi-lobed protein with a deep cleft interposed between the lobes (Figure 15.3). The substrate, MgATP, binds at the base of the cleft; a Gly–X–Gly–X–X–Gly

Figure 15.3 Model of the bi-lobed structure of the catalytic subunit of PKA. The smaller amino-terminal lobe is rich in β-sheet structure (broad arrows) and provides several contacts for MgATP. The larger lobe is rich in α-helices (coiled structures) and provides multiple contacts for protein/peptide substrates. Solid circles indicate the glycine-rich loop involved in anchoring ATP and diamonds indicate the catalytic loop. [Model adapted with permission from Knighton, D.R., Zheng, J., Ten Eyck, L.F. *et al.* (1991) *Science*, 253, 407–14.]

sequence (residues 50–55) and a lysine (Lys72) in the small lobe anchors the ATP phosphate moiety and directs the γ-phosphate for transfer to proteins. A 'catalytic loop' (residues 165–171) extends into the cleft and is suggested to provide for catalytic transfer of phosphate.

The structure for C was determined using C that was complexed with an inhibitory peptide containing a pseudosubstrate site (Arg–Arg–Asn–Ala–Ile). Amino acids in this peptide interact with sites within, or near, the catalytic cleft of C. Each arginine contacts a pair of glutamates (Glu127/Glu331 and Glu170/Glu230), thus accounting for the importance of two basic residues in these positions in the PKA consensus substrate sequence. The alanine replaces the serine/threonine phosphorylation site and is near the putative catalytic base, Asp166. The isoleucine occupies a position that is usually hydrophobic in PKA substrates; it sits in a hydrophobic pocket in the larger lobe of C. Thus, the smaller lobe of C provides several contacts for MgATP and the larger lobe forms numerous contacts with the protein substrate. However, the interactions between R and C and the mechanism for activation of PKA are still unknown.

Isoforms of cAMP-dependent protein kinases

There are two major classes of mammalian R subunit isoforms, RI and RII (~45 kDa), that are highly homologous, and each isoform has subtypes that are products of different genes.[26] RI and RII expression varies according to species and tissue, and they complex with the same isoforms of C (Cα, Cβ and Cγ), which are products of different genes. Many studies suggest that activation of either isoform can elicit a particular physiological response.[27] In some tissues, a portion of the type II PKA is bound via the amino terminus of RII to particulate fraction proteins known as A Kinase Anchoring Proteins, some of which are PKA substrates.[31,32] Co-localization of PKA with substrates could enhance the response to stimuli.

Effects of cAMP on gene transcription

Studies suggest that prolonged elevation of intracellular cAMP causes C to translocate to the nucleus, but PKA holoenzyme is excluded.[33] Transcription of genes containing cAMP-responsive elements (CREs) in their promoters is increased by overexpression of C, an effect that is blocked by co-expression of R. Whether C translocates by diffusing through the nuclear membrane or by another process is unknown. Diffusion of cytosolic cAMP into the nucleus and activation of nuclear PKA could also account for cAMP effects on transcription. Either process could increase nuclear protein phosphorylation (e.g. transcription factors such as CREB).[34,35]

Cyclic GMP signaling via cGMP-dependent protein kinases

Cyclic GMP action in mammalian tissues is mediated in large part by activation of PKG which is associated with lowering of cytosolic Ca^{2+}.[9,36,37] PKG and cGMP concentrations range from 0.1 to 1 μM in tissues such as lung, cerebellum, platelets, neutrophils, intestinal epithelium and smooth muscle, but are very low in other tissues, such as liver and skeletal muscle. PKG and PKA are closely related, but in PKGs, regulatory and catalytic components reside in a single protein (Figure 15.4). There are two major classes of PKGs. The cytosolic type I PKGs (types Iα and Iβ) form homodimers of subunits of 76 and 78 kDa, respectively, and are products of alternative mRNA splicing near the amino terminus. Type II PKG is homologous to type I PKGs, although somewhat larger (87 kDa subunits), and is associated with the particulate fraction in some tissues.[9,26,36,37] PKGs dimerize through a leucine zipper motif near the amino terminus, but whether this region 'anchors' PKGs to subcellular structures, as in the case of type II PKA, is unclear.[38] Binding of two cGMP molecules to dimeric type Iα PKG partially activates the enzyme, but full activation requires binding of four cGMP molecules.[37]

The consensus sequence for PKG and PKA substrates is similar (–Arg–Arg–X–Ser–X–), but there are some differences.[21,26] Generally, PKGs and PKAs phosphorylate different substrates *in situ*, but in some tissues the same substrate is phosphorylated by both PKGs and PKAs. Suggested physiological substrates for PKG include cGMP-binding, cGMP-specific PDE, Ca^{2+}/ATPases from plasma membrane and the sarcoplasmic reticulum, the InsP_3 receptor, phospholamban, certain G proteins, the cerebellar 'G substrate' and vasodilator-stimulated phosphoprotein.[9,36,37]

Determinants of cyclic nucleotide specificity and cross-activation

Cyclic nucleotide-binding domains of PKG and PKA have an approximately 50-fold specificity for cGMP and cAMP, respectively[6,9,23,26,38] but each can be activated by high concentrations of the other nucleotide.

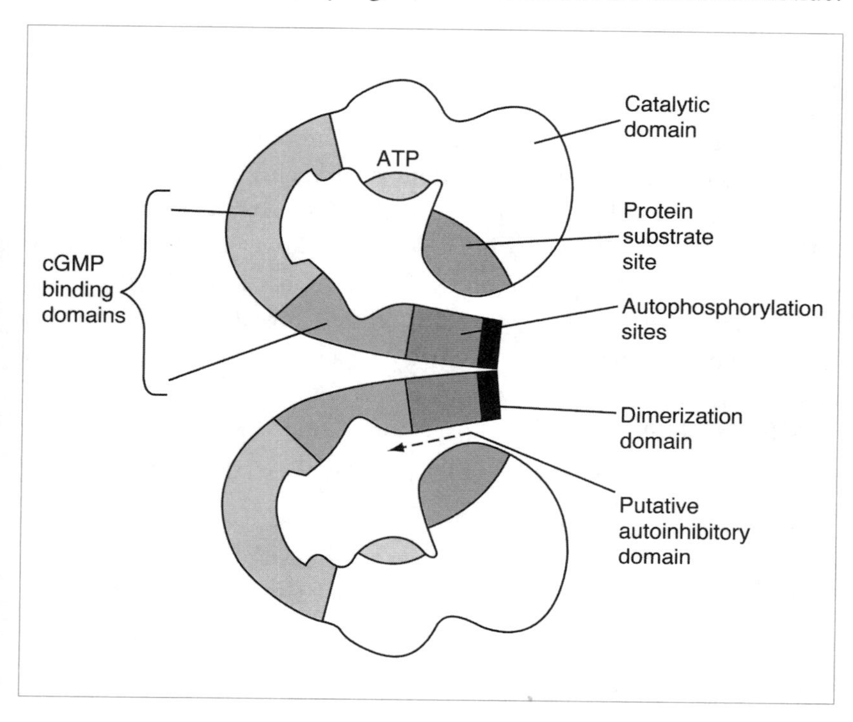

Figure 15.4 Working model of PKG. PKGs have their regulatory and catalytic components located in a single polypeptide chain. The regulatory domain is located in the amino-terminal half of the protein and the catalytic domain is located in the carboxy-terminal portion. Sequences of the amino-terminal ~100 amino acids in PKGs are very divergent and contain the leucine zipper dimerization motif and the autoinhibitory/autophosphorylation domain.[36,38]

Sequences of the cyclic nucleotide-binding sites in PKGs and PKAs have many differences, but a threonine/serine at a single position in each site in PKG (Thr177 and Thr301 in type Iα PKG) accounts for high affinity binding of cGMP, compared with cAMP.[6] PKAs have alanine at this location. The threonine/serine hydroxyl is proposed to form a hydrogen-bond with the C2-amino of cGMP (Box 15.1).[6] Cyclic nucleotide-gated channels (see below), members of the same cyclic nucleotide receptor family as PKA and PKG, preferentially bind cGMP and have threonine in this position.[4–7,39]

PKG and PKA are thought to mediate separate pathways in most cells, but cross-talk can occur where cAMP activates PKG (e.g. in smooth muscle) or cGMP activates PKA (e.g. in intestinal cells).[40,41] The latter may have only pathological importance, but cross-activation in smooth muscle may be physiological. Cross-activation is possible when concentrations of cGMP or cAMP are sufficient to overcome the approximately 50-fold specificity difference between cGMP and cAMP for activation of the respective kinase. Selectivity in kinase substrates may also discriminate between cGMP or cAMP effects; significant differences in protein/peptide substrate specificities are known.[21,26] Subcellular localization (i.e. compartmentalization) of PKA and PKG could also affect which enzyme elicits an effect.[31–33,36–38]

Autophosphorylation of PKA and PKG

PKA and PKG autophosphorylate serines/threonines that are near or within their competitive autoinhibitory domains (Figures 15.2 and 15.4).[29] Autophosphorylation of type II PKA increases its affinity for cAMP so that the PKA can be activated by lower concentrations of cAMP. Autophosphorylation of PKG increases basal (no cGMP) kinase activity and also produces a higher affinity for cGMP and cAMP. Increased basal activity could also produce a sustained increase in PKG activity after cGMP has declined. Whether factors other than cAMP/cGMP alter the phosphorylation state of PKA or PKG *in vivo* is not known, and the protein phosphatases that dephosphorylate the kinases have not been determined.

Signaling through cyclic nucleotide-gated channels

Physiological effects of cAMP and cGMP are also produced by their direct gating of a diverse family of channels that conduct both

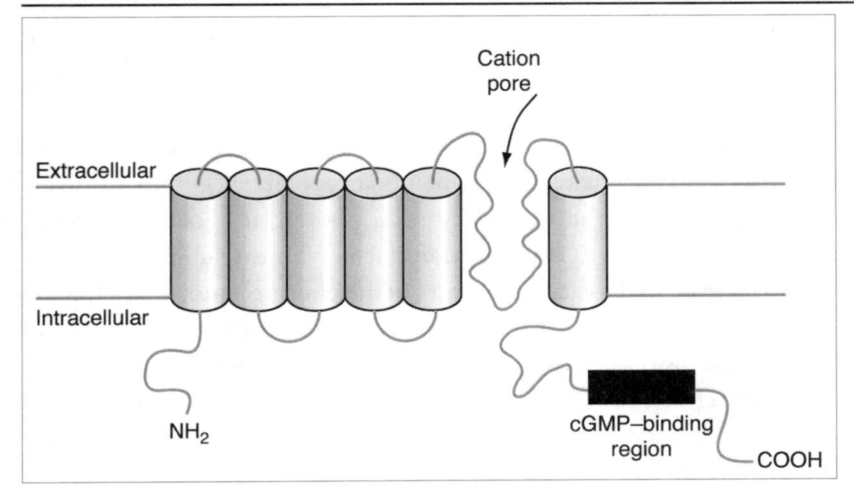

Figure 15.5 Overall structure of the cyclic nucleotide-binding subunit of the cyclic nucleotide-gated cation channel. These proteins contain three functional regions; the amino terminus and the carboxyl terminus face the cytosol and the central region has a transmembrane topography, including a segment that forms the ion pore. The cyclic nucleotide-binding domain is located in the carboxy-terminal segment.

monovalent and divalent cations. Although lacking voltage sensitivity, they are structurally homologous to voltage-gated channels. The single cGMP/cAMP binding site is a ~100 residue segment near the carboxyl terminus and is homologous in structure to the CAP/kinase family of cAMP/cGMP-binding sites (Figures 15.1 and 15.5).[4–7,39]

One channel is thought to contain four or five of the cyclic nucleotide-binding proteins (ranging from 63 to 76 kDa), and upon binding cyclic nucleotide the channels are maintained in an open state. This provides for increased cation influx[4,7;39,42] and relative depolarization of the membrane potential. The number of cyclic nucleotide molecules that are bound to the channel is suggested to determine the facility with which cations traverse the pore.[4–7,39]

Cyclic nucleotide-gated channels occur in many tissues.[4,5,7,39,43] These channels from rod, cone and olfactory epithelium bind cGMP in preference to cAMP, but in olfactory cilia there is evidence that cAMP functions as the signal.[4–7] Cyclic nucleotide-gated channels from different tissues have unique electrical properties and the concentration of cGMP/cAMP required to activate conductance varies.[4,39] Significantly lower cyclic nucleotide concentrations will activate olfactory channels than will activate photoreceptor channels. Features of the primary structure of the channels, as well as the presence of a second type of subunit,[43] alter the affinity of the channels for cGMP/cAMP. A hetero-oligomeric structure is similar to that of other ligand-gated channels, but whether this is common to all cyclic nucleotide-regulated channels remains to be determined.

In the visual system, cGMP-gated cation channels provide a critical component of a highly complex signaling pathway (Figure 15.6).[39,42] In dark-adapted rods and cones, a significant portion of the cellular cGMP is bound to cGMP-binding sites in the photoreceptor PDEs (see below). The high concentration of unbound cGMP (~4 μM) provides for cGMP binding by the cGMP-gated cation channels, and the inward conductance of Na^{+}/Ca^{2+} is facilitated. The membrane potential is thus relatively depolarized. When rhodopsin absorbs a photon, it becomes activated and, in turn, activates a G protein, transducin, whose α-subunit (α_t) causes a rapid increase in the catalytic activity of the photoreceptor PDEs (Figure 15.6). Consequently, cGMP concentrations are rapidly reduced, and cGMP that dissociates from the channel is not replaced. The channel pore then closes, thus abolishing the depolarizing influx of cations into the photoreceptor. However, the activity of a Na^{+}/Ca^{2+}, K^{+} exchanger persists, so that the membrane potential of the cell becomes hyperpolarized. This hyperpolarization provides the signal that is perceived as light in higher integrative centers (Figure 15.7).

The light-induced closing of the channel causes a decrease in intracellular Ca^{2+} which, in turn, activates a negative feedback loop to increase photoreceptor guanylyl cyclase activity.[39,42] The decrease in Ca^{2+} does not directly affect the cyclase but alters the activity of a Ca^{2+}-binding regulatory protein, recoverin. When the cGMP level recovers, the photoreceptor cell regains its sensitivity to light.

Olfaction also utilizes a cyclic nucleotide-gated channel.[4,7,43,44] Odorants bind to membrane receptors on the cilia of olfactory epithelial cells and stimulate adenylyl cyclase activity by activating a G protein. Increased cAMP binding causes the olfactory cyclic nucleotide-gated channel to open, and the influx of Na^{+}/Ca^{2+} depolarizes the membrane

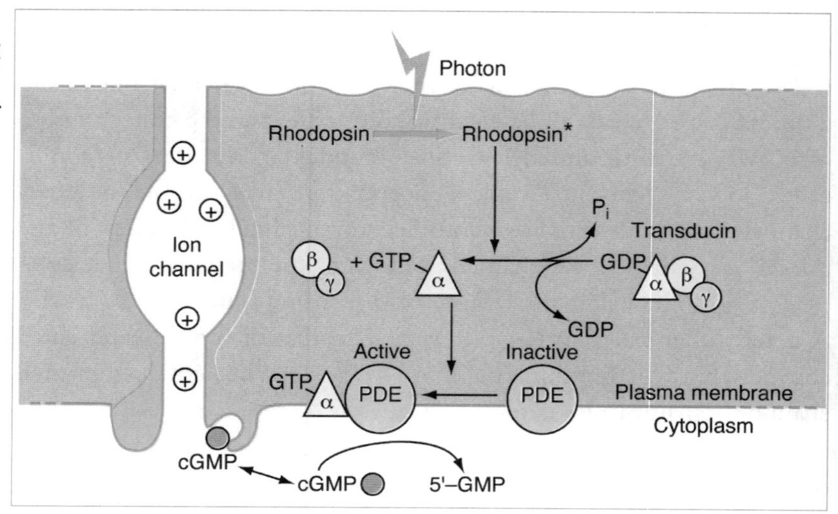

Figure 15.6 Cyclic GMP signaling pathway in the visual system. PDE = cGMP–PDE; α, β, γ = subunits of transducin; ⊕ = cations; • = cGMP.

potential of the olfactory cell. This provides a signal to higher centers for detection of an odor (Figure 15.7).

Signaling through cGMP-binding phosphophodiesterase

cGMP-binding PDEs are a diverse subgroup of PDEs that form another family of cGMP intracellular receptors[6,8,9,19] (Figure 15.1). These PDEs bind cGMP with high specificity at sites in their amino-terminal regulatory domains; no cAMP-binding PDEs have been identified. This group of PDEs includes the rod and cone PDEs (PDE6), a cGMP-binding, cGMP-specific PDE (PDE5) and the cGMP-stimulated PDEs (PDE2) (Figure 15.8). Most of these are dimeric, and each monomer has two allosteric cGMP-binding sites of ~100 amino acids each that are homologous. The role of these sites is best understood in PDE2, where cGMP binding increases catalytic activity by lowering the K_m for cAMP with little effect on the V_{max}. Little is known about the function of the sites in other cGMP-binding PDEs. PDE cGMP-binding sites differ from those of the kinase/CAP family in their amino acid sequences and in their cyclic nucleotide analog specificities, and are likely to be distinct in their overall structure.[6,8]

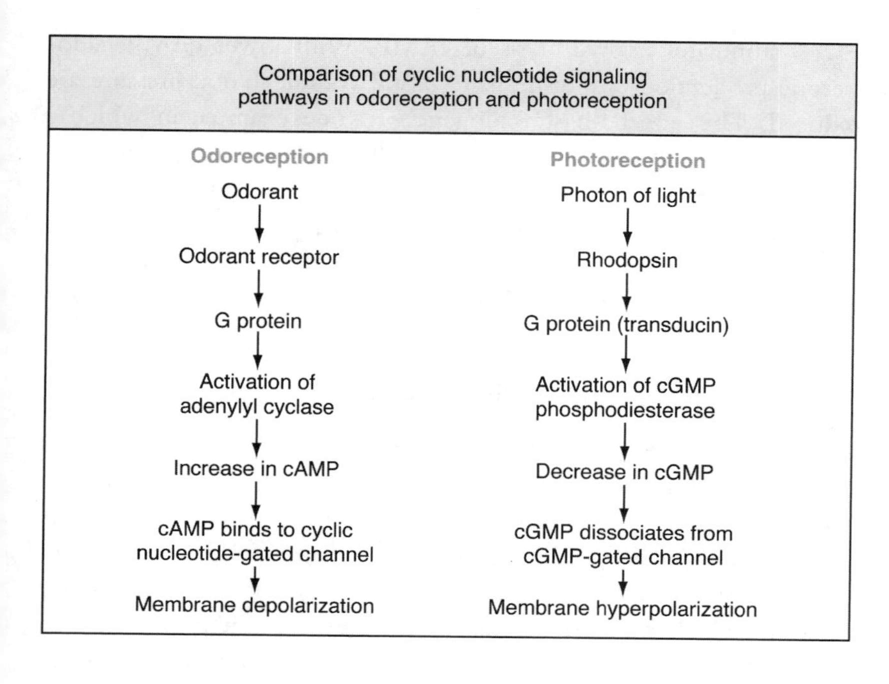

Figure 15.7 Comparison of cyclic nucleotide signaling pathways in odoreception and photoreception.

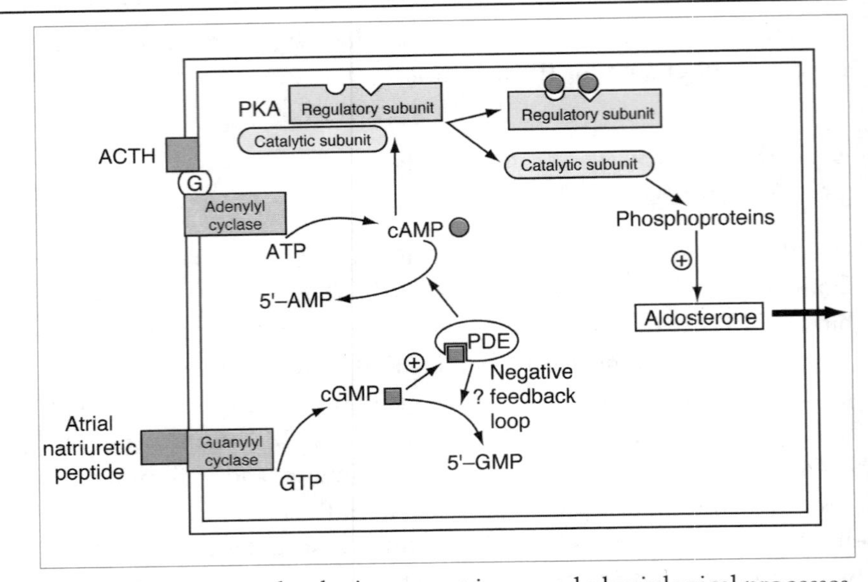

Figure 15.8 The role of PDE2, a cGMP-binding PDE, in regulating aldosterone production. Cyclic GMP binding at allosteric sites on PDE2 stimulates the catalytic activity which hydrolyzes both cAMP and cGMP. ACTH = adrenal corticotropic hormone; ⊕ = stimulatory effect.

PDE2 is suggested to be important in several physiological processes, including regulation of Ca^{2+} channels and regulation of aldosterone secretion (Figure 15.8).[19,45] In adrenal glands, where PDE2 is abundant, an adrenocorticotropic hormone-induced increase in cAMP promotes mineralocorticoid synthesis (e.g. aldosterone), thereby increasing blood volume. Atrial natriuretic peptide activation of a guanylyl cyclase increases cGMP levels, and cGMP binding to cGMP-binding sites in PDE2 stimulates its hydrolysis of cAMP. With lower cAMP, aldosterone production diminishes, and blood volume/blood pressure are reduced. This effect on steroidogenesis is one example in which a cGMP-binding PDE may regulate the cell's response to cyclic nucleotide.[19,45] Whether this effect is exclusive to cyclic nucleotide-gated channels or PKG participation is not known. PDE2 degrades both cGMP and cAMP *in vitro* and could also mediate negative feedback regulation of the cGMP signal (Figure 15.8).

Regulation of cyclic nucleotide levels

The sensitivity of physiological processes to cAMP/cGMP signals requires that their levels be maintained within an optimal range for responsiveness. Two- to three-fold increases in cyclic nucleotides (basal level ~10^{-7} M) produce maximum responses in most tissues and involve very limited changes in PKA or PKG activity.[26] Agents acting through cAMP or cGMP typically cause only a transient rise in cyclic nucleotide, despite the continued presence of hormone. Several mechanisms are likely to contribute to the rapid decline, including desensitization of adenylyl cyclases, receptor sequestration, receptor phosphorylation by PKA/PKG and other kinases and increased PDE activity.[11,46–48]

Cellular levels of cAMP or cGMP rarely approach the K_ms that have been determined for the PDEs *in vitro*,[19,26] so that increased cyclic nucleotide synthesis is accompanied by increased rates of hydrolysis. Thus, the rule of mass action dampens increases in cyclic nucleotide (Figure 15.1). In addition, PDE activities are regulated by cGMP concentrations (Figure 15.8), Ca^{2+}/calmodulin and phosphorylation mechanisms. The latter, which involves phosphorylation and activation of PDEs, enhances cAMP/cGMP degradation in tissues (Figure 15.9).

Thyroid-stimulating hormone causes an increase in cAMP levels in thyroid cells in addition to the phosphorylation and activation of a cAMP-specific PDE (PDE4).[46] In adipose tissue, liver, platelets and other tissues, another PDE (PDE3) that hydrolyzes either cAMP or cGMP is phosphorylated and activated by PKA[47,48] (Figure 15.9). By this negative feedback control, the PKA down-regulates its activity and accelerates termination of a cyclic nucleotide signal after the extracellular stimulus declines. PDE3 is also activated by insulin (Figure 15.9), which causes it to be phosphorylated by an uncharacterized protein kinase, insulin-stimulated kinase (ISK).[48] This could explain insulin's action in opposing effects of several cAMP-elevating hormones, including the antagonism of gene transcription that is mediated by CREB phosphorylation. Other PDEs are also phosphorylated by PKA and PKG and may be similarly regulated. Persistent cAMP elevation increases PDE synthesis which would contribute further to down-regulation of cyclic nucleotide signals.[49]

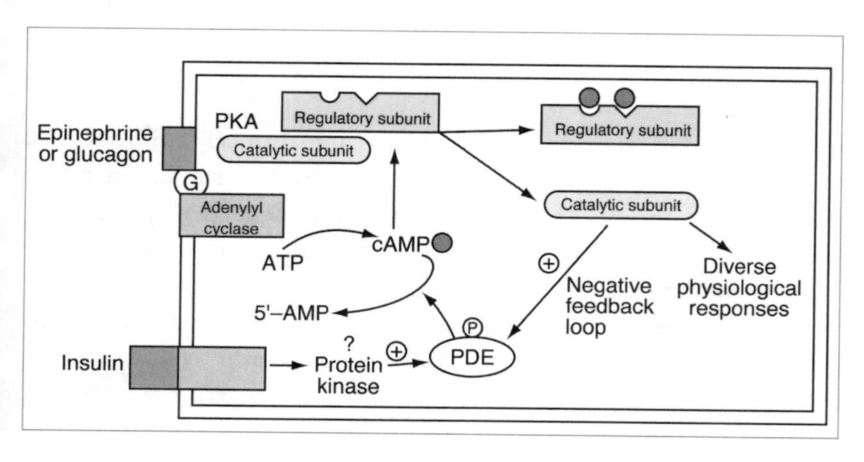

Figure 15.9 Depiction of negative feedback control of cAMP levels. PDE3 is phosphorylated and activated by both the C subunit of the activated PKA and by an unknown insulin-stimulated kinase (ISK). The ⊕ indicates that these phosphorylations increase the catalytic activity of the PDE3 which hydrolyzes both cAMP and cGMP.

Amplification in cyclic nucleotide signaling pathways

Amplification of the initial stimulus is a central theme to signaling cascades.[22] In cyclic nucleotide signaling, amplification differs somewhat depending on the intracellular receptors that are used. Amplification first occurs in these pathways when hormone binds to a single receptor and activates either adenylyl or guanylyl cyclase. The adenylyl cyclase pathway includes an intervening amplification at the G-protein level which then activates the adenylyl cyclase. The increased catalytic activity of the cyclases produces many molecules of extra cAMP or cGMP which then activate the PKAs and PKGs. These kinases further amplify the signal by phosphorylating many protein substrate(s) molecules. This may increase the catalytic functions of some proteins. Other effects, such as phosphorylation of protein phosphatase-1 inhibitor, produce potent phosphatase inhibitors that block removal of phosphates that have been introduced into substrates.[24,25]

In the visual pathway, amplification occurs at the levels of the G protein, the PDE and the cGMP-gated channel.[39,42] A single photon converts rhodopsin into photoactivated rhodopsin (Figure 15.6) which promotes GDP/GTP exchange on many transducins. Hundreds of α_t-GTPs dissociate from transducin complexes, and a single α_t-GTP activates a PDE which then hydrolyzes perhaps 1000 cGMP molecules to 5′-GMP. The lowered cGMP leads to closure of the cGMP-gated channel and the influx of thousands of cations is blocked. In olfaction, amplification is likely to occur at the level of the G protein, the adenylyl cyclase and opening of the channel. Numerous mechanisms dampen amplification in these pathways, thus constraining the cellular response and enhancing recovery of the unstimulated state.

Summary

Cyclic nucleotide signaling pathways in mammalian cells utilize at least two evolutionarily distinct and apparently unrelated families of cyclic nucleotide receptors: (1) the cyclic nucleotide-dependent protein kinases (PKA and PKG)/cyclic nucleotide-gated channels and (2) the cGMP-binding PDEs. Mechanisms used by these receptors are diverse and involve at least three modes of action: protein kinase pathways, the ligand-gated channels pathway and/or the PDE pathway. The contributions of these signaling cascades to physiological processes and the interplay between these receptors are complex and not fully understood. Regulation of the kinases, channels and PDEs by other cellular components and feedback mechanisms undoubtedly impacts further on the functioning of these pathways.

References

1 Butcher, R.W. and Sutherland, E.W. (1967) The effects of the catecholamines, adrenergic blocking agents, prostaglandin E_1, and insulin on cyclic AMP levels in the rat epididymal fat pad *in vitro*. *Ann. N.Y. Acad. Sci.*, **139**, 849–59.

2 Robison, G.A., Butcher, R.W. and Sutherland, E.W. (1971) Cyclic AMP and hormone action, in *Cyclic AMP* (eds Robison, G.A., Butcher, R.W. and Sutherland, E.W.), Academic Press, New York, pp. 17–47.

3 Doskeland, S.O., Boe, R., Bruland, T. *et al.* (1991) *Cell Signaling: Experimental Strategies*, in the series *Methodological Surveys in Biochemistry and Analysis*, Vol. 21, Special Publication No. 92 (eds Reid, E., Cook, G.M.W. and Luzio, J.P.), Royal Society of Chemistry, Cambridge pp.103–4.

4 Kaupp, U.B. (1991) The cyclic nucleotide-gated channels of vertebrate photoreceptors and olfactory epithelium. *Trends Neurosci.*, **14**, 150–7.

5 Bonigk, W., Altenhofen, W., Muller, F. *et al.* (1993) Rod and cone photoreceptor cells express distinct genes for cGMP-gated channels. *Neuron*, **10**, 865–77.

6 Shabb, J.B. and Corbin, J.D. (1992) Cyclic nucleotide-binding domains in proteins having diverse functions. *J. Biol. Chem.*, **267**, 5723–6.

7 Dhallan, R.S., Yau, K.-W., Schrader, K.A. and Reed, R.R. (1990) Primary structure and functional expression of a cyclic nucleotide-activated channel from olfactory neurons. *Nature*, **347**, 184–7.

8 Charbonneau, H., Prusti, R.K., LeTrong, H. *et al.* (1990) Identification of a non-catalytic cGMP-binding domain conserved in both the cGMP stimulated and photoreceptor cyclic nucleotide phosphodiesterase. *Proc. Natl. Acad. Sci. USA*, **87**, 288–92.

9 Lincoln, T.M. and Cornwell, T.L. (1993) Intracellular cyclic GMP receptor proteins. *FASEB J.*, **7**, 328–38.

10 Taussig, R. and Gilman, A.G. (1995) Mammalian membrane-bound adenylyl cyclases. *J. Biol. Chem.*, **270**, 1–4.

11 Collins, S., Caron, M.G. and Lefkowitz, R.J. (1992) From ligand binding to gene expression: new insights into the regulation of G-protein-coupled receptors. *Trends Biol. Sci.*, **17**, 37–9.

12 Frey, U., Huang, Y.-Y. and Kandel, E.R. (1993) Effects of cAMP simulate a late stage of LTP in hippocampal CA1 neurons. *Science*, **260**, 1661–4.

13 Kandel, E. and Abel, T. (1995) Neuropeptides, adenylyl cyclase, and memory storage. *Science*, **268**, 825–6.

14 Garbers, D.L. and Lowe, D.G. (1994) Guanylyl cyclase receptors. *J. Biol. Chem.*, **269**, 30741–4.

15 Garbers, D.L. (1992) Guanylyl cyclase receptors and their endocrine, paracrine and autocrine ligands. *Cell*, **71**, 1–4.

16 Rudolph, S.A., Johnson, E.M. and Greengard, P. (1971) The enthalpy of hydrolysis of various 3′,5′- and 2′,3′-cyclic nucleotides. *J. Biol. Chem.*, **246**, 1271–73.

17 Pasternak, T. (1971) Chemistry of cyclic nucleoside phosphates and synthesis of analogs, in *Cyclic AMP* (eds Robison, G.A., Butcher, R.W. and Sutherland, E.W.), Academic Press, New York, pp 48–71.

18 Charbonneau, H. (1990) Structure–function relationships among cyclic nucleotide phosphodiesterases, in *Cyclic Nucleotide Phosphodiesterases: Structure, Regulation and Drug Action*, (eds Beavo, J. and Houslay, M.) J. Wiley, New York, pp.267–96.

19 Sonnenburg, W.K. and Beavo, J.A. (1994) Cyclic GMP and regulation of cyclic nucleotide hydrolysis, in *Cyclic GMP: Synthesis, Metabolism, and Function* (ed. Murad, F.), *Advances in Pharmacology*, Vol. 26, Academic Press, San Diego, pp. 87–114.

20 Davis, R.L. and Dauwalder, B. (1991) The *Drosophila dunce* locus: learning and memory genes in the fly. *Trends Genet.*, **7**, 224–9.

21 Kennelley, P.J. and Krebs, E.G. (1991) Consensus sequences as substrate specificity determinants for protein kinases and protein phosphatases. *J. Biol. Chem.*, **266**, 15555–8.

22 Shacter, E., Stadtman, E.R., Jurgensen, S.R. and Chock, P.B. (1988) Role of cAMP in cyclic cascade regulation, in *Initiation and Termination of Cyclic Nucleotide Action* (eds Corbin, J.D. and Johnson, R.A.), *Methods in Enzymology*, Vol 159, Academic Press, San Diego, pp. 3–19.

23 Gettys, T.W. and Corbin, J.D. (1989) The protein kinase family of enzymes, in *Receptor Phosphorylation* (ed. Moudgil, V.K.), CRC Press, Boca Raton, FL, pp. 40–88.

24 Cohen, P. and Cohen, P.T.W. (1989) Protein phosphatases come of age. *J. Biol. Chem.*, **263**, 21435–8.

25 Shenolikar, S. (1994) Protein serine/threonine phosphatases – new avenues for cell regulation. *Annu. Rev. Cell. Biol.*, **10**, 55–86.

26 Francis, S.H. and Corbin, J.D. (1994) Structure and function of cyclic nucleotide-dependent protein kinases. *Annu. Rev. Physiol.*, **56**, 237–72.

27 Doskeland, S.O., Maronde, E. and Gjertsen, B.T. (1993) The genetic subtypes of cAMP-dependent protein kinase – functionally different or redundant? *Biochim. Biophys. Acta*, **1178**, 249–58.

28 Soderling, T.R. (1990) Protein kinases. Regulation by autoinhibitory domains. *J. Biol. Chem.*, **265**, 12111–14.

29 Smith, J.A., Francis, S.H. and Corbin, J.D. (1993) Autophosphorylation: a salient feature of protein kinases. *Mol. Cell. Biochem.*, **127/128**, 51–70.

30 Knighton, D.R., Zheng, J., Ten Eyck, L.F. *et al.* (1991) Crystal structure of the catalytic subunit of cyclic adenosine monophosphate-dependent protein kinase. *Science*, **253**, 407–14.

31 Bregman, D.B., Bhattacharyya, N. and Rubin, C.S. (1989) High affinity binding

protein for the regulatory subunit of cAMP-dependent protein kinase II-B. *J. Biol. Chem.*, **264**, 4648–56.

32 Scott, J.D., Stofko, R.E., McDonald, J.R. *et al.* (1990) Type II regulatory subunit dimerization determines the subcellular localization of the cAMP-dependent protein kinase. *J. Biol. Chem.*, **265**, 21561–6.

33 Meinkoth, J.L., Ji, Y., Taylor, S.S. and Feramisco, J.R. (1990) Dynamics of the distribution of cyclic AMP-dependent protein kinase in living cells. *Proc. Natl. Acad. Sci. USA*, **87**, 9565–9.

34 Mellon, P.L., Clegg, C.H., Correll, L.A. and McKnight, G.S. (1989) Regulation of transcription by cyclic AMP-dependent protein kinase. *Proc. Natl. Acad. Sci. USA*, **86**, 4887–91.

35 Zalli, E. and Sassone-Corsi, P. (1994) Signal transduction and gene regulation: the nuclear response to cAMP. *J. Biol. Chem.*, **269**, 17359–62.

36 Butt, E., Geiger, J., Jarchau, T. *et al.* (1993) The cGMP-dependent protein kinase – gene, protein and function. *Neurochem. Res.*, **18**, 27–42.

37 Francis, S.H. and Corbin, J.D. (1994) Progress in understanding the mechanism and function of cyclic GMP-dependent protein kinase, in *Cyclic GMP: Synthesis, Metabolism, and Function* (ed. Murad, F.), *Advances in Pharmacology*, Vol. 26, Academic Press, San Diego, pp. 115–70.

39 Kaupp, U.B. and Koch, K.-W. (1992) Role of cGMP and Ca^{2+} in vertebrate photoreceptor excitation and adaptation. *Annu. Rev. Physiol.*, **54**, 153–75.

40 Forte, L., Thorne, P.K., Eber, S.L. *et al.* (1992) Stimulation of intestinal Cl^- transport by heat-stable enterotoxin: activation of cAMP-dependent protein kinase by cGMP. *Am. J. Physiol.*, **263**, C607–15.

41 Jiang, H., Colbran, J.L., Francis, S.H. and Corbin, J.D. (1992) Direct evidence for cross-activation of cGMP-dependent protein kinase by cAMP in pig coronary arteries. *J. Biol. Chem.*, **267**, 1015–9.

42 Yarfitz, S. and Hurley, J.B. (1994) Transduction mechanisms of vertebrate and invertebrate photoreceptors. *J. Biol. Chem.*, **269**, 14329–32.

43 Snyder, S.H., Sklar, P.B. and Pevsner, J. (1988) Molecular mechanism of olfaction. *J. Biol. Chem.*, **263**, 13971–4.

44 Liman, E. and Buck, L.B. (1994) A second subunit of the olfactory cyclic nucleotide-gated channel confers high sensitivity to cAMP. *Neuron*, **13**, 611–21.

45 MacFarland, R.T., Zelus, B.D. and Beavo, J.A. (1991) High concentrations of a cGMP-stimulated phosphodiesterase mediate ANP induced decreases in cAMP and steroidogenesis in adrenal glomerulosa cells. *J. Biol. Chem.*, **266**, 136–42.

46 Sette, C., Iona, S. and Conti, M. (1994) The short-term activation of a rolipram-sensitive cAMP-specific phosphodiesterase by thyroid-stimulating hormone in thyroid FRTL-5 cells is mediated by a cAMP-dependent phosphorylation. *J. Biol. Chem.*, **269**, 9245–52.

47 Gettys, T.W., Blackmore, P.F., Redmon, J.B. *et al.* (1987) Short term feedback regulation of cAMP by accelerated degradation in rat tissues. *J. Biol. Chem.*, **262**, 333–9.

48 Degerman, E., Smith, C.J., Tornqvist, H. *et al.* (1990) Evidence that insulin and isoprenaline activate the cGMP-inhibited low K_m cAMP phosphodiesterase in rat fat cells by phosphorylation. *Proc. Natl. Acad. Sci. USA*, **87**, 533–7.

49 Morena, A.R., Boitani, C., de Grossi, S. *et al.* (1995) Stage and cell-specific expression of the adenosine 3′,5′ monophosphate–phosphodiesterase genes in rat seminiferous epithelium. *Endocrinology*, **136**, 687–95.

50 Su, Y., Dostmann, R.G., Herberg, F.W. *et al.* (1995) Regulatory subunit of protein kinase A: structure of deletion mutant with cAMP binding domains. *Science*, **269**, 807–13.

16 The protein kinase C gene family

Peter J. Parker

The transduction of signals from plasma membrane receptors to cellular processes is effected by a number of different strategies. One of these involves the generation of diffusable second messengers that interact with specific intracellular proteins altering their function and so triggering a cellular response. Protein kinases play a key role as targets for second messengers as evidenced by the first discovered and best understood, the cAMP-dependent protein kinase. Within this class of second-messenger-dependent protein kinases is protein kinase C (PKC). Members of the PKC family are activated by the second messenger diacylglycerol (DAG). This lipid is produced in response to a variety of cellular agonists operating through tyrosine-kinase-linked receptors and seven-transmembrane, serpentine receptors.

The reversible activation of PKC through the binding of DAG in the presence of phospholipid was established initially *in vitro*; the agonist-dependent turnover of inositol lipids was already known and provided the basis of a signaling pathway.[1] The use of cell-permeable DAG species has provided strong evidence for the activation of PKC by DAG *in vivo* (e.g. ref. 2). DAG from sources other than the action of PLC (e.g. phospholipase D + phosphatidic acid phosphohydrolase, or *de novo* biosynthesis) has recently grown in importance and may also be responsible for PKC activation (reviewed in ref. 3).

The present family of 11 PKC isotypes (α, β_{I}, β_{II}, γ, δ, ε, ζ, θ, ζ, ι/λ and μ) are grouped on the basis of a combination of structural and functional considerations (Figure 16.1). Most PKC isotypes display typical DAG (or phorbol ester; see Box 16.1) dependence for kinase activation – only the atypical isotypes (aPKC) are insensitive and it has yet to be determined which if any lipid(s) can activate the aPKCs *in vivo*. At a functional level, the catalytic domains of the c, n and a subgroups all have very similar (and distinctive) features which implies overlapping specificity (see below). By contrast, the catalytic domain of PKC-μ (also termed PKD) shows features characteristic of the Ca^{2+}-calmodulin-de-

Box 16.1 Phorbol esters
The tumor-promoting phorbol esters activate PKC *in vitro* and have been widely used as a pharmacological tool to investigate the function of PKC in cells. Esters such as phorbol dibutyrate (PDBU) are relatively hydrophilic and display excellent binding properties, competing for the same binding site on PKC as DAG.[12]

Signal Transduction. Edited by Carl-Henrik Heldin and Mary Purton. Published in 1996 by Chapman & Hall. ISBN 0 412 70810 8

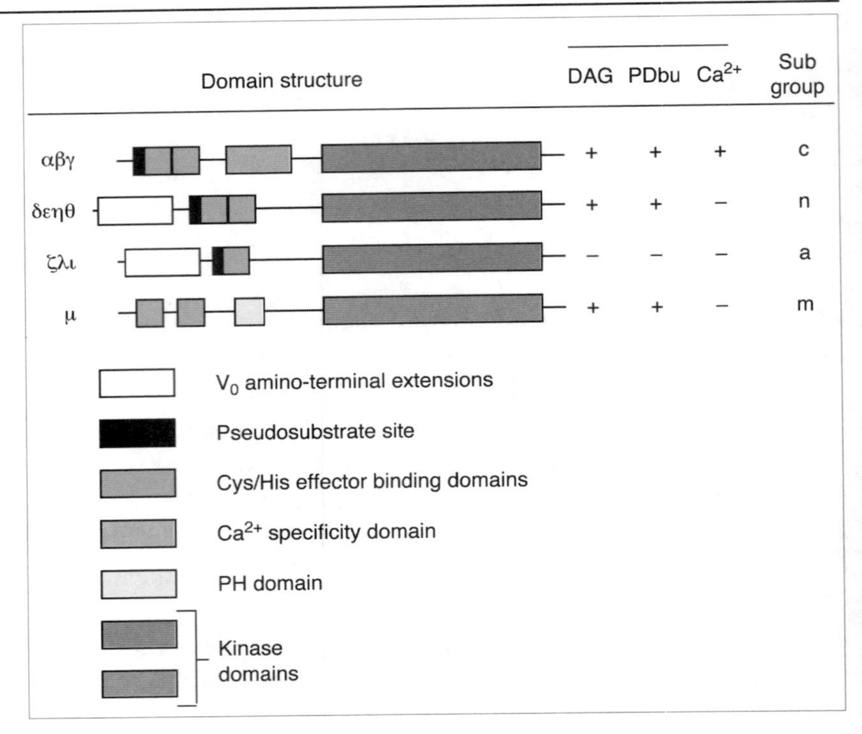

Figure 16.1 The PKC gene family. The domain structure and effector dependences of members of the PKC gene family are shown. The individual isotypes are indicated by Greek letters and the subgroups as shown (see text). The variable V_0 domains of δ and Θ are related, as are those of ε and η and ζ and ι. No pseudosubstrate site has been identified in PKC-μ; the PH domain in this isotype has been ascribed based upon sequence conservation.

pendent protein kinase family and has a distinct substrate specificity.[4,5]

In general, most PKC isotypes are DAG-activated (without any apparent preference for DAG species) and for the purposes of much of the discussion the behavior of these kinases will be considered in this light.

Inhibition and activation

The reversible activation of PKC by DAG implies that PKC, as isolated, exists in a latent state that is maintained through domain interactions and relieved on DAG binding. This situation parallels the latent (R_2C_2) state of the cyclic, AMP-dependent protein kinase (see Chapter 15), the difference being that for PKC the regulatory and catalytic domains reside on one polypeptide chain. Indeed, limited proteolysis of PKC also causes activation through the untethering of these two domains.[6]

The region of the regulatory domain responsible for the tonic inhibition of catalytic activity has been termed the pseudosubstrate site[7] and resides immediately amino-terminal of the cysteine-rich motif (see below). At least for the c, n and a subgroups of PKC, the pseudosubstrate site is readily recognized as a basic region surrounding an alanine residue. Peptides based upon these sequences but with serine substituted for alanine are good (but not exclusive) substrates for these PKC isotypes[8] (which are serine/threonine kinases). That these sequences indeed confer inhibition is clearly evidenced by mutagenesis; even the single substitution of glutamic acid for alanine within these sequences serves to switch the mutant PKC to a constitutively active form.[9] Such mutants have been useful in comparing selective functions.

Inhibition of the catalytic domain by the pseudosubstrate site is relieved when DAG binds to the proximal cysteine-rich (C1) domain. This domain of PKC is characterized by a C_6H_2 motif that binds two Zn^{2+} ions[10,11] and in those PKC isotypes responsive to DAG (or phorbol esters) is repeated in a tandem fashion (see Figure 16.1). Mutagenesis and deletion analysis indicate that one of these two repeats is sufficient for phorbol ester binding. Deletion analysis has further refined the core binding domain to a linear sequence of 53 amino acids (based upon PKC-γ) that binds a single Zn^{2+} ion with high affinity. Interestingly, a recent NMR study on a core C_6H_2 binding motif from PKC-β has demonstrated that the two Zn^{2+} ions present are not complexed by sequential His and Cys residues.[13] The determined structure has, as yet, not defined the specific DAG/phorbol ester binding site.

It is not evident why the single C_6H_2 motif in PKC-ζ is insufficient for phorbol ester binding while single motifs based upon other PKC isotypes are fully functional (e.g. ref. 14). However, the highly conserved core residues suggest conserved topology for this ζ domain and the presumption that it binds a related lipid molecule.

The C1 motif is also found in otherwise unrelated proteins including DAG kinase (which does not bind phorbol esters; D. Schaap, personal communication), unc-13 (which does bind phorbol esters)[15] and N-chimaerin (which shows pharmacological properties essentially indistinguishable from PKC).[16] Whether these proteins like PKC contribute to acute cell signaling downstream of DAG accumulation is unclear.

While the binding of DAG to PKC C1 domains leads to activation, this is not a simple on/off switch but involves a shift in the equilibrium between the pseudosubstrate bound and unbound states of the catalytic domain.[17] This is evidenced by the distinct kinetic properties of protease-activated and DAG/lipid-activated PKC. The possibility that secondary constraints act to 'lock' the protein in a fully active conformation remains an open issue. For example, interactions between the amino-terminal VO domain and other proteins(s) may contribute to the activation of the n isotypes, for which the disparity in kinetics between the effector and protease-activated states is greatest. The

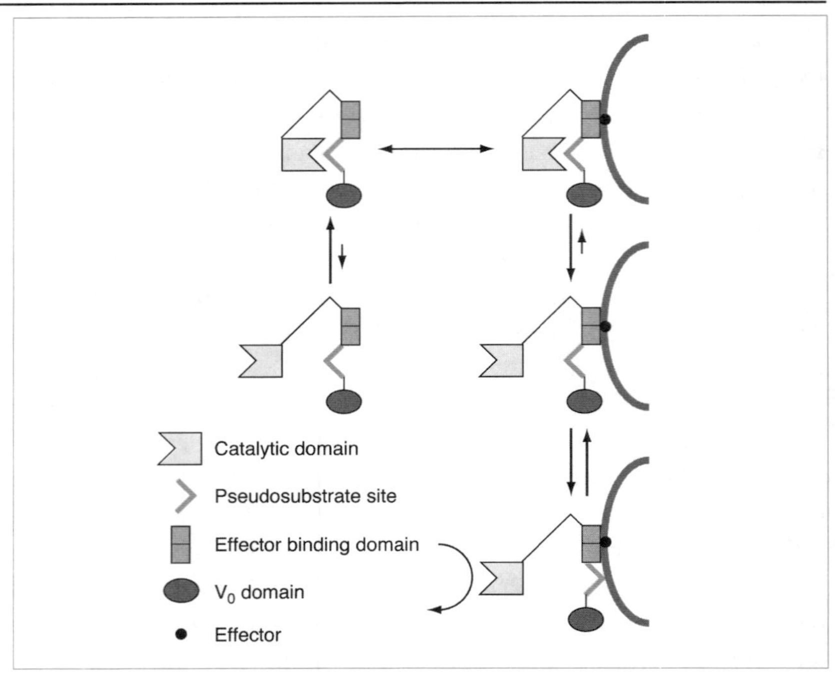

Figure 16.2 Effector activation of PKC. A generic PKC isotype is shown on the lower left-hand side, in its effector-free state, where the equilibrium between pseudosubstrate bound and pseudosubstrate free strongly favors the bound inactive state. On effector binding at the membrane, this equilibrium is shifted towards the unbound state. Sequestration of the pseudosubstrate site, for example through phospholipid interaction as shown, shifts the equilibrium further towards the unbound active state. The domains are as keyed in Figure 16.1.

extended variable regions (VO) in the n and aPKC isotypes are expected to confer particular properties on these proteins. That these sequences may indeed do so is most clearly evidenced by the finding that the PKC-ε and -η VO domains are related to the HR2 domains of the PRK gene family.[18] However, with respect to activation it may transpire that direct interactions between the phospholipid bilayer and the positively charged pseudosubstrate site[19] may also contribute to shift the equilibrium towards a fully 'deinhibited' state (Figure 16.2).

In addition to DAG, there is evidence that phosphorylated inositol phospholipids will also activate PKC isotypes.[20–22] This would provide an important downstream target for the phosphoinositide 3-kinase (PI3K; see Chapter 13) which is responsible for the generation of phosphoinositol 3,4,5 trisphosphate [PtdIns(3,4,5)P_3] (reviewed in refs 23 and 24). However, while an attractive hypothesis, no *in vivo* evidence for a link is established and the *in vitro* data leave little evidence of specificity for PKC since another lipid-activated kinase can also be stimulated by PtdIns(3,4,5)P_3 *in vitro*.[25]

Catalytic potential

While expression of PKCα (or γ) in *Escherichia coli* can produce a soluble ligand binding protein, no kinase activity is obtained (e.g. ref. 26). This and other considerations have led to the view that phosphorylation of PKCα is needed to prime the normal latent kinase activity.[27] This does not appear to be an autophosphorylation event since phosphatase treatment of purified PKCα leads to inactivation that is not reversible *in vitro*.[27] The critical site of phosphorylation has been mapped to Thr497 in PKCα[28] and to an homologous site in PKCβ.[29] This site is in the 'hot lip' or activation loop in the predicted PKC kinase domain structure and is predicted to be located at precisely that site occupied by Thr197 in the cAMP-dependent protein kinase (cAK).[29,30] Equivalent activation loops also bear responsibility for activation of cdc2, MAP kinase and Mek.[31] It is likely that the threonine residues homologous to that in PKCα and present in all PKC isotypes, is also phosphorylated and necessary for activity (Figure 16.3).

The phosphorylation in PKCα described appears to be sufficient for activity, since a proximal substitution of a threonine residue for a glutamic acid residue confers some activity on the *E. coli* expressed mutant,[28] although at only a fraction of the activity of wild-type PKC in eukaryotic cells. The kinase that is responsible for the normal phosphorylation of this threonine has not been identified although there is clear evidence for a common component acting upon both PKCα and PKCβ.[28]

In their activated states, most PKC isotypes display a similar broad specificity, phosphorylating a variety of proteins and peptides. Consistent with this, the primary structural requirements of individual isotypes are not stringent and sites with basic residues amino-terminal to or carboxy-terminal to the target serine or threonine residue can act as

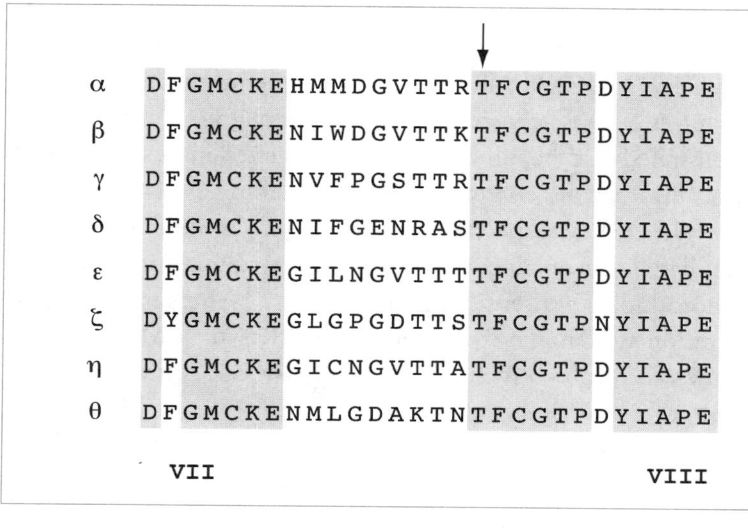

Figure 16.3 Sequence comparison of the PKC activation loops. The sequences between kinase subdomains VII and VIII are shown. The conserved threonine residue that plays an essential role in PKCα regulation is indicated by an arrow; other wholly conserved residues are also indicated.

effective substrates.[32] While conditions for the activation of PKC-ζ have not been clarified, it is evident from the generation of a regulated chimeric δ/ζ construct that the ζ catalytic domain has a similar broad specificity and preference for basic residues surrounding the phosphorylation site.[33] In line with the divergent catalytic domain of PKC-μ this isotype displays a distinctive specificity *in vitro*.[34,35]

Biological potential

While protein isotypes *in vitro* show significant promiscuity and overlapping functions, the differential patterns of expression and growing evidence for distinct subcellular location indicate specialized functions *in vivo*.

Much of what has been ascribed to PKC action *in vivo* has derived from the ability of the membrane permeant phorbol esters (and short-chain diacylglycerols) to activate the cPKC and nPKC isotypes. However, it is now well established that these agents can interact with other proteins (such as N-chimaerin and unc-13 as noted above). Furthermore, the decompartmentalization caused by these membrane permeant agents destroys the specificity afforded by differential subcellular localization. Thus conclusions of cause and effect must be tempered.

The most direct assessment of biological potential has come from the (over) expression and deletion of specific isotypes. For example, PKC-γ knock-out mice appear healthy but suffer defects in long-term potentiation (PKC-γ is normally expressed only in the brain).[36,37] Functional deletion has also been achieved through the use of antisense oligonucleotides, for example providing evidence for specific roles in intercellular adhesion molecule (ICAM) induction by phorbol esters in A549 cells[38] and phospholipase D (PLD) activation in Madin–Darby canine kidney (MDCK) cells.[39] In the A549 study, suppression of PKC-α expression was achieved with antisense oligonucleotides and it was demonstrated that the inhibition of phorbol ester-induced ICAM-1 mRNA occurred in the context of normal levels of PKC-δ,-ε and -ζ protein, i.e. the effect was specific for this isotype.[38] Inhibition of PKC-α expression has also been achieved *in vivo* with systemic administration of phosphorothioate antisense oligonucleotides.[40] The suppression of PLD activation was shown to be specific to PKC-α in MDCK cells. Using antisense constructs to either PKC-α or PKC-β, only the former blocked phorbol ester-induced PLD activation.[39]

The (selective) down-regulation (i.e. loss) of PKC isotypes caused by chronic exposure to phorbol esters has also been employed to implicate particular PKC isotypes in biological responses. However, the use of this method is limited by the complex nature of cellular responses to phorbol esters.

The availability of purified PKC isotypes has permitted their study in permeabilized cells to assess selectivity of function. In particular, this approach has been used to investigate the regulation of receptor-dependent secretory responses in rodent basophilic leukemia cells.[41–43] Specifically, it has been shown that PKC-β or -δ will restore exocytosis to permeabilized RBL cells while PKC-α and -ε appear to inhibit phosphoinositide hydrolysis.[42] Similar studies coupled to transfection experiments also provide evidence for an inhibitory role of PKC-γ.[43]

The ectopic expression of PKC isotypes has been most effective in demonstrating their differential effects. Thus the overexpression of PKC-α, -β, -δ and -ε in fibroblasts has been shown to induce distinct phenotypes ranging from growth arrest (δ in NIH3T3 cells) to transformation (ε in NIH3T3 cells).[44] Similar distinctions have been made for overexpression of PKC-γ and PKC-δ in a human glioblastoma cell line.[45] Here, PKC-γ is found to confer an increased growth rate in addition to increased colony-forming efficiency in soft agar. By contrast, glioblastoma cells (U251MG) over-expressing PKC-δ displayed reduced growth rate and colony-forming capacity.[45] Selective roles for PKC-α and -δ (compared with PKC-β_{II}, -ε, -ζ and -η) have been noted on their introduction into the 32D myeloid cell. Thus, both PKC-α and PKC-δ can participate in the differentiation of these cells.[46] These distinct biological effects are presumably due to differential protein phosphorylation. One direct assessment of this issue has come from studies on the insulin receptor, where it has been shown that overexpression of any one of three cPKCs (i.e. α, β, γ) but not nPKC-ε will lead to an increase in insulin receptor phosphorylation.[47] A similar array of responses has been observed for promoter–reporter activation on transient overexpression of PKC isotypes, in particular point mutated (pseudosubstrate site) constitutively active forms.[9,48,49] Such studies have demonstrated differential sensitivity of TRE, NFAT, NFκB promoter elements to PKC isotypes in T cells[49] and similar sensitivities of TRE and ANF promoter elements in cardiomyocytes.[50] It has been reported that PKC-ζ is involved in a pathway leading to the phosphorylation of IκB and hence to NFκB activation,[51] although the inability of a constitutively active form of PKC-ζ to stimulate NFκB *in vivo* questions this link.[49]

Summary

The studies described above illustrate both the potential for PKC isotype specificity and also the cell-specific nature of this specificity. Ultimately, however, the problem of 'actual' as opposed to 'potential' function will need to be tackled. Introduction of proteins into permeabilized cells and the ectopic expression of specific PKC isotypes in cells may well break down the normal compartmentalization, in a sense much as phorbol ester treatment does. As for any class of lipid-activated protein kinases, a major objective will be to determine where individual PKC isotypes are in relation to the substrates they may control.

Progress in defining the subcellular localization of PKC has come with the identification of binding proteins.[52–55] Many of these are also substrates for PKC[56] and their potential importance is underscored by their altered expression on transformation.[55] However, little specificity has been demonstrated for these proteins and, in respect of the selective localization of individual PKC isotypes, this issue is likely to be the focus of future work.

References

1 Hokin, L.E. (1985) Receptors and phosphoinositide-generated second messengers. *Annu. Rev. Biochem.*, **54**, 205–35.

2 Kaibuchi, K., Takai, Y., Sawamura, M. *et al.* (1983) Synergistic functions of protein phosphorylation and calcium mobilisation in platelet activation. *J. Biol. Chem.*, **258**, 6701–4.

3 Liscovitch, M. (1992) Crosstalk among multiple signal-activated phospholipases. *Trends Biochem. Sci.*, **17**, 393–9.

4 Valverde, A.M., Sinnett-Smith, J., Van Lint, J. and Rozengurt, E. (1994) Molecular cloning and characterisation of protein kinase D: a target for diacylglycerol and phorbol esters with an unusual catalytic domain. *Proc. Natl. Acad. Sci. USA*, **91**, 8572–6.

5 Johannes, F.J., Prestle, J., Eis, S. *et al.* (1994) PKC is a novel, atypical member of the protein-kinase-C family. *J. Biol. Chem.*, **269**, 6140–8.

6 Kishimoto, A., Mikawa, K., Hashimoto, K. *et al.* (1989) Limited proteolysis of protein kinase C subspecies by calcium-dependent neutral protease (calpain). *J. Biol. Chem.*, **264**, 4088–92.

7 House, C. and Kemp, B.E. (1987) Protein kinase C contains a pseudosubstrate prototope in its regulatory domain. *Science*, **238**, 1726–8.

8 Schaap, D., Parker, P.J., Bristol, A. *et al.* (1989) Unique substrate specificity and regulatory properties of PKC-ε: a rationale for diversity. *FEBS Lett.*, **243**, 351–7.

9 Pears, C.J., Kour, G., House, C. *et al.* (1990) Mutagenesis of the pseudosubstrate site of protein kinase C leads to activation. *Eur. J. Biochem.*, **194**, 89–94.

10 Hubbard, S.R., Bishop, W.R., Kirschmeier, P. *et al.* (1991) Identification and characterization of zinc binding sites in protein kinase C. *Science*, **254**, 1776–9.

11 Quest, A.F.G., Bloomenthal, J., Bardes, E.S.G. and Bell, R.M. (1992) The regulatory domain of protein kinase C coordinates four atoms of zinc. *J. Biol. Chem.*, **267**, 10193–7.

12 Sharkey, N. and Blumberg, P. (1985) Kinetic evidence that 1,2-diolein inhibits phorbol ester binding to protein kinase-C via a competitive mechanism. *Biochem. Biophys. Res. Commun.*, **133**, 1051–4.

13 Hommel, U., Zurini, M. and Luyten, M. (1994) Solution structure of a cysteine rich domain of rat protein kinase C. *Struct. Biol.*, **1**, 383–7.

14 Quest, A.F.G., Bardes, E.S.G. and Bell, R.M. (1994) A phorbol ester binding domain of protein kinase Cγ. *J. Biol. Chem.*, **269**, 2961–70.

15 Maruyama, I. and Brenner, S. (1991) A phorbol ester diacylglycerol-binding protein encoded by the unc-13 gene of caenorhabditis-elegans. *Proc. Natl. Acad. Sci. USA*, **88**, 5729–33.

16 Areces, L.B., Kazanietz, G. and Blumberg, P.M. (1994) Close similarity of baculovirus-expressed *amino* chimaerin and protein kinase Cα as phorbol ester receptors. *J. Biol. Chem.*, **269**, 19553–8.

17 Dekker, L.V., McIntyre, P. and Parker, P.J. (1993) Mutagenesis of the regulatory domain of rat protein C-η a molecular basis for restricted histone kinase activity. *J. Biol. Chem.*, **268**, 19498–504.

18 Palmer, R.H., Ridden, J. and Parker, P.J. (1995) Cloning and expression patterns of two members of a novel protein-kinase-C-related kinase family. *Eur. J. Biochem.*, **227**, 344–51.

19 Moisor, M. and McLaughlin, S. (1991) Peptides that mimic the pseudosubstrate region of protein-kinase-C bind to acidic lipids in membranes. *Biophys. J.*, **60**, 149–59.

20 Nakanishi, H., Brewer, K.A. and Exton, J.H. (1993) Activation of the ζ isozyme of protein kinase C by phosphatidylinositol 3,4,5-trisphosphate. *J. Biol. Chem.*, **266**, 13–6.

21 Singh, S.S., Chauhan, A., Brockerhoff, H. and Chauhan, V.P.S. (1993) Activation of protein kinase C by phosphatidylinositol 3,4,5,-trisphosphate. *Biochem. Biophys. Res. Commun.*, **195**, 104–12.

22 Toker, A., Meyer, M., Reddy, K.K. *et al.* (1994) Activation of protein-kinase-C family members by the novel polyphosphoinositides ptdins-3,4-p-2 and ptdins-3,4,5-p-3. *J. Biol. Chem.*, **269**, 32358–67.

23 Carpenter, C.L. and Cantley, L.C. (1990) Phosphoinositide kinases. *Biochemistry*, **29**, 11147–56.

24 Parker, P.J. and Waterfield, M.D. (1992) Phosphatidylinositol 3-kinase: a novel effector. *Cell Growth Diff.*, **3**, 747–52.

25 Palmer, R.H., Dekker, L.V., Woscholski, R. *et al.* (1995) Activation of PRK1 by

phosphatidylinositol 4,5-bisphosphate and phosophatidylinositol 3,4,5-trisphosphate. A comparison with protein kinase C isotypes. *J. Biol. Chem.*, **270**, 22412–16.

26 Cazaubon, S., Webster, C., Camoin, L. *et al.* (1990) Effector dependent conformational changes in protein kinase Cγ through epitope mapping with inhibitory monoclonal antibodies. *Eur. J. Biochem.*, **194**, 799–804.

27 Pears, C., Stabel, S., Cazaubon, S. and Parker, P.J. (1992) Studies on the phosphorylation of protein kinase C-α. *Biochem. J.*, **283**, 515–18.

28 Cazaubon, S., Bornancin, F. and Parker, P.J. (1994) Threonine-497 is a critical site for permissive activation of protein kinase Cα. *Biochem. J.*, **301**, 443–8.

29 Orr, J.W. and Newton, A.C. (1994) Requirement for negative charge on 'activation loop' of protein kinase C. *J Biol. Chem.*, **269**, 27715–8.

30 Srinivasan, N., Bax, B., Blundell, T.L. and Parker, P.J. (1996) Structural aspects of the functional modules in human protein kinase Cα deduced from comparative analyses. *Proteins* (in press).

31 Taylor, S.S. and Radzio-Andzelm, E. (1994) Three protein kinase structures define a common motif. *Structure*, **2**, 345–55.

32 Marais, R.M., Nguyen, O., Woodgett, J.R. and Parker, P.J. (1990) Studies on the primary sequence requirements for PKC-α, β_1 and γ peptide substrates. *FEBS Lett.*, **277**, 151–5.

33 Goode, N.T. and Parker, P.J. (1994) A phorbol ester-responsive PKC-zeta generated by fusion with the regulatory domain of PKC-delta. *FEBS Lett.*, **340**, 145–50.

34 Van Lint, J., Sinnett-Smith, J. and Rozengurt, E. (1995) Expression and characterization of PKD, a phorbol ester and diacylglycerol-stimulated serine protein kinase. *J. Biol. Chem.*, **270**, 1455–61.

35 Johannes, F.-J., Prestle, J., Dieterich, S. *et al.* (1995) Characterization of activators and inhibitors of protein kinase Cμ. *Eur. J. Biochem.*, **227**, 303–7.

36 Abeliovich, A., Chen, C., Goda, Y. *et al.* (1993) Modified hippocampal long-term potentiation in PKCγ-mutant mice. *Cell*, **75**, 1253–62.

37 Abeliovich, A., Paylor, R., Chen, C. *et al.* (1993) PKCγ mutant mice exhibit mild deficits in spatial and contextual learning. *Cell*, **75**, 1263–71.

38 Dean, N.M., McKay, R., Condon, T.P. and Bennett, C.F. (1994) Inhibition of protein-kinase C-alpha expression in human a549 cells by antisense oligonucleotides inhibits induction of intercellular-adhesion molecule-1 (icam-1) messenger-RNA by phorbol esters. *J. Biol. Chem.*, **269**, 16416–24.

39 Balboa, M.A., Firestein, B.L., Godson, C. *et al.* (1994) Protein-kinase C-alpha mediates phospholipase-d activation by nucleotides and phorbol ester in Madin–Darby canine kidney-cells – stimulation of phospholipase-d is independent of activation of polyphosphoinositide-specific phospholipase-c and phospholipase a(2). *J. Biol. Chem.*, **269**, 10511–6.

40 Dean, N.M. and McKay, R. (1994) Inhibition of protein-kinase C-alpha expression in mice after systemic administration of phosphorothioate antisense oligodeoxynucleotides. *Proc. Natl. Acad. Sci. USA*, **91**, 11762–6.

41 Ozawa, K., Szallasi, Z., Kazanietz, M.G. *et al.* (1993) Ca^{2+}-dependent and

Ca^{2+}-independent isozymes of protein-kinase-c mediate exocytosis in antigen-stimulated rat basophilic rbl-2h3 cells – reconstitution of secretory responses with Ca^{2+} and purified isozymes in washed permeabilized cells. *J. Biol. Chem.*, **268**, 1749–56.

42 Ozawa, K., Yamada, K., Kazanietz, M.G. *et al.* (1993) Different isozymes of protein-kinase-c mediate feedback inhibition of phospholipase-C and stimulatory signals for exocytosis in rat rbl-2h3 cells. *J. Biol. Chem.*, **268**, 2280–3.

43 Baumgartner, R.A., Ozawa, K., Cunhamelo, J.R. *et al.* (1994) Studies with transfected and permeabilized rbl-2h3 cells reveal unique inhibitory properties of protein-kinase C-gamma. *Mol. Biol. Cell*, **5**, 475–84.

44 Mischak, H., Goodnight, J., Kolch, W. *et al.* (1993) Overexpression of protein kinase C-δ and -ε in NIH 3T3 cells induces opposite effects on growth, morphology, anchorage dependence, and tumorigenicity. *J. Biol. Chem.*, **268**, 6090–6.

45 Mishima, K., Ohno, S., Shitara, N. *et al.* (1994) Opposite effects of the overexpression of protein-kinase C-gamma and c-delta on the growth-properties of human glioma cell-line u251 mg. *Biochem. Biophys. Res. Commun.*, **201**, 363–72.

46 Mischak, H., Pierce, J.H., Goodnight, J. *et al.* (1993) Phorbol ester-induced myeloid differentiation is mediated by protein kinase-C-alpha and -delta. *J. Biol. Chem.* **268**, 20110–5.

47 Chin, J.E., Dickens, M., Tavare, J.M. and Roth, R.A. (1993) Overexpression of protein-kinase-c isoenzymes-alpha, beta-i, gamma, and epsilon in cells overexpressing the insulin-receptor – effects on receptor phosphorylation and signaling. *J. Biol. Chem.*, **268**, 6338–47.

48 Wotton, D., Ways, D.K., Parker, P.J. and Owen, M.J. (1993) Activity of both *Raf* and *Ras* is necessary for activation of transcription of the human T-cell receptor β gene by protein kinase C; *Ras* plays multiple roles. *J. Biol. Chem.*, **268**, 17975–82.

49 Genot, E., Parker, P.J. and Cantrell, D.A. (1995) Analysis of the role of PKC-α, -ε and -ζ in T-cell activation. *J. Biol. Chem.*, **270**, 9833–39.

50 Decock, J.B.J., Gillespie-Brown, J., Parker, P.J. *et al.* (1994) Classical, novel and atypical isoforms of PKC stimulate ANF- and TRE/AP-1-regulated-promoter activity in ventricular cardiomyocytes. *FEBS Lett.*, **356**, 275–8.

51 Diazmeco, M.T., Dominguez, I., Sanz, L. *et al.* (1994) Zeta-PKC induces phosphorylation and inactivation of i-kappa-b-alpha in-vitro. *EMBO J.*, **13**, 2842–8.

52 Mochlyrosen, D., Khaner, H. and Lopez, J. (1991) Identification of intracellular receptor proteins for activated protein-kinase-C. *Proc. Natl. Acad. Sci. USA*, **88**, 3997–4000.

53 Ron, D., Chen, C.H., Caldwell, J. *et al.* (1994) Cloning of an intracellular receptor for protein-kinase-C – a homolog of the beta-subunit of G-proteins. *Proc. Natl. Acad. Sci. USA*, **91**, 839–43.

54 Chapline, C., Ramsay, K., Klauck, T. and Jaken, S. (1993) Interaction cloning of protein-kinase-C substrates. *J. Biol. Chem.*, **268**, 6858–61.

55 Hyatt, S.L., Liao, L., Chapline, C. and Jaken, S. (1994) Identification and

characterization of alpha-protein kinase-C binding-proteins in normal and transformed ref52 cells. *Biochemistry*, **33**, 1223–8.

56 Hyatt, S.L., Liao, L., Aderem, A. *et al.* (1994) Correlation between protein-kinase-C binding-proteins and substrates in ref52 cells. *Cell Growth Diff.*, **5**, 495–502.

17 Protein tyrosine phosphatases

N.K. Tonks

Since the initial observation of tyrosine phosphorylation, the family of protein tyrosine kinases (PTKs) has expanded rapidly to encompass approximately 100 members, about half of which are receptor-like in their configuration. Stimulation of PTKs controls a diverse array of fundamental cellular responses including growth and proliferation, migration, differentiation and metabolism as well as cell cycle and cytoskeletal function. However, it is important to realize that *in vivo* protein tyrosine phosphorylation is a reversible, dynamic process in which the net level of phosphate in a target substrate reflects not only the activity of the PTKs that phosphorylate it, but also the competing action of the protein tyrosine phosphatases (PTPs) that catalyze the dephosphorylation reaction. This chapter focuses on a discussion of recent advances in our understanding of the structure, regulation and function of members of the PTP family.

The PTP family

Since the early studies of tyrosine phosphorylation, it has been clear that PTPs must exist, but their number and properties were, until recently, poorly defined. PTP1B was the first of these enzymes to be isolated in homogeneous form.[1,2] Its primary sequence illustrated two important points.[3] First, PTP1B did not show significant overall sequence similarity to the serine/threonine, acid or alkaline phosphatases, indicating that unlike the protein kinases, which are all derived from a common ancestor, the phosphatases have evolved in separate families. However, a high degree of similarity was noted between PTP1B and each of the two tandemly repeated intracellular domains of the leukocyte common antigen, CD45. CD45 is a transmembrane protein with a highly glycosylated, cysteine-rich extracellular segment that displays hallmarks of a ligand-binding motif. Thus CD45 may be regarded as a prototype for receptor-linked PTPs with the capacity to play a direct role in modulating cellular signaling responses. This concept was further

Signal Transduction. Edited by Carl-Henrik Heldin and Mary Purton. Published in 1996 by Chapman & Hall. ISBN 0 412 70810 8

strengthened by the demonstration of intrinsic PTP activity in CD45.[4]

These observations have been followed by an explosion of interest in the PTP family, examples of which have now been identified in sources as diverse as mammals, *Drosophila*, *Caenorhabditis elegans*, *Dictyostelium*, yeast, viruses and even prokaryotes. In fact, it has recently been shown that an essential virulence determinant of many strains of the pathogenic bacterium *Yersinia*, the causative agent of plague, is a PTP termed Yop. Apparently, following infection, the Yop PTP dephosphorylates tyrosine residues in proteins of the eukaryotic host cell, thus abrogating normal cell function.[5] Some 50 PTPs have been identified to date and current predictions suggest that there will be ~500 in humans. Each PTP contains at least one conserved catalytic domain of ~240 residues, which typically displays 30–40% identity between individual enzymes. Within this domain is the signature motif [I/V]HCXAGXXR[S/T]G which uniquely defines the PTP family of enzymes. The cysteine residue in this motif is invariant and is involved in forming a thiophosphate intermediate as part of the catalytic mechanism.[6] Individual PTPs can be distinguished on the basis of the noncatalytic segments fused to either end of the catalytic domain. As will be described, these segments are frequently of regulatory significance. Most strikingly, the PTPs, like the PTKs, can be subdivided into receptor-like and nontransmembrane, cytoplasmic forms. In addition, a subfamily of dual-specificity phosphatases which also contain this signature motif, has been described recently (Figure 17.1).

The receptor-like PTPs

The observation of receptor-like PTPs (RPTPs) is important because it predicts that signaling events may be initiated at the plasma membrane through ligand-modulated **dephosphorylation** of tyrosine residues in proteins. Generally, RPTPs consist of an intracellular segment containing one or two phosphatase domains, a single transmembrane domain and a variable extracellular segment. Why most RPTPs have two phosphatase domains is unclear.[7] There is some evidence to suggest that interactions between the two domains may determine overall activity. Nevertheless, in some cases (e.g. PTPα) both domains I and II are active independently and display distinct substrate specificities. There may also be differential regulation of each domain; for example, in CD45 it has been suggested that domain II requires phosphorylation for activity.[8] However, whether or not each domain is active in all RPTPs remains

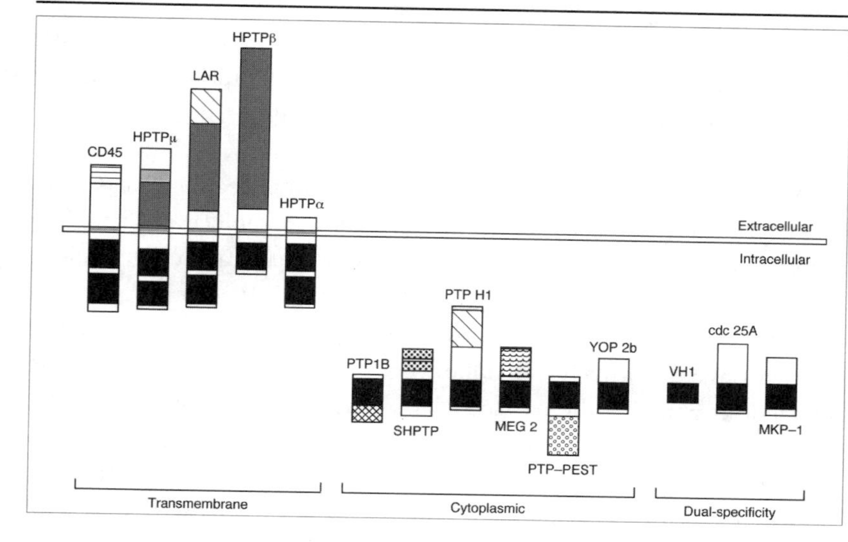

Figure 17.1 Structural organization of representatives of the PTP family of enzymes. The PTPs can be classified as (i) transmembrane, receptor-like, (ii) nontransmembrane, cytoplasmic or (iii) dual-specificity enzymes. The conserved catalytic domains are shown in black. The receptor-like enzymes are distinguished by the structure of their extracellular segments. Multiple isoforms of CD45 arise from a single gene through differential splicing of three exons encoding sequences at the extreme amino terminus (horizontal lines). The extracellular segments of LAR and PTPμ contain immunoglobulin (diagonal lines) and fibronectin type III (FNIII) motifs (stippled pattern), whereas PTPβ possesses arrays of multiple FNIII motifs alone. The cytoplasmic PTPs are characterized by their noncatalytic, potentially regulatory segments. Examples are illustrated by various shaded sections and include ER-targeting motifs in PTP1B, two SH2 domains in the SHPTPs, a segment of homology to the band 4.1 superfamily in PTPH1, a domain with homology to lipid-binding proteins in MEG2 and a segment that is rich in the amino acids Pro, Glu/Asp and Ser/Thr in PTP-PEST.

a point of controversy. Interestingly, if one constructs a dendrogram to analyze structural relationships among PTP sequences, for the most part the nontransmembrane PTPs and domains I and II of RPTPs form three different clusters, i.e. domain I from one RPTP shares higher sequence similarity with domain I from another RPTP than with domain II of the same gene product.[9] Thus, whereas domain II may serve a related function from enzyme to enzyme, this may be distinct from the function of domain I. The identification of physiological substrates for these enzymes should resolve some of these issues.

RPTPs may be distinguished on the basis of their extracellular segments. The diversity of these segments presumably reflects an equivalent diversity in ligands to which they may respond. The search for such ligands is one area in which further progress needs to be made. Although it was originally suggested that CD22, a B-cell surface protein, bound specifically to the smallest, 180 kDa form of CD45, it now appears that CD22 is a lectin that binds generally to sialoglycoproteins and its significance to cell signaling remains unclear. More recently, an alternatively spliced form of the extracellular segment of PTPζ has been identified as phosphacan, a chondroitin sulfate proteoglycan that binds to the extracellular matrix (ECM) protein tenascin.[10] The adhesive interactions that occur at focal contacts between integrins and ECM proteins such as fibronectin have been shown to initiate tyrosine phosphorylation-dependent signaling events through the activation of PTKs such as focal adhesion kinase (FAK).[11] Tenascin can mediate anti-adhesive effects,[12] and so its potential interaction with PTPζ, which may promote net dephosphorylation of tyrosine residues in proteins, would provide an intriguing counterpoint to the integrin-induced activation of FAK. However, it remains to be determined whether the full-length form of PTPζ displays similar binding properties to phosphacan and what the effects of such an interaction would be on phosphatase activity.

A common structural theme in the extracellular segment of RPTPs is the presence of motifs found in cell adhesion molecules, including repeated arrays of fibronectin type III (FNIII) domains and combinations of FNIII and immunoglobulin (Ig)-like domains. Recently, one such enzyme (PTPμ) was shown to promote aggregation when expressed in cells that were normally nonadhesive.[13,14] This is due to homophilic interactions between the extracellular segments of molecules of PTPμ. Furthermore, PTPμ, as it is normally expressed on the surface of Mink lung cells, retains the capacity for such homophilic binding interactions.[13] Similar observations have been reported for PTKκ, a close homolog of PTPμ.[15] When cells expressing PTPμ were mixed with cells expressing PTPκ, they formed separate aggregates, demonstrating that the homophilic binding interaction is highly specific and indicating that a ligand for PTPμ (or PTPκ) is another molecule of the same enzyme expressed on the surface of an apposing cell. By means of such homophilic binding interactions, these PTPs may be able to sense directly cell–cell contact and, through their PTP domains, transduce this signal into an intracellular response. In addition, the expression of a novel receptor PTP, termed DEP-1 (high cell density enhanced PTP) is dramatically enhanced as cells in culture approach confluence.[16]

These observations may be particularly important to the phenomenon of contact inhibition of cell growth. As normal cells in culture approach confluence, and adjacent cells touch each other, growth is inhibited. In cancer cells such contact inhibition is disrupted. Tyrosine phosphorylation has been implicated in promoting cell growth and proliferation and so PTPs may exert a negative effect on such growth-promoting signals by triggering net dephosphorylation of tyrosine residues in membrane proteins. Engagement of the extracellular segment of an RPTP may modulate activity directly or target it to a particular junctional complex so as to trigger dephosphorylation of a defined subset of phosphotyrosyl proteins that are important for growth or adhesion. For example, recent data suggest that PTPμ is targeted to adherens junctions, which are defined points of cell–cell contact at which the actin cytoskeleton is anchored. At these junctions PTPμ associates directly with members of the cadherin superfamily of cell adhesion molecules and regulates their phosphorylation state.[17] Tyrosine phosphorylation of cadherins and catenins, which are proteins that link cadherins to the actin cytoskeleton, interferes with normal adhesive function, potentially contributing to transformation and metastasis. Thus PTPμ may provide the regulatory balance to such phosphorylation events by maintaining the cadherin–catenin adhesion complex in its functional, dephosphorylated state.

Nontransmembrane, cytoplasmic PTPs

As with the RPTPs, one of the striking features of the cytoplasmic PTPs is the structural diversity of the noncatalytic segments that are fused to individual catalytic domains. These noncatalytic segments serve a regulatory function, particularly in subcellular compartmentalization. Such regulatory motifs include, among others, (1) the carboxy-terminal hydrophobic segment of PTP1B that targets the enzyme to the cytoplasmic face of membranes of the endoplasmic reticulum,[18] (2) SH2 domains in HCP/SH-PTP1 and Syp/SH-PTP2 which direct the association of these enzymes with particular sites of tyrosine phosphorylation in signaling receptors,[19] (3) amino-terminal segments of homology to the cytoskeletal-associated proteins of the band 4.1 superfamily, consistent with a co-localization with cortical actin at interfaces between the plasma membrane and the cytoskeleton,[20] and (4) amino-terminal segments of homology to lipid-binding proteins, again consistent with targeting to membranes.[21] Thus, by restricting the subcellular distribution of these enzymes, the spectrum of substrates with which they may interact is similarly restricted. In some cases, this phenomenon may be further fine-tuned through alternative splicing. For example, the *Drosophila* enzyme DPTP61F exists in two alternatively spliced forms. In one form the extreme carboxyl terminus is hydrophobic and directs the enzyme to intracellular membranes, whereas in the other this hydrophobic stretch is missing and the enzyme is targeted to the nucleus.[22] Thus, differential splicing may represent a regulatory mechanism by which a particular PTP may be targeted to different regions of the cell, possibly in response to a signaling event, thereby exerting its effects on discrete pools of substrates.

Regulation of PTPs by phosphorylation

There is evidence to indicate that many of the PTPs are phosphoproteins *in vivo*, modified on serine, threonine and tyrosine residues, and that such phosphorylation events exert a regulatory function. For example, PTP-PEST, a cytosolic PTP that is expressed ubiquitously in mammalian tissues, is phosphorylated by protein kinase A (PKA) and protein kinase C (PKC) on two sites both *in vitro* and *in vivo*. Phosphorylation of one of these sites, Ser39, is directly inhibitory, reducing the affinity of PTP-PEST for its substrate.[23] PTP1B is also phosphorylated and its activity is modulated in a cell-cycle dependent manner.[24] In addition, following TPA-induced differentiation of HL-60 promyelocytic leukemia cells into macrophages, phosphorylation of serine residues in HCP/SH-PTP1 was enhanced and this was accompanied by translocation of the phosphatase to the membrane.[25] Such observations are not restricted to cytoplasmic PTPs. Treatment of T cells with ionomycin, which mobilizes Ca^{2+} from intracellular stores, resulted in a dephosphorylation of serine residues in CD45 coincident with a decrease in its activity.[26] These data suggest that Ser/Thr kinases are capable of phosphorylating and modifying the activity of PTPs *in vivo*, thus offering a mechanism whereby signal transduction pathways acting through Ser/Thr kinases may influence cellular processes involving reversible tyrosine phosphorylation. This illustrates another point of 'crosstalk' between Ser/Thr and Tyr phosphorylation events in signal transduction *in vivo*.

There have also been reports of the phosphorylation of tyrosine residues in PTPs *in vivo*. Following recruitment of the SH2 domain-containing PTP, Syp/SH-PTP2, into the platelet-derived growth factor (PDGF) receptor signaling complex, through interaction between its SH2 domains and the Tyr1009 autophosphorylation site in the receptor, the phosphatase becomes phosphorylated on Tyr542 near the carboxyl terminus.[27] The sequence surrounding this residue matches the consensus for interaction with the SH2 domain of the adaptor protein Grb2, which in turn binds Sos, a guanine nucleotide exchange factor for Ras which converts Ras from an inactive GDP bound form to an active form in which GTP is bound (see review in Chapter 10). Binding of Grb2 to Syp that is phosphorylated on Tyr542 has now been shown both *in vitro* and *in vivo*, with Sos also recovered in the complex *in vivo*.[27] Hence it is possible that Syp may also function as an adaptor molecule to link the PDGF receptor to the activation of Ras, thus facilitating the signaling response to PDGF. In addition, constitutive tyrosine phosphorylation of the receptor-like enzyme PTPα has been observed in NIH3T3 cells. Again, the site of phosphorylation (Tyr789) matches the consensus for interaction with Grb2 and association between tyrosine phosphorylated PTPα and Grb2 has been demonstrated.[28] These observations suggest that tyrosine phosphorylation of PTPs may play a role in the assembly of multiprotein complexes in the membrane. Perhaps PTPs modulate signaling responses *in vivo* not only through dephosphorylation of key tyrosine residues in proteins but also through assembling complexes.

Dual-specificity phosphatases

In addition to the classical PTPs described above, a rapidly expanding subfamily of dual-specificity phosphatases has recently been described. These enzymes, which dephosphorylate Ser/Thr in addition to Tyr residues, contain the unique signature motif that defines the PTP family but otherwise show little, if any, similarity to the bona-fide phosphotyrosine-specific enzymes. The dual-specificity phosphatases, which have been detected in pox viruses, yeast, mammalian cells and the cyanobacterium *Nostoc commune*, tend to display a restricted substrate specificity. In at least two cases, they have been shown to catalyze dephosphorylation events that are of fundamental importance to the control of cell function.

Dephosphorylation of Thr14 and Tyr15 in cdc2 is the key step in activating this protein kinase and driving the transition from the G_2 to M phase of the eukaryotic cell cycle. Both of these dephosphorylation events are catalyzed by a dual-specificity phosphatase, cdc25, and this mechanism of initiation of mitosis is highly conserved across all eukaryotic species.[29] Interestingly there is a positive feedback loop in which phosphorylation of cdc25 by cdc2 stimulates the phosphatase and serves to amplify rapidly the initial increase in cdc2 activity. In addition, a link has been suggested recently between growth factor signaling pathways and activation of cdc25; at least one of the three forms of human cdc25, cdc25A, associates with Raf and phosphorylation of cdc25A by Raf *in vitro* results in an increase in phosphatase activity.[30] However, whether cdc25A is phosphorylated by Raf *in vivo* and the identity of the substrates for cdc25A at the G_0/G_1 phase of the cell cycle remain to be established.

Dual-specificity phosphatases have also been shown to function directly in the control of signaling pathways initiated by mitogenic stimuli. Growth factor stimulation of receptor-PTKs sets in train a series of signaling events (see Figure 17.2) that lead to the activation of MAP kinase through phosphorylation of two regulatory sites, Thr183 and Tyr185. The MAP kinase family of enzymes has been implicated as common and essential components of signaling pathways induced by diverse mitogenic stimuli. Once activated, MAP kinases can phosphorylate a number of substrates, including transcription factors, that are essential for triggering the expression of genes that are required for the mitogenic response (see Chapter 11). Therefore, growth factor binding initiates a complex network of protein phosphorylation events that lead a quiescent cell to enter the cell cycle, undergo DNA replication and ultimately divide. One of these immediate early genes that is activated rapidly and transiently in quiescent fibroblasts treated with serum growth factors is 3CH134, which encodes a dual-specificity (Tyr/Thr) phosphatase termed MKP-1. MKP-1 possesses intrinsic phosphatase activity that is highly specific for both the Tyr and Thr regulatory sites in MAP kinases. It thus has the potential to feed back on the growth factor signaling pathway to prevent uncontrolled growth and proliferation (Figure 17.2). MKP-1 is also induced by a wide variety of stimuli, including agents that elevate cAMP, raising the possibility that it may play a role in crosstalk between different signaling pathways as well as in feedback control. Several other dual-specificity phosphatases have been identified that have the potential to dephosphorylate MAP kinases but which differ from MKP-1 in both their tissue distribution and time course of expression.[31–35] This apparent multiplicity of dual-specificity phosphatases raises the possibility that they may each be responsible for the dephosphorylation of distinct MAP kinase isoforms specific for different cell types and signaling pathways. However, there are also situations in which inactivation of MAP kinases occurs too rapidly to invoke the involvement of an inducible dual-specificity phosphatase. Under such conditions, it has been proposed that the inactivation is catalyzed by the co-ordinated action of two single-specificity enzymes.[36]

PTPs as both positive and negative regulators of signal transduction

PTPs have the potential to exert a negative influence on PTK-dependent signaling pathways (Figure 17.3). One of the best characterized examples of a growth suppressive function for a cytoplasmic PTP involves the SH2 domain-containing enzyme HCP/SH-PTP1, which is expressed exclusively in hematopoietic cells. Mutations in the gene for HCP, which result in aberrant splicing of the transcript, have been shown to be the cause of the *motheaten* phenotype in mice (see review[19]). In one mutation, a single base deletion in the HCP gene results in the

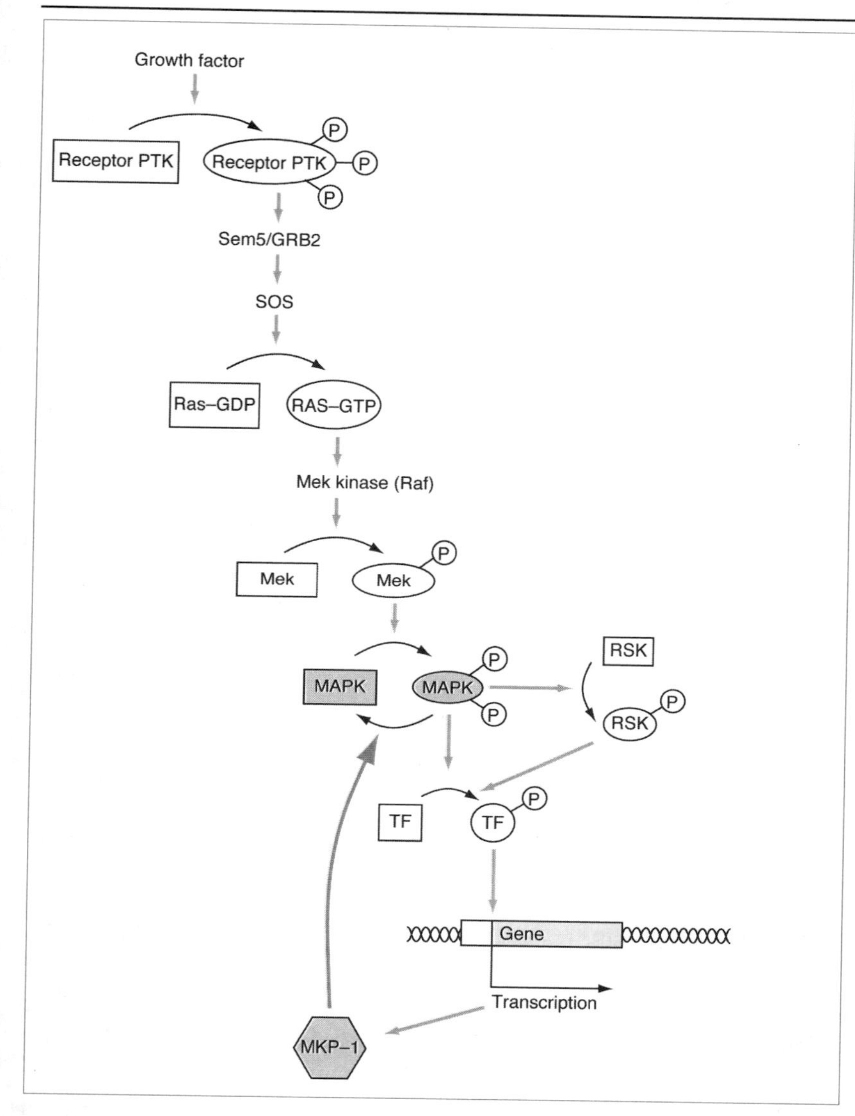

Figure 17.2 A signaling pathway triggered by growth factor receptor PTKs. Following activation of a growth factor receptor PTK by binding to its cognate ligand, autophosphorylation of the receptor creates docking sites that recruit the adaptor protein Grb2, leading to activation of Sos, the GDP–GTP exchanger for Ras. Then a cascade of phosphorylation is triggered, leading to activation of mitogen-activated protein kinase (MAPK). MAPK in turn phosphorylates transcription factors (TF), either directly or through activation of another Ser/Thr kinase, Rsk, thus promoting the transcription of genes required for the growth response. One of the genes induced by growth factor stimulation, presumably through the MAPK pathway, encodes a MAPK phosphatase (MKP-1). As denoted by the filled arrow, MKP-1 may feed back on the pathway by dephosphorylating and inactivating MAPK, thus attenuating the signaling response. [Reproduced with permission from Sun, H. and Tonks, N.K. (1994) *Trends Biochem. Sci.*, 19, 480–5.]

production of a severely truncated polypeptide that is completely devoid of the PTP domain and most of the SH2 sequences. Such *motheaten* mice display severe immunodeficiency and systemic autoimmune disease and generally live for only 2–3 weeks. There is a less severe form of the disease, *motheaten viable* (*me^v*), in which the animals may survive a few months. This results from a distinct mutation that generates a protein with sequences either inserted into or deleted from the catalytic domain. The activity of *me^v* mutant HCP is reduced by ~80% relative to wild type. Mice displaying the *motheaten* phenotype present a panoply of hematopoietic abnormalities in which cells proliferate at a much faster

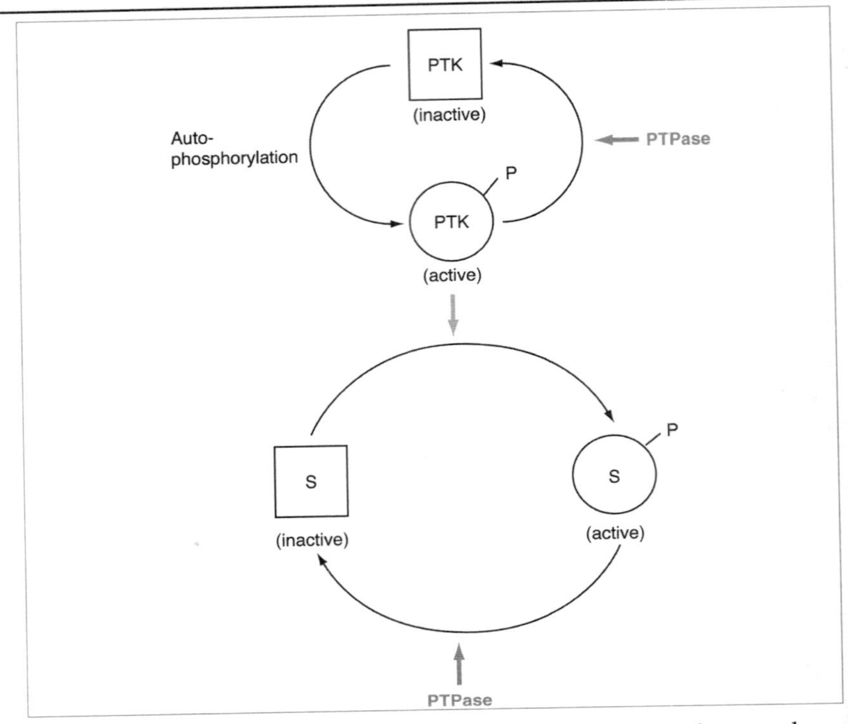

Figure 17.3 PTPs may act negatively to antagonize PTK-induced signaling events *in vivo*. PTKs catalyze the phosphorylation of tyrosyl residues in proteins and PTPs dephosphorylate these residues. Many PTKs require autophosphorylation of tyrosyl residues in the kinase itself for optimal activity. An inhibitory effect of a PTP may be exerted through dephosphorylation and inactivation of either the PTK or its target substrate(s). [Reproduced with permission from Sun, H. and Tonks, N.K. (1994) *Trends Biochem. Sci.*, 19, 480–5.]

rate than normal. Thus, HCP may be an important negative regulator of cytokine signaling in general, its loss resulting in sustained tyrosine phosphorylation with consequent enhanced proliferation. The net effect is similar to that observed with constitutively activated oncogenic PTKs. One of the best defined effects of HCP concerns its role in modulating signaling through the erythropoietin (EPO) receptor.[37] Binding of EPO to its receptor activates the associated Jak2 PTK and induces the signaling response. One of the substrates of Jak2 is Tyr429 of the EPO receptor itself, which serves as a docking site for the SH2 domains of HCP. Once recruited into the EPO receptor signaling complex, HCP can dephosphorylate phosphotyrosyl proteins in that complex, in particular Jak2, and thus terminate the signal. The importance of HCP in down-regulation of the EPO signal is seen by the fact that cells expressing EPO receptors in which Tyr429 is mutated to Phe proliferate in 10–20% of the normal amount of EPO required for growth.

This growth-suppressive effect of PTPs raises the possibility that they may be the products of tumor suppressor genes. A considerable focus of effort in many laboratories is the determination of the chromosomal localization of PTP genes with the aim of correlating map positions with sites of abnormality in human disease states. New and tantalizing results are being reported frequently. One of the more notable examples is the gene for RPTPγ, which is found on 3p21, a segment frequently altered in renal and small cell lung carcinomas. Recent studies indicate that mutations may occur in the extracellular segment of RPTPγ and involve deletions in a putative ligand-binding domain related to carbonic anhydrase, thus creating an RPTP that can no longer respond to external signals.[38] It is also interesting that deletions and translocations of human chromosome 12 p12–13, the position of the HCP gene, are found in patients with acute lymphocytic leukemia.[19]

It is now clear that the PTPs cannot simply be regarded as antagonists of PTK function. Several have recently been shown to act positively in mediating signaling responses (Figure 17.4). Perhaps the best characterized example is CD45. A major breakthrough in the determination of its function was the generation of T cells that do not express CD45 (see review[39]). These cells failed to proliferate in response to antigen or to crosslinking of CD3, but did respond normally to interleukin-2. Subsequently, a number of CD45-deficient T-cell lines were also developed that were shown to be defective in T-cell receptor (TCR) signaling responses. Perhaps most striking was the observation that in $CD45^-$ Jurkat cells there was no TCR-induced increase in cellular pTyr. These signaling defects could be rescued by restoring expression of the PTP by transfecting back cDNA for CD45. Thus, CD45 is required for coupling stimulation of the TCR to activation of these signaling pathways, indicating that it plays a positive role in mediating signal transduction in response to T-cell activation. A similar

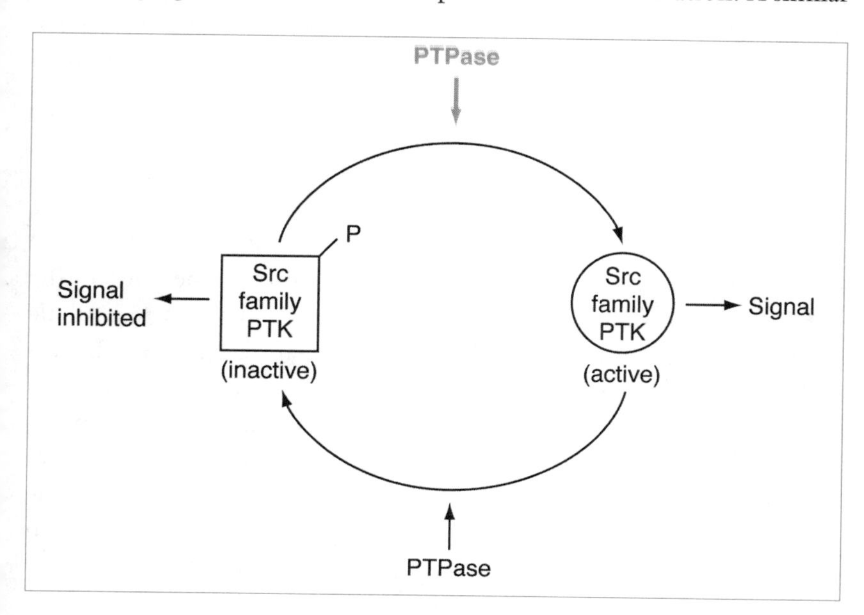

Figure 17.4 PTPs may act positively to promote signaling events *in vivo*. One mechanism by which PTPs may promote signaling responses is through the activation of members of the Src family of PTKs. Src family PTKs have an inhibitory site of phosphorylation in their carboxyl termini. Dephosphorylation of this by a PTP promotes kinase activity and triggers the signaling cascade. Such a system requires strict subcellular compartmentalization or precise specificity of the PTP to prevent it from also acting on the downstream targets of the Src family PTK. [Reproduced with permission from Sun, H. and Tonks, N.K. (1994) *Trends Biochem. Sci.*, 19, 480–5.]

situation is now known to be true for B cells. These data have been extended by gene knock-out experiments in which ablation of CD45 was shown to inhibit thymic development profoundly and to impair signal transduction through the B-cell receptor. This presents something of a paradox – ablation of the expression of a PTP results in the failure of the TCR to elicit an enhancement in the level of pTyr. The resolution of this paradox serves as a classic illustration of the interrelationship between PTK and PTP function. It was observed that the Src family kinases such as Lck (found in association with the T-cell surface accessory proteins CD4 or CD8) and Fyn (found in association with the CD3 complex that interacts with the TCR) are abnormally regulated in $CD45^-$ cells, although the relative importance of each to a particular pathway varies with cell type. Members of the Src family of PTKs possess a site of tyrosine phosphorylation at their carboxyl termini, equivalent to Tyr527 in c-Src itself, which when phosphorylated exerts an inhibitory effect on kinase activity. In $CD45^-$ cells, these kinases accumulate in a form in which the carboxy-terminal site is hyperphosphorylated and thus activity is repressed. Therefore, the signaling defect in $CD45^-$ cells appears to be a failure to activate Src family kinases efficiently. Under normal conditions, CD45 can actually promote tyrosine phosphorylation through the dephosphorylation and activation of Src family PTKs. Such a concept reinforces the importance of subcellular compartmentalization in the control of PTP activity to prevent inappropriate activation of Src family PTKs and inappropriate dephosphorylation of the downstream targets of the activated Src PTK. The results of similar studies aimed at determining physiological roles for other PTPs through ablation of expression are anticipated in the near future.

Three-dimensional structure of PTPs

In order to understand more fully the catalytic mechanism and function of PTPs, the X-ray crystallographic structures of some members of the family have been determined. PTP1B is composed of a single catalytic domain in which the polypeptide chain is organized into 8 α-helices and 12 β-strands with a 10-stranded mixed β-sheet which adopts a highly twisted conformation, spanning the entire length of the molecule.[40] The unique signature motif that defines the PTP family of enzymes forms the phosphate recognition site. In PTP1B the sequence CSAGIGR (residues 215–221) adopts a rigid cradle structure in which the main-chain nitrogen atoms and the side chain of Arg221 encircle

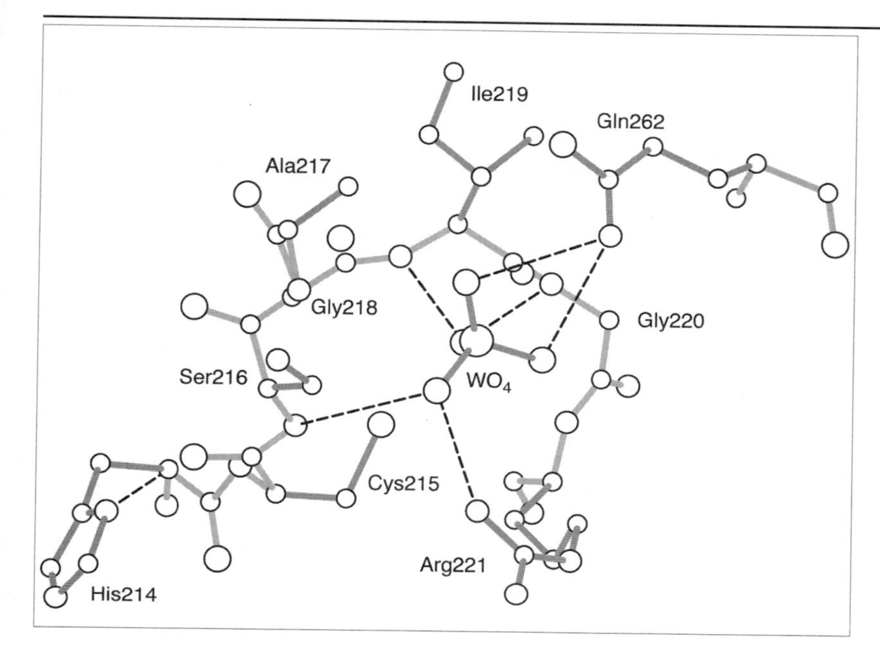

Figure 17.5 View of the catalytic site of PTP1B to illustrate binding of tungstate (phosphate). The main-chain bonds are shown as open lines and side-chain bonds as solid lines. The hydrogen bonding interactions are illustrated by the dotted lines. [Reprinted with permission from Barford, D., Flint, A.J. and Tonks, N.K. (1994) *Science*, 263, 1397–404. Copyright 1994 American Association for the Advancement of Science.]

the phosphate and hold it in place through hydrogen bonding interactions, with the Sγ atom of Cys215 ideally positioned to act as a nucleophile toward the phosphate moiety of bound substrate (Figure 17.5). Similar conclusions were drawn from the structure of the *Yersinia* PTP, Yop[41] and the low-M_r acid phosphatase/tyrosine phosphatase.[42–44]

The phosphate binding loop is located at the base of a cleft on the surface of the protein, leading to the proposal that the absolute specificity of PTP1B for phosphotyrosine-containing substrates resulted from the depth of the cleft, since the structure suggested that the smaller phosphoserine and phosphothreonine residues would not reach down to the phosphate binding site.[40] This issue has now been examined by the determination of the structure of PTP1B complexes with a high-affinity peptide substrate (DADEpYL) representing an autophosphorylation site of the epidermal growth factor receptor.[45] The analysis made use of the fact that a catalytically inactive mutant of PTP1B, in which the essential Cys from the signature motif, Cys215, was replaced by Ser, retained the ability to bind substrate even though it could not dephosphorylate it. Peptide binding to the protein was accomplished by a large conformational change in a surface loop at one end of the substrate binding cleft, which created a recognition pocket that surrounded the phosphotyrosine in the substrate. In fact, PTP1B represents an example of Koshland's concept of 'induced fit' – substrate binding induces a conformational change that creates the catalytically competent form of the enzyme. As initially postulated, the depth of the pTyr binding pocket, which at ~9 Å exactly matches the length of a pTyr residue, is a major determinant of specificity. A similar mechanism has been proposed to explain the specificity of SH2 domains for pTyr residues.[46] Another determinant of specificity for PTP1B is the presence of non-polar side chains lining the cleft that form hydrophobic interactions with the phenyl ring of the pTyr residue. Engagement of pTyr by the binding pocket anchors the peptide to the peptide binding site. Hydrogen bonds between peptide main-chain atoms and the protein contribute to binding affinity and specific interactions of acidic residues of the peptide with basic residues on the surface of the enzyme confer sequence specificity. However, the peptide binding site has a relatively open structure, consistent with the ability of PTP1B to dephosphorylate a variety of pTyr substrates.

These data are exciting because the structural information can be used as a guide to prepare mutants of PTP1B in which key residues that are important for substrate recognition have been altered. Preliminary studies suggest that some of these mutants retain a high affinity for substrate but display an exceedingly poor ability to catalyze dephosphorylation. Expression of such mutants may allow the trapping and identification of physiological substrates for PTP1B.

Acknowledgements

Work in the author's laboratory is supported by grants from NIH (CA53840 and CA64593), the Tobacco Research Council, the Mellam Family Foundation and Pew Scholars Program in the Biomedical Sciences. Carol Marcincuk is thanked for typing the manuscript.

Summary

The PTPs have been defined as a large, structurally diverse family of receptors, with the ability to transmit a signal directly across the membrane, and cytoplasmic enzymes. They have the ability to act both positively and negatively in the control of cell function. These observations further emphasize the important concept that it is the coordinated actions of both PTKs and PTPs that are responsible for the regulatory effects of reversible tyrosine phosphorylation in cell signaling. The PTPs represent crucial targets for future research to define precisely the physiological roles of tyrosine phosphorylation.

References

1 Tonks, N.K., Diltz, C.D. and Fischer, E.H. (1988) Purification of the major protein tyrosine phosphatases of human placenta. *J. Biol. Chem.*, **263**, 6722–30.

2 Tonks, N.K., Diltz, C.D. and Fischer, E.H. (1988) Characterization of the major protein tyrosine phosphatases of human placenta. *J. Biol. Chem.*, **263**, 6731–7.

3 Charbonneau, H., Tonks, N.K., Walsh, K.A. and Fischer, E.H. (1988) The leukocyte common antigen (CD45): a putative receptor-linked protein tyrosine phosphatase. *Proc. Natl. Acad. Sci. USA*, **85**, 7182–6.

4 Tonks, N.K., Charbonneau, H., Diltz, C.D. *et al.* (1988) Demonstration that the leukocyte common antigen CD45 is a protein tyrosine phosphatase. *Biochemistry*, **27**, 8695–701.

5 Guan, K.-L. and Dixon, J.E. (1993) Bacterial and viral protein tyrosine phosphatases. *Semin. Cell Biol.*, **4**, 389–96.

6 Guan, K.L. and Dixon, J.E. (1991) Evidence for protein-tyrosine-phosphatase catalysis proceeding via a cysteine-phosphate intermediate. *J. Biol. Chem.*, **266**, 17026–30.

7 Pallen, C.J. (1993) The receptor-like protein tyrosine phosphatase α: a role in cell proliferation and oncogenesis. *Semin. Cell Biol.*, **4**, 403–8.

8 Stover, D.R. and Walsh, K.A. (1994) Protein-tyrosine phosphatase activity of CD45 is activated by sequential phosphorylation by two kinases. *Mol. Cell. Biol.*, **14**, 5523–32.

9 Yang, Q. and Tonks, N.K. (1993) Structural diversity within the protein tyrosine phosphatase family. *Adv. Protein Phosphatases*, **7**, 353–66.

10 Barnea, G., Grumet, M., Milev, P. *et al.* (1994) Receptor tyrosine phosphatase β is expressed in the form of proteoglycan and binds to the extracellular matrix protein tenascin. *J. Biol. Chem.*, **269**, 14349–52.

11 Schaller, M.D. and Parsons, J.T. (1993) Focal adhesion kinase: an integrin-linked protein tyrosine kinase. *Trends Cell Biol.*, **3**, 258–62.

12 Spring, J., Beck, K. and Chiquet-Ehrismann, R. (1989) Two contrary functions of Tenascin: dissection of the active sites by recombinant Tenascin fragments. *Cell*, **59**, 325–34.

13 Brady-Kalnay, S.M., Flint, A.J. and Tonks, N.K. (1993) The receptor-type protein tyrosine phosphatase PTPμ mediates cell–cell aggregation. *J. Cell Biol.*, **122**, 961–72.

14 Gebbink, M.F.B.G., Zondag, G.C.M., Wubbolts, R.W. *et al.* (1993) Cell–cell adhesion mediated by a receptor-like protein tyrosine phosphatase. *J. Biol. Chem.*, **268**, 16101–4.

15 Sap, J., Jiang, Y.-P., Friedlander, D. *et al.* (1994) Receptor tyrosine phosphatase R-PTP-κ mediates homophilic binding. *Mol. Cell. Biol.*, **14**, 1–9.

16 Östman, A., Yang, Q. and Tonks, N.K. (1994) Expression of DEP-1, a receptor-like protein-tyrosine-phosphatase, is enhanced with increasing cell density. *Proc. Natl. Acad. Sci. USA*, **91**, 9680–4.

17 Brady-Kalnay, S.M., Rimm, D.L. and Tonks, N.K. (1995) The receptor protein tyrosine phosphatase PTPμ associates with cadherins and catenins *in vivo*. *J. Cell Biol.*, **130**, 977–86.

18 Frangioni, J.V., Beahm, P.H., Shifrin, V. *et al.* (1992) The nontransmembrane tyrosine phosphatase PTP-1B localizes to the endoplasmic reticulum via its 35 amino acid C-terminal sequence. *Cell*, **68**, 545–60.

19 Neel, B.G. (1993) Structure and function of SH2-domain containing tyrosine phosphatases. *Semin. Cell Biol.*, **4**, 419–32.

20 Yang, Q. and Tonks, N.K. (1991) Isolation of a cDNA clone encoding a human protein-tyrosine phosphatase with homology to the cytoskeletal-associated proteins band 4.1, ezrin, and talin. *Proc. Natl. Acad. Sci. USA*, **88**, 5949–53.

21 Del Vecchio, R.L. and Tonks, N.K. (1994) Characterization of two structurally related *Xenopus laevis* protein tyrosine phosphatases with homology to lipid-binding proteins. *J. Biol. Chem.*, **269**, 19639–45.

22 McLaughlin, S. and Dixon, J.E. (1993) Alternative splicing gives rise to a nuclear protein tyrosine phosphatase in *Drosophila*. *J. Biol. Chem.*, **268**, 6839–42.

23 Garton, A.J. and Tonks, N.K. (1994) PTP-PEST: a protein tyrosine phosphatase regulated by serine phosphorylation. *EMBO J.*, **13**, 3763–71.

24 Flint, A.J., Gebbink, M.F.B.G., Franza, J. *et al.* (1993) Multi-site phosphorylation of the protein tyrosine phosphatase, PTP1B: identification of cell cycle regulated and phorbol ester stimulated sites of phosphorylation. *EMBO J.*, **12**, 1937–46.

25 Zhao, Z., Shen, S.-H. and Fischer, E.H. (1994) Phorbol ester-induced expression, phosphorylation, and translocation of protein-tyrosine-phosphatase 1C in HL-60 cells. *Proc. Natl. Acad. Sci. USA*, **91**, 5007–11.

26 Ostergaard, H.L. and Trowbridge, I.S. (1991) Negative regulation of CD45 protein tyrosine phosphatase activity by ionomycin in T cells. *Science*, **253**, 1423–25.

27 Bennett, A.M., Tang, T.L., Sugimoto, S. *et al.* (1994) Protein-tyrosine-phosphatase SHPTP2 couples platelet-derived growth factor receptor β to Ras. *Proc. Natl. Acad. Sci. USA*, **91**, 7335–9.

28 den Hertog, J., Tracy, S. and Hunter, T. (1994) Phosphorylation of receptor protein-tyrosine phosphatase α on Tyr789, a binding site for the SH3–SH2–SH3 adaptor protein GRB-2 *in vivo*. *EMBO J.*, **13**, 3020–32.

29 Atherton-Fessler, S., Hannig, G. and Piwnica-Worms, H. (1993) Reversible tyrosine phosphorylation and cell cycle control. *Semin. Cell Biol.*, **4**, 433–42.

30 Galaktionov, K., Jessus, C. and Beach, D. (1995) Raf1 interaction with Cdc25 phosphatase ties mitogenic signal transduction to cell cycle activation. *Genes Dev.*, **9**, 1046–58.

31 Doi, K., Gartner, A., Ammerer, G. *et al.* (1994) MSG5, a novel protein phosphatase promotes adaptation to pheromone response in *S. cerevisiae*. *EMBO J.*, **13**, 61–70.

32 Kwak, S.P. and Dixon, J.E. (1995) Multiple dual specificity protein tyrosine phosphatases are expressed and regulated differentially in liver cell lines. *J. Biol. Chem.*, **270**, 1156–60.

33 Ward, Y., Gupta, S., Jensen, P. *et al.* (1994) Control of MAP kinase activation by the mitogen-induced threonine/tyrosine phosphatase PAC1. *Nature*, **367**, 651–4.

34 Ishibashi, T., Bottaro, D.P., Michieli, P. *et al.* (1994) A novel dual specificity phosphatase induced by serum stimulation and heat shock. *J. Biol. Chem.*, **269**, 29897–902.

35 Guan, K.-L. and Butch, E. (1995) Isolation and characterization of a novel dual

specific phosphatase, HVH2, which selectively dephosphorylates the mitogen-activated protein kinase. *J. Biol. Chem.*, **270**, 7197–203.

36 Alessi, D.R., Gomez, N., Moorhead, G. *et al.* (1995) Inactivation of p42 MAP kinase by protein phosphatase 2A and a protein tyrosine phosphatase, but not CL100, in various cell lines. *Curr. Biol.*, **5**, 283–95.

37 Klingmüller, U., Lorenz, U., Cantley, L.C. *et al.* (1995) Specific recruitment of SH-PTP1 to the Erythropoietin receptor causes inactivation of Jak2 and termination of proliferative signals. *Cell*, **80**, 729–38.

38 Wary, K.K., Lou, Z., Buchberg, A.M. *et al.* (1993) A homologous deletion within the carbonic anhydrase-like domain of the Ptprg gene in murine L-cells. *Cancer Res.*, **53**, 1498–1502.

39 Woodford-Thomas, T. and Thomas, M.L. (1993) The leukocyte common antigen, CD45 and other protein tyrosine phosphatases in hematopoietic cells. *Semin. Cell Biol.*, **4**, 409–18.

40 Barford, D., Flint, A.J. and Tonks, N.K. (1994) Crystal structure of human protein tyrosine phosphatase 1B. *Science*, **263**, 1397–404.

41 Stuckey, J.A., Schubert, H.L., Fauman, E.B. *et al.* (1994) Crystal structure of *Yersinia* protein tyrosine phosphatase at 2.5 Å and the complex with tungstate. *Nature*, **370**, 571–5.

42 Logan, T.M., Zhou, M.-M., Nettesheim, D.G. *et al.* (1994) Solution structure of a low molecular weight protein tyrosine phosphatase. *Biochemistry*, **33**, 11087–96.

43 Zhang, M., Van Etten, R.L. and Stauffacher, C.V. (1994) Crystal structure of bovine heart phosphotyrosyl phosphatase at 2.2-Å resolution. *Biochemistry*, **33**, 11097–105.

44 Su, X.-D., Taddei, N., Stefani, M. *et al.* (1994) The crystal structure of a low-molecular-weight phosphotyrosine protein phosphatase. *Nature*, **370**, 575–8.

45 Jia, Z., Barford, D., Flint, A.J. and Tonks, N.K. (1995) Structural basis for phosphotyrosine peptide recognition by protein tyrosine phosphatase 1B. *Science*, **268**, 1754–8.

46 Pawson, T. and Gish, G.D. (1992) SH2 and SH3 domains: from structure to function. *Cell*, **71**, 359–62.

18 Protein serine/threonine phosphatases

Stanislaw Zolnierowicz and Brian A. Hemmings

Many cellular processes are regulated by reversible phosphorylation through the concerted action of protein kinases and protein phosphatases. Until recently, only protein kinases were considered as important regulators of protein phosphorylation with the protein phosphatases considered as passive bystanders. However, it is now apparent that the phosphorylation state of a target protein may be actively regulated by phosphatases (Figure 18.1). Therefore, reversible phosphorylation–dephosphorylation of 'target' proteins can be viewed as a binary switch with the kinases and phosphatases pushing the 'target' protein into the 'on' or 'off' state. Obviously, to achieve fine regulation the interconverting kinases/phosphatases must receive, process and respond to incoming signals. The mechanisms whereby a phosphatase is able to respond to signals are the focus of much current research.

Moreover, information from the *Caenorhabditis elegans* genome sequencing project indicates that the number of protein phosphatases is much higher than previously anticipated. Approximately 500 genes are predicted to encode protein serine/threonine phosphatases (PSTPs) in the human genome.[1] Of this number, around 35 catalytic subunits have been identified by a number of techniques and studied in some detail. Eventually the total number of PSTP holoenzymes is likely to be higher than 500 because the catalytic subunits associate with different regulatory or targeting subunits to make divergent heteromeric proteins.

In this chapter, we describe the structure of various PSTPs and some aspects of their role in the regulation of cell function. To comprehend and appreciate fully the complexity of the regulation of cellular processes by protein phosphatases, the reader is also referred to Chapter 17.

Signal Transduction. Edited by Carl-Henrik Heldin and Mary Purton. Published in 1996 by Chapman & Hall. ISBN 0 412 70810 8

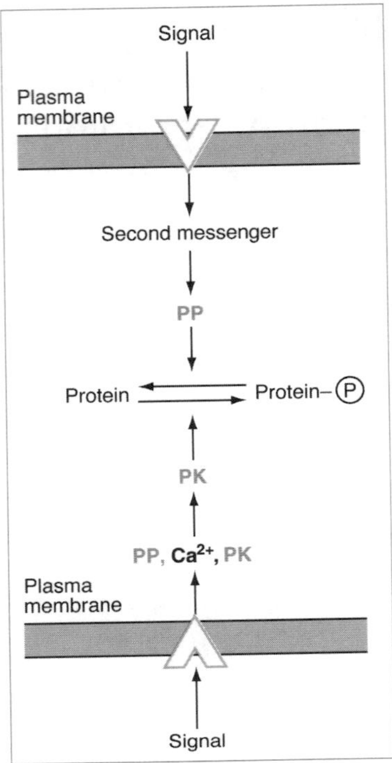

Figure 18.1 Phosphorylation state of a target protein is regulated by both protein kinases (PKs) and protein phosphatases (PPs).

Classification

Biochemical identification of PSTPs was initiated in laboratories working on the regulation of glycogen metabolism. Initially, PSTPs were divided into two groups based on their preference to dephosphorylate either β (type 1) or α (type 2) subunits of phosphorylase kinase. Further classification of type 2 protein phosphatases (PP2) into PP2A, 2B and 2C was based on their different requirements for divalent cations. PP2A does not require divalent cations to be active, whereas PP2B is Ca^{2+}-dependent and PP2C requires Mg^{2+}. A further characteristic of PP2A holoenzymes is the stimulation of activity by certain polycations such as protamine, polylysine and histone H1.

Initially, only four PSTPs, PP1 and PP2A, -2B and -2C were identified in mammalian tissue.[2–4] Subsequent sequence analysis of cDNA clones revealed that PP1, PP2A and PP2B are closely related, and that PP2C displays a limited homology to the other PSTPs. The number of PSTPs has since been extended considerably through the application of molecular cloning techniques (see Table 18.1). It appears that all PSTPs are derived from the same ancestral gene, a prototype of which is bacteriophage λ phosphatase encoded by open reading frame 221.[5]

The early analysis of PSTPs focused on the catalytic subunits of PP1, PP2A, PP2B and PP2C. Subsequently PP1, PP2A and PP2B were isolated as heteromeric proteins consisting of a catalytic subunit associated with regulatory subunits. The structure of the major PSTPs is discussed below and gives ample evidence of the diversity of these enzymes.

Protein phosphatase type 1

Five isoforms of PP1 catalytic subunit (PP1c) encoded by three distinct genes have been identified. These 37 kDa proteins complex with several different regulatory (or targeting) subunits and inhibitory proteins (Table 18.2; see reviews[6,7]). Significant free PP1c has not been identified in cells.

The first characterized targeting subunit was a protein of 124 kDa termed G subunit. The G subunit directs PP1c to glycogen phosphorylase and glycogen synthase located on glycogen particles (see Figure 18.2). The same or a very similar protein directs PP1c to the sarcoplasmic reticulum, whereas the myofibrillar-bound PP1c complexes to 130 and 20 kDa regulatory subunits. These regulatory subunits target and tether PP1c to intracellular locations and impose substrate specificity.

Table 18.1 PSTPs

Name	Source/accession number[a]	Major characteristics
PP1	Many species/X70848	IC_{50} for okadaic acid 10–20 nM
PP2A	Many species/X12656	IC_{50} for okadaic acid 0.05–0.1 nM
PP2B	Many species/M29551	Ca^{2+}-dependent
PP2C	Many species/D28117	Mg^{2+}-dependent
PDP	Mammals/L18966	Mg^{2+}-dependent, Ca^{2+}-stimulated
PPX/PP4	Mammals/S57412	Centrosomal localization
PPT/PP5	Mammals/X72237	Nuclear localization
SIT4	*S. cerevisiae*/M24393	Involved in cyclin transcription
PPH21	*S. cerevisiae*/X58856	Amino-terminal acidic domain
PPG	*S. cerevisiae*/M94269	Carboxy-terminal extension of 50 amino acids
PPZ	*S. cerevisiae*/M86242	Amino-terminal domain rich in Ser/Arg
PPQ	*S. cerevisiae*/X75485	Amino-terminal domain rich in Ser/Asn
rdgC	*Drosophila*/M89628	Carboxy-terminal calcium-binding domain
ABI1	*A. thaliana*/X78886	Carboxy-terminal calcium-binding domain

[a] For phosphatases identified in many species an accession number for one of the mammalian isoforms is listed.

Soluble cytoplasmic forms of PP1c also complex with two inhibitory proteins (I-1 and I-2), while in the nucleus the inhibitor NIPPI-1 controls the activity of PP1.[8] These inhibitory proteins are potential targets of protein kinases and other PSTPs (see below).

Protein phosphatase type 2A

The catalytic subunit of PP2A also associates with a number of different regulatory subunits (Table 18.2).[9] The basic structure of PP2A[9] holoenzymes is a heterotrimeric configuration consisting of a 36 kDa catalytic (PP2Ac) complexed with a 65 kDa regulatory subunit (PR65 or A subunit), to which a 'variable' third subunit associates. So far, three unrelated families of variable regulatory subunits have been characterized: the 55 kDa regulatory subunit (PR55 or B subunit); the 72 kDa regulatory subunit (PR72/PR130; the PR130 subunit is produced by the alternative splicing from the same gene as PR72); and the 54–57 kDa regulatory subunit (PR61 or B′) family.

The PR65/A subunit provides binding sites for both PP2Ac and the 'variable' subunits and is a potent inhibitor of catalytic activity. In addition, two low-molecular-mass protein inhibitors of PP2Ac, termed I_1^{PP2A} and I_2^{PP2A}, have been isolated.[10] PP2Ac also displays phosphotyrosine phosphatase activity that can be up-regulated by the association of the PP2A core subunits with a protein termed phosphotyrosine phosphatase activator (PTPA).[11] Furthermore, PP2Ac is modified by carboxymethylation of its carboxy-terminal leucine.[12,13] The precise role of this modification is still under investigation but it may be

Table 18.2 Regulatory subunits and inhibitory proteins interacting with PSTPs

PSTP	Regulatory subunit	Inhibitory protein
PP1	G subunit M subunits	I-1/DARPP-32 I-2 NIPP-1
PP2A	PR65 PR55 PR72/PR130 PR61 PTPA	I_1^{PP2A} I_2^{PP2A}
PP2B	B subunit Calmodulin	

important for the assembly of holoenzymes or association with membranes.

Protein phosphatase type 2B

PP2B (also termed calcineurin) was initially identified as an abundant brain protein (approximately 1% of total) that inhibited cAMP-phosphodiesterase, owing to its Ca^{2+}- and calmodulin-binding properties (see review[14]). PP2B is a heterodimer consisting of a catalytic subunit of 57–61 kDa (A subunit) complexed to a regulatory subunit (B subunit) which is a homolog of calmodulin (Table 18.2). However, to obtain fully active PP2B requires the association of Ca^{2+}-calmodulin to the AB complex. The activity of the A catalytic subunit is stimulated fivefold by the addition of B subunit, 50-fold by calmodulin and 600-fold by the simultaneous addition both regulatory subunits. PP2B activity is inhibited by an intramolecular interaction of the carboxyl terminus with the catalytic site. The role of B regulatory subunit and calmodulin is to relieve this inhibition.

Protein phosphatase type 2C

Mammalian PP2C is a monomeric protein of 42–45 kDa and requires Mg^{2+} for maximal activity. PP2C shows homology to the other PSTP catalytic subunits only in its putative catalytic domain[15] (e.g. residues 190–270 in PP2Cα1). Interestingly, both PP2Cα and -β contain functional nuclear localization signals. Putative PEST degradation signals present in PP2Cα suggest that these proteins can be down-regulated by proteolysis. Pyruvate dehydrogenase phosphatase, which displays Mg^{2+}-dependent and Ca^{2+}-stimulated activity, is apparently related to PP2C (Table 18.1).

Novel serine/threonine phosphatases

Many phosphatases that escaped characterization by biochemical methods have recently been identified by molecular genetic techniques. PP4 (PPX) and PP5 were identified by the polymerase chain reaction (PCR) technique using primers corresponding to the most conserved regions in PP1/2A/2B. At present we can only speculate about their physiological roles. PP4 localizes both to centrosomes and the nucleus and so it has been proposed that it may take part in microtubule nucleation. PP5 is found predominantly in the nucleus and contains a domain with three repeated elements of 43 amino acids localized amino-terminally to the catalytic domain. Other proteins containing these tetratricopeptide repeat motifs are apparently involved in mitosis, transcription and RNA splicing.

A *Drosophila* PSTP (rdgC) was isolated in a genetic screen designed to identify proteins that prevent light-induced retinal degeneration. The rdgC possesses a regulatory domain with a Ca^{2+}-binding motif at the carboxyl terminus of the molecule, implying that it participates in Ca^{2+}-dependent signaling pathways. The *Saccharomyces cerevisiae* phosphatases *PPG* and *PPZ* also have regulatory domains attached to the catalytic part. *PPG* contains a carboxy-terminal domain rich in acidic amino acids and its function is apparently linked with glycogen metabolism. *PPZ* has an amino-terminal domain rich in serine and basic amino acids and its function is connected with osmotic regulation.

Inhibitors

In addition to the physiological protein inhibitors of PP1 and PP2A mentioned above, several environmental toxins act by binding and potently inhibiting these PSTPs.[16–18] This diverse group of chemicals includes the polyether fatty acid okadaic acid (produced by *Dinoflagellates*), the polyketides tautomycin (*Streptomyces*) and calyculin (marine sponges) and the cyclic peptides, microcystin and nodularin (*Cyanobacteria*). PP2B requires approximately 1000-fold higher concentrations than PP1 for inhibition and PP2C is insensitive to these inhibitors. PP4 and PP5 are also inhibited by okadaic acid. Interestingly, in tests on laboratory animals PSTP inhibitors displayed tumor-promoting properties. Other compounds identified as PP1/PP2A inhibitors include terpenoids, cantharidin (a product of blister beetles) and its derivative endothall.[19] Both cantharidin and endothall inhibit PP2A with higher potency than PP1. Since most of these inhibitors are products of either marine or freshwater organisms they are possible contaminants of food and water. Okadaic acid-contaminated seafood is known to cause diarrheal shellfish poisoning.

Regulation of cellular processes by protein serine/threonine phosphatases

PSTPs are implicated in the regulation of several metabolic processes, including glycogen metabolism, fatty acid and cholesterol biosynthesis, glycolysis and gluconeogenesis (see reviews[2,3,6]). Regulation of muscle glycogen metabolism by the second-messenger promoted phosphorylation cascade has served as a paradigm for other signaling cascades (Figure 18.2).

Glycogen phosphorylase, a glycogen-degrading enzyme, is activated upon phosphorylation catalyzed by phosphorylase kinase. Phosphorylase kinase itself is activated by phosphorylation through adrenalin-activated cAMP-dependent protein kinase (PKA). At the same time, PKA phosphorylates and inactivates glycogen synthase. The action of adrenalin also inhibits the dephosphorylation of glycogen phosphorylase through phosphorylation of two proteins that regulate the function of PP1c, G subunit and I-1 (see review[20]). Upon phosphorylation of G subunit, PP1c is released from the vicinity of glycogen particles and is trapped by the phosphorylated I-1. This situation is reversed when the G subunit is phosphorylated at a distinct site by an insulin stimulated kinase (p90 S6 kinase) (Figure 18.2(b)). PP1c associates with the G subunit with a concomitant increase in phosphatase activity towards glycogen synthase and

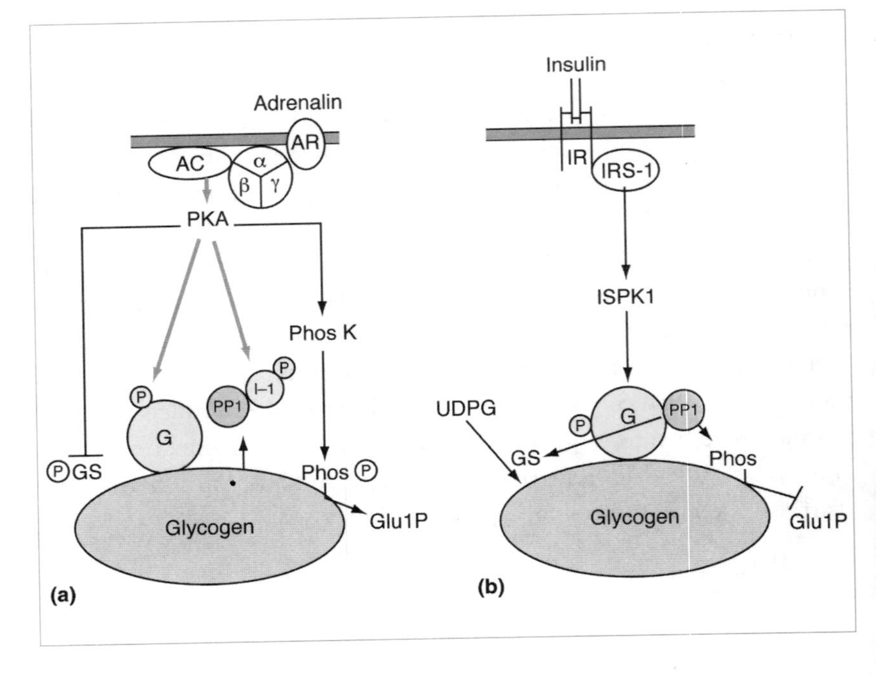

Figure 18.2 Regulation of glycogen metabolism in skeletal muscle. (a) Glycogenolysis stimulated by adrenalin binding to the adrenergic receptor (AR) involves a phosphorylation cascade of cAMP-stimulated protein kinase (PKA), phosphorylase kinase (Phos K) and phosphorylase (Phos). Adrenalin also promotes the phosphorylction of the regulatory (G subunit) and inhibitory (I-1) subunits of PP1c. (b) Insulin promotes glycogenogenesis by a cascade involving insulin receptor (IR), insulin receptor substrate-1 (IRS-1) and insulin-stimulated protein kinase-1 (ISPK1). The G subunit is phosphorylated on a distinct site and reactivates PP1c. GS, glycogen synthase; UDPG, uridine diphosphate glucose; Glu1P, glucose 1-phosphate.

glycogen phosphorylase. This, in turn, leads to an increase in glycogenesis. Thus, the G subunit functions as a molecular switch which recruits PP1c to stimulate glycogen catabolism and releases it to allow glycogen synthesis.

Signaling cascades

Recently, several kinase cascades involved in various cellular responses including growth stimuli, heat shock, UV irradiation and hyperosmolarity have been described. In these cascades, an initial signal detected by a membrane receptor is passed down a chain of activating phosphorylations involving several protein kinases, resulting in increased gene transcription through the modification of the appropriate transcription factor (see examples below). The mitogen-activated protein kinase (MAP kinase) cascade is the prototype (see Chapter 11). A potential role of PP2A in controlling the mammalian MAP kinase cascade is described below.

The best documented example is the MAP-kinase-like cascade involved in response to osmolarity changes in yeast.[21,22] Interestingly, this eukaryotic pathway starts with a signaling module analogous to the prokaryotic two-component system, consisting of an osmosensor protein termed SLN1 that in turn activates the response regulator protein SSK1 (Figure 18.3). SLN1 is a histidine kinase which phosphorylates both itself and SSK1, the latter on aspartate residue. This results in the stimulation of PBS2 and HOG1 and, ultimately, increased glycerol production. *Saccharomyces cerevisiae* PP2C-like PTC1 and PTC3 together with a tyrosine phosphatase PTP2 regulate this pathway by dephosphorylating PBS2 and HOG1 (Figure 18.3). Similarly, in *Schizosaccharmyces pombe ptc1*, *ptc2* and *ptc3* PP2C-like phosphatases are involved in osmoregulatory pathway involving *wis1* protein kinase (MEK homolog).

Translational control

Stimulation of protein synthesis correlates with changes in the phosphorylation state of ribosomal proteins and several transcription factors (see below). Phosphorylation of S6 protein, a constituent of 40S ribosomal complex, represents one of the first changes that link mitogenic stimulation of cells with increased protein synthesis. Both PP1 and PP2B have been implicated in S6 dephosphorylation.[23] Phosphorylation of S6 protein is catalyzed by p70 and p90 S6 kinases (see Chapter 11). Early studies on p70 S6 kinase established that protein phosphatase inhibitors are required

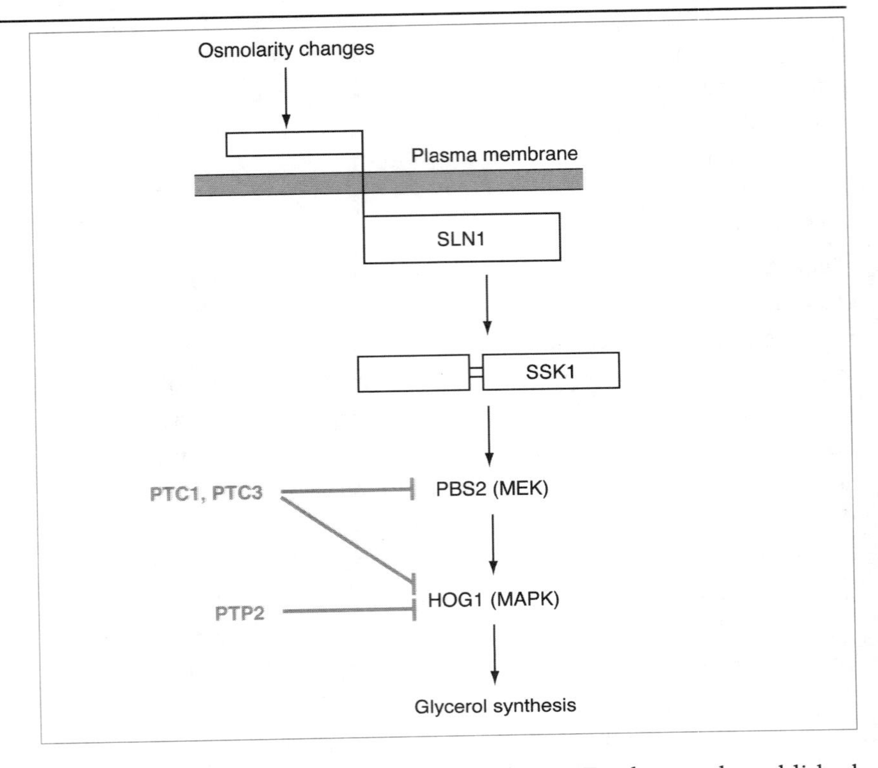

Figure 18.3 Regulatory role of the protein phosphatases PTC1, PTC3 and PTP2 in the osmoregulatory MAP-kinase-like cascade in yeast.

to maintain full activity during enzyme isolation. Further work established that p70 S6 kinase is preferentially dephosphorylated and inactivated by PP2Ac *in vitro*.[24] Subsequently, many other protein kinases in various signaling cascades have been shown to be regulated by PP2A (see below). Interestingly, activation of p70 S6 kinase but not p90 S6 kinase is inhibited by immunosuppressive macrolide rapamycin, by an as yet unknown mechanism.

In yeast, the PSTP PPQ appears to participate in another aspect of translational regulation. Its amino-terminal region is rich in serine and asparagine residues (Table 18.1) and deletion of *PPQ* gene causes hypersensitivity to protein synthesis inhibitors.

Transcriptional regulation

The phosphorylation state of many transcription factors regulates their nuclear transport, DNA binding and trans-activation activities. One of the best studied transcription factors in terms of phosphatase action is the cAMP response element binding protein (CREB). CREB is phosphorylated on multiple residues but phosphorylation of Ser133 by PKA seems to be crucial for trans-activation (see Chapter 15). PP1 was shown to dephosphorylate Ser133 and subsequently inactivate CREB[25] in

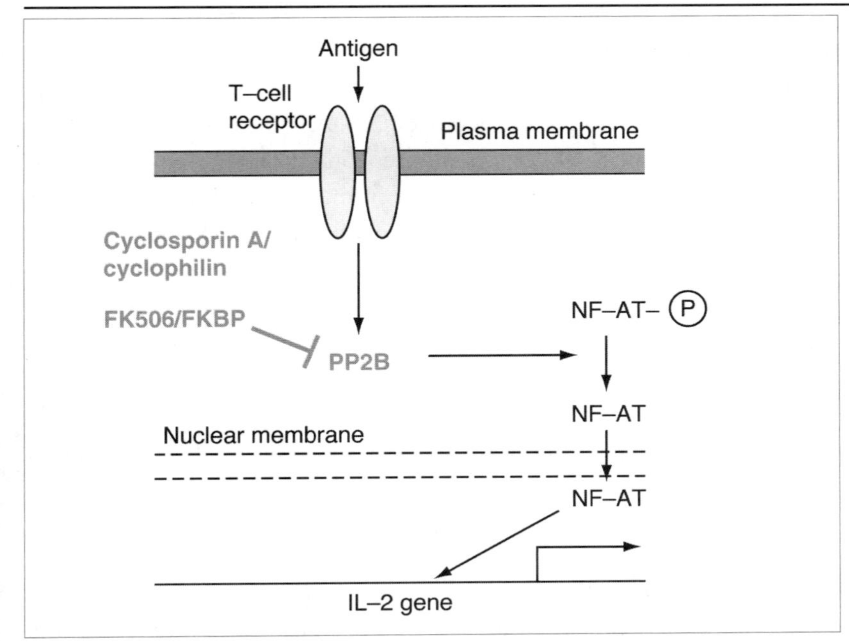

Figure 18.4 Inhibitory action of cyclosporin A and FK506 on interleukin-2 (IL-2) gene expression in T cells.

experiments, conducted using both PC12 and NIH3T3 cells, that utilized overexpression and microinjection of PP1c, PP2Ac and the constitutively active form of I-1. However, using human HepG2 cells as starting material, PP2A was purified from nuclear extracts as a major CREB phosphatase.[26]

Some viruses utilize interaction with PSTPs to interfere with transcription of their eukaryotic host. For instance, the transcription factor EBNA2 of Epstein–Barr virus associates with nuclear PP1,[27] which results in phosphatase inhibition and subsequent activation of viral transcription via a cAMP-responsive element. Other examples include the interaction of adenoviral and papillomavirus-encoded proteins with PP2A (see reviews[4,28]).

T-cell signaling

T cells constitute an important part of a defense machinery towards foreign antigens. Binding of antigen to T-cell receptors initiates a signal transduction cascade that leads to activation of interleukin 2 (IL-2) gene transcription and thus to T-cell activation. This signaling pathway is disrupted by the immunosuppressant drugs cyclosporin A (a cyclic undecapeptide) and FK506 (a macrocyclic lactone), which block T-cell proliferation and differentiation. Cyclosporin A and FK506 bind to intracellular receptor proteins termed cyclophillins and FKBPs (FK506 binding proteins), respectively, and these complexes inhibit PP2B

(Figure 18.4; see review[14]). The inhibition of PP2B blocks the dephosphorylation of cytoplasmic pools of the transcription factor NF-AT. The phosphorylated form of NF-AT is unable to enter the nucleus and stimulate IL-2 gene expression. In agreement with this, overexpression of PP2B A subunit in human T lymphocytes, makes them more resistant to the action of cyclosporin and FK506.

PP2B also stimulates IL-2 gene expression by another mechanism involving transcription factor NF-κB.[29] NF-κB is anchored in the cytoplasm by an inhibitory protein termed IκB (see Chapter 21). NF-κB activation involves phosphorylation and subsequent proteolysis of IκB. The role played by PP2B in NF-κB activation is indirect and apparently involves IκB protein kinase which is activated by PP2B-mediated dephosphorylation.

Neuronal-specific processes

Many neuronal processes involve reversible phosphorylation,[30] including release of neurotransmitters, control of neurotransmitter receptors and ion channels. The importance of phosphorylation and dephosphorylation in neuronal processes is reflected by high concentration of protein kinases and phosphatases in brain tissue. As mentioned before, PP2B represents up to 1% of total brain protein. Other PSTPs are also highly enriched in brain, including the catalytic subunits of PP1 and PP2A and some of their regulatory subunits and inhibitory proteins. For instance, both PR55β/Bβ and PR55γ/Bγ regulatory subunits of PP2A are specifically expressed in neural tissue.[31,32] Also, a protein related to I-1 and designated DARPP-32 (dopamine and cAMP-regulated phosphoprotein) is enriched in brain (see review[30]).

An example of the role of PSTPs in brain function is a phosphatase cascade involved in the regulation of synaptic membrane polarization in the hippocampus.[33] This cascade has been shown recently to regulate a process termed long-term depression (LTD). LTD, together with long-term potentiation (LTP), plays an important role in the storage of information in the brain involved in learning and memory. The phosphatase cascade that functions in LTD starts from the activation of PP2B by Ca^{2+} influx through the NMDA (*N*-methyl-D-aspartate) receptor–ionophore. Subsequently, PP2B dephosphorylates I-1, thus activating PP1. The proteins that are dephosphorylated by PP1 and important for LTD have still to be identified.

Cell cycle control

Mutants that are deficient in both catalytic and regulatory subunits of PSTPs and display various cell cycle defects have been isolated (see review[34]). These include the *bimG* (blocked in mitosis) mutant of the fungus *Aspergillus nidulans*. BimG is a homolog of mammalian PP1c, which apparently regulates chromosome condensation and spindle function. A homolog of PP1 from *Schizosaccharomyces pombe* and termed dis2 (defective in sister chromatid disjoining) was isolated as a cold-sensitive mutation while looking for defects in chromosome segregation. *SIT4* was cloned from *Saccharomyces cerevisiae* as a suppressor of mutation that repressed transcription of the *HIS4* gene.[35] SIT4 represents a PP2A-like PSTP and it is necessary for the cell cycle progression from G1 to S phase. It has been established that SIT4 activates expression of G1 yeast cyclin genes termed *CLN1* and *CLN2*.

Mutations in *S. cerevisiae CDC55* and *TPD3* genes which encode proteins that correspond to PR55/B and PR65/A regulatory subunits of PP2A, respectively, have also been isolated.[36,37] These mutants display a similar phenotype, with cells becoming elongated, multinuclear and affected in cytokinesis. In *Drosophila*, two mutations which result from insertion of P-elements in a gene encoding the PR55/B regulatory subunit of PP2A have been described[38,39]. Mutants termed *aar* (abnormal anaphase resolution) and *tws* (twins) display two different phenotypes depending on the severity of the mutation. The conclusion from analysis of these *Drosophila* mutants is that the PP2A holoenzyme containing the 55 kDa regulatory subunit is involved in both cell division and cell fate determination. The precise mechanisms that link a phosphatase regulatory subunit to these cellular events remain to be elucidated.

Studies conducted with *Xenopus* oocyte extracts resulted in the isolation of a factor termed INH, which was able to inhibit cell cycle progression. INH corresponds to PP2A comprising PR65/A, PR55α/Bα and PP2Ac. Its action involves inhibition of cdc2 cyclin B complex by blocking the phosphorylation of Thr161 on cdc2.[40]

The above examples indicate that both PP1 and PP2A are important for the regulation of cell cycle. However, it is likely that other phosphatases will be discovered that also participate in nuclear events.

Cell transformation

Recent results indicate that inhibition of PP1 and PP2A can lead to cell transformation. The best studied example of this is okadaic acid-induced tumor promotion in mice.[41] The transforming properties of okadaic acids have been extensively studied in different tissue culture cells. It is important to note that the effect of okadaic acid is concentration dependent. At very high concentrations (up to 1 μM) okadaic acid induces either programmed cell death (apoptosis) or growth arrest (cells round up and detach from a tissue culture dish), whereas low concentrations (<10 nM) applied for longer periods (several weeks) cause cell transformation (see review[4]).

Transformation of cells by DNA viruses involves deregulation of the function of PSTPs. Among the best-studied examples are polyoma middle T and small t and SV40 small t antigens which associate with PP2A and its regulatory subunit PR65/A.[42] In the case of polyoma middle T antigen, this association is required for cell transformation. Experiments with SV40 small t antigen[43] show that this protein stimulates cell proliferation by inhibiting PP2A-dependent dephosphorylation of MAP kinases and MAP kinase kinases (Mek). (Figure 18.5).

Studies conducted on the retinoblastoma protein (Rb) imply that PSTPs may act as tumor suppressors. It is known that the phosphorylation state of Rb correlates with growth suppressor activity, with the hypophosphorylated form being the strongest inhibitor of cell cycle

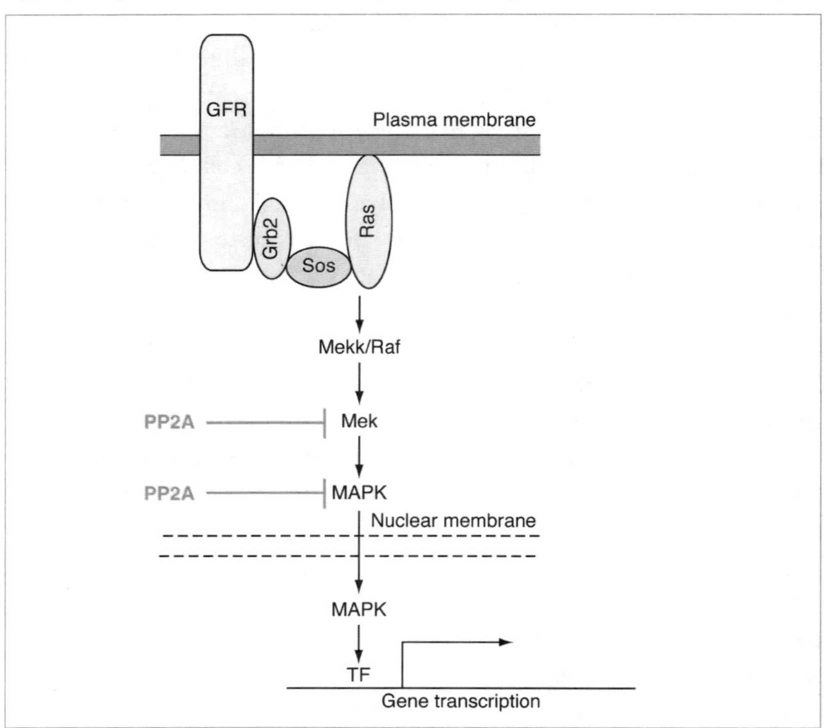

Figure 18.5 Regulatory role of protein phosphatase 2A (PP2A) in growth-factor-induced MAP kinase (MAPK) cascade. GFR, growth factor receptor; Grb2, adaptor molecule consisting of an SH2 and two SH3 domains; Sos, guanine-nucleotide-releasing protein; Ras, low-molecular-mass GTP binding protein; TF, transcription factor.

progression. Several reports have pointed out that PP1 acts as an Rb phosphatase. By applying the yeast two-hybrid system, a novel isoform of PP1, termed PP1α2, was isolated that contains an insertion of 11 amino acids close to its amino terminus. The interaction of PP1α2 with Rb occurs during mitosis and in early G1.[44]

Summary

After the initial characterization and classification of PSTPs in the early 1980s, we now realize that the number of these enzymes is higher than previously anticipated. Many PSTPs discovered in yeast and *Drosophila* are unique but one can predict that mammalian homologs will be soon identified. Further clarification of the roles that regulatory subunits play in targeting and regulating phosphatase function will be an important step forward in our understanding of the function of PSTPs. An important issue will be to establish which PSTPs dephosphorylate a particular target phosphoprotein *in vivo*. In this case, the yeast two-hybrid system, utilizing both catalytic and regulatory subunits of PSTPs as bait, will be instrumental in identifying important physiological substrates. Recently, the three-dimensional structures of both PP1 and PP2B were solved, and the results indicate that they are similar in overall structures[45,46]. As other structures are solved it is likely to emerge that all PSTPs have a similar catalytic domain but have different regulatory strategies. The elucidation of the structure of the regulatory subunits will provide important insights into activity regulation and substrate specificity. These advances allied with transgenic and gene knockout techniques will provide significant advances in understanding the role of PSTPs in regulating cellular processes.

References

1 Hunter, T. (1995) Protein kinases and phosphatases: the yin and yang of protein phosphorylation and signaling. *Cell*, **80**, 225–36.

2 Cohen, P. (1989) The structure and regulation of protein phosphatases. *Annu. Rev. Biochem.*, **58**, 453–508.

3 Shenolikar, S. and Nairn, A.C. (1991) Protein phosphatases: recent progress, in *Advances in Second Messengers and Phosphoprotein Research* (eds Greengard, P. and Robinson, G.A.) Vol. 25, Raven Press, New York, pp. 1–121.

4 Walter, G. and Mumby, M. (1993) Protein serine/threonine phosphatases and cell transformation. *Biochim. Biophys. Acta*, **1155**, 207–26.

5 Zuo, S., Clemens, J.C., Stone, R.L. and Dixon, J.E. (1994) Mutational analysis of

Ser/Thr phosphatase. Identification of residues important in phosphoesterase substrate binding and catalysis. *J. Biol. Chem.*, **269**, 26234–8, and references cited therein.

6 Bollen, M. and Stalmans, W. (1992) The structure, role, and regulation of type 1 protein phosphatases. *Crit. Rev. Biochem. Mol. Biol.*, **27**, 227–81.

7 Hubbard, M.J. and Cohen, P. (1993) On target with a new mechanism for the regulation of protein phosphorylation. *Trends Biochem. Sci.*, **18**, 172–7.

8 Beullens, M., Van Eynde, A., Stalmans, W. and Bollen, M. (1992) The isolation of novel inhibitory polypeptides of protein phosphatase 1 from bovine thymus nuclei. *J. Biol. Chem.*, **267**, 16538–44.

9 Wera, S. And Hemmings, B.A. (1995) Serine/threonine phosphatases. *Biochem. J.*, **311**, 17–29.

10 Li, M.L., Guo, H. and Damuni, Z. (1995) Purification and characterization of two potent heat-stable protein inhibitors of protein phosphatase 2A from bovine kidney. *Biochemistry*, **34**, 1988–96.

11 Cayla, X., Goris, J., Hermann, J. *et al.* (1990) Isolation and characterization of a tyrosyl phosphatase activator from rabbit skeletal muscle and *Xenopus laevis* oocytes. *Biochemistry*, **29**, 658–67.

12 Favre, B., Zolnierowicz, S., Turowski, P. and Hemmings, B.A. (1994) The catalytic subunit of protein phosphatase 2A is carboxyl-methylated *in vivo*. *J. Biol. Chem.*, **269**, 16311–7, and references cited therein.

13 Zolnierowicz, S., Favre, B. and Hemmings, B.A. (1994) Carboxyl methylation – a new regulatory mechanism for an old enzyme, protein phosphatase 2A. *Adv. Protein Phosphatases*, **8**, 355–69, and references cited therein.

14 Kincaid, R.L. and O'Keefe, S.J. (1993) Calcineurin and immunosuppression: a calmodulin-dependent protein phosphatase acts as the 'gatekeeper' to interleukin-2 gene transcription. *Adv. Protein Phosphatases*, **7**, 543–83.

15 Peruski, L.F., Wadzinski, B.E. and Johnson, G.L. (1993) Analysis of the multiplicity, structure, and function of protein serine/threonine phosphatases. *Adv. Protein Phosphatases*, **7**, 9–30.

16 Bialojan, C. and Takai, A. (1988) Inhibitory effect of a marine-sponge toxin, okadaic acid, on protein phosphatases. *Biochem. J.*, **256**, 283–90.

17 Holmes, C.F.B. and Boland, M.P. (1993) Inhibitors of protein phosphatase-1 and -2A; two of the major serine/threonine protein phosphatases involved in cellular regulation. *Curr. Opinion Struct. Biol.*, **3**, 934–43.

18 MacKintosh, C. and MacKintosh, R. (1994) Inhibitors of protein kinases and phosphatases. *Trends Biochem. Sci.*, **19**, 444–8.

19 Li, Y.-M. and Casida, J.E. (1992) Cantharidin-binding protein: identification as protein phosphatase 2A. *Proc. Natl. Acad. Sci. USA*, **89**, 11867–70.

20 Cohen, P. (1992) Signal integration at the level of protein kinases, protein phosphatases and their substrates. *Trends Biochem. Sci.*, **17**, 408–13.

21 Maeda, T., Wurgler-Murphy, S.M. and Saito, H. (1994) A two-component system that regulates an osmosensing MAP kinase cascade in yeast. *Nature*, **369**, 242–5.

22 Shiozaki, K. and Russell, P. (1995) Counteractive roles of protein phosphatase 2C (PP2C) and a MAP kinase kinase homolog in the osmoregulation of fission yeast. *EMBO J.*, **14**, 492–502.

23 Andres, J.L., Johansen, J.W. and Maller, J.L. (1987) Identification of protein phosphatases 1 and 2B as ribosomal protein S6 phosphatases *in vitro* and *in vivo*. *J. Biol. Chem.*, **262**, 14389–93.

24 Ballou, L.M., Jeno, P. and Thomas, G. (1988) Protein phosphatase 2A inactivates the mitogen stimulated S6 kinase from Swiss mouse 3T3 cells, *J. Biol. Chem.*, **263**, 1188–94.

25 Hagiwara, M., Alberts, A., Brindle, P. *et al.* (1992) Transcriptional attenuation following cAMP induction requires PP-1 mediated dephosphorylation of CREB. *Cell*, **70**, 105–13.

26 Wadzinski, B.E., Wheat, W.H., Jaspers, S. *et al.* (1993) Nuclear protein phosphatase 2A dephosphorylates protein A-phosphorylated CREB and regulates CREB transcriptional stimulation. *Mol. Cell. Biol.*, **13**, 2822–34.

27 Fahraeus, R., Palmqvist, L., Nerdstedt, A. *et al.* (1994) Response to cAMP levels of the Epstein–Barr virus EBNA2-inducible LMP1 oncogene and EBNA2 inhibition of a PP1-like activity. *EMBO J.*, **13**, 6041–51.

28 Mayer-Jaekel, R.E. and Hemmings, B.A. (1994) Protein phosphatase 2A – a 'menage a trois.' *Trends Cell Biol.*, **4**, 287–91.

29 Frantz, B., Nordby, E., Bren, G. *et al.* (1994) Calcineurin acts in synergy with PMA to inactivate IκB/MAD3, an inhibitor of NF-κB. *EMBO J.*, **13**, 861–70.

30 Nairn, A.C. and Shenolikar, S. (1992) The role of protein phosphatases in synaptic transmission, plasticity and neuronal development. *Curr. Opinion Neurobiol.*, **2**, 296–301.

31 Mayer, R.E., Hendrix, P., Cron, P. *et al.* (1991) Structure of the 55 kDa regulatory subunit of protein phosphatase 2A: evidence of a neuronal specific isoform. *Biochemistry*, **30**, 3589–97.

32 Zolnierowicz, S., Csortos, C., Bondor, J. *et al.* (1994) Diversity in the regulatory B-subunits of protein phosphatase 2A: identification of a novel isoform highly expressed in brain. *Biochemistry*, **33**, 11858–67.

33 Mulkey, R.M., Endo, S., Shenolikar, S. and Malenka, R. (1994) Involvement of a calcineurin/inhibitor-1 phosphatase cascade in the hippocampal long-term depression. *Nature*, **369**, 486–8.

34 Cyert, M.S. and Thorner, J. (1989) Putting it on and taking it off: phosphoprotein phosphatase involvement in cell cycle regulation. *Cell*, **57**, 891–3.

35 Sutton, A., Immanuel, D. and Arndt, K.T. (1991) The SIT4 protein phosphatase functions in late G_1 for progression into S phase. *Mol. Cell. Biol.*, **11**, 2133–48.

36 Healy, A.M., Zolnierowicz, S., Stapleton, A.E. *et al.* (1991) *CDC55*, a *Saccharomyces cerevisiae* gene involved in cellular morphogenesis: identification, characterization and homology to the B subunit of Mammalian type 2A protein phosphatase. *Mol. Cell. Biol.*, **11**, 5767–80.

37 Van Zyl, W.H. (1992) Inactivation of the protein phosphatase 2A regulatory

subunit A results in morphological and transcriptional defects in *Saccharomyces cerevisiae*. *Mol. Cell. Biol.*, **12**, 4946–59.

38 Mayer-Jaekel, R.E., Ohkura, H., Gomes, R. *et al.* (1993) The 55 kd regulatory subunit of *Drosophila* protein phosphatase 2A is required for anaphase. *Cell*, **72**, 621–33.

39 Shiomi, K., Takeichi, M., Nishida, Y. *et al.* (1994) Alternative cell fate choice induced by low-level expression of a regulator of protein phosphatase 2A in the *Drosophila* peripheral nervous system. *Development*, **120**, 1591–9.

40 Lee, T.H., Turck, C. and Kirschner, M.W. (1994) Inhibition of cdc2 activation by INH/PP2A. *Mol. Biol. Cell*, **5**, 323–38.

41 Fujiki, H. and Suganuma, M. (1993) Tumor promotion by inhibitors of protein phosphatases 1 and 2A: the okadaic acid class of compounds. *Adv. Cancer Res.*, **61**, 143–94.

42 Pallas, D.C., Shahrik, L.K., Martin, B.L. *et al.* (1990) Polyoma small and middle t antigens and SV40 t antigen form stable complexes with protein phosphatase 2A. *Cell*, **60**, 167–76.

43 Sontag, E., Fedorov, S., Kamibayashi, C. *et al.* (1993) The interaction of SV40 small tumor antigen with protein phosphatase 2A stimulates the MAP kinase pathway and induces cell proliferation. *Cell*, **75**, 887–97.

44 Durfee, T., Becherer, K., Chen, P.-L. *et al.* (1993) The retinoblastoma protein associates with the protein phosphatase type 1 catalytic subunit. *Genes Dev.*, **7**, 555–69.

45 Goldberg, J., Huang, H., Kwon, Y. *et al.* (1995) Three-dimensional structure of the catalytic subunit of protein serine/threonine phosphatase-1. *Nature*, **376**, 745–53.

46 Kissinger, C.R., Parge, H.E., Knighton, D.R., *et al.* (1995) Crystal structures of human calcineurin and the human FKBP12-FK506-calcineurin complex. *Nature*, **378**, 641–4.

19 G proteins in signal transduction

Paul C. Sternweis

To respond appropriately to their environment, cells need to sense the extracellular milieu, integrate the various signals detected and initiate regulation of intracellular metabolism. One mechanism at the disposal of cells involves receptors that function through heterotrimeric, GTP-dependent regulatory proteins (G proteins). This large class of G-protein-coupled receptors detects a wide variety of stimuli (such as hormones, neurotransmitters, light and odorants) and was discussed in Chapter 7. The structure and function of their downstream signaling proteins will be described in this chapter. The Ras-like GTP binding proteins (see Chapter 10) have also frequently been called G proteins, but do not appear to interact directly with receptors and can function as monomers.

The initial molecular dissection of a hormone-linked G protein system dates back to the elucidation of cAMP as a second messenger for epinephrine and glucagon and the first description of this messenger's synthesis by a membrane-associated enzyme.[1,2] This was followed several years later by the key observation that GTP was required for effective stimulation of adenylyl cyclase by hormones.[3] The discovery that guanine-nucleotide-sensitive adenylyl cyclase actually consisted of two components, a GTP-sensitive protein and a catalytic moiety,[4] then set the stage for isolation and characterization of the G_s protein. The subsequent discovery of many additional G proteins and identification of their roles in a broad spectrum of biology have demonstrated the importance of this regulatory mechanism. A discussion of the early work in this field[5] and some recent reviews[6–9] supply abundant details and references.

G protein systems

Signaling via G proteins is achieved by vectorial pathways consisting of three basic components (Figure 19.1). Receptors act as specific detectors for signals in the environment. The receptor spans the plasma membrane

Signal Transduction. Edited by Carl-Henrik Heldin and Mary Purton. Published in 1996 by Chapman & Hall. ISBN 0 412 70810 8

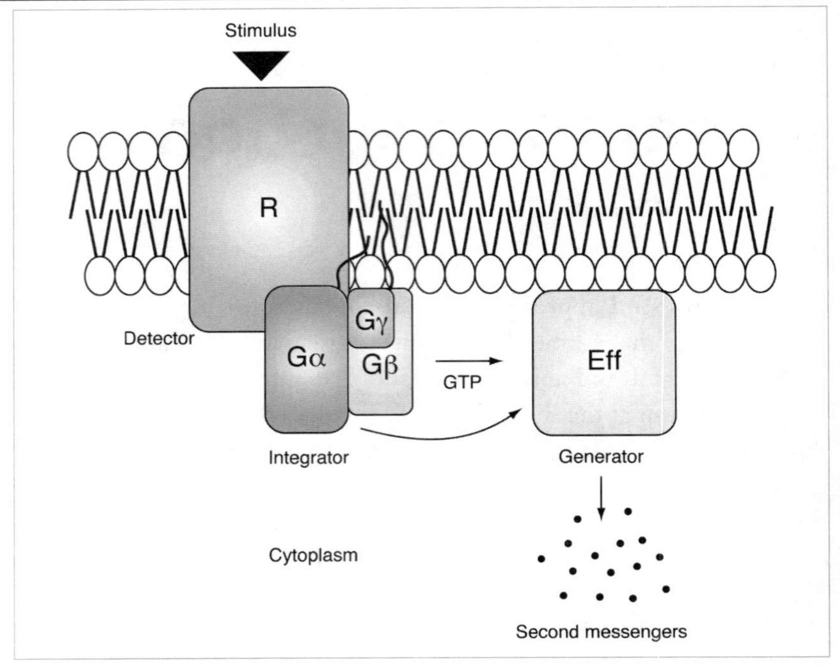

Figure 19.1 Basic pathway for G-protein-dependent signal transduction. Stimulation of receptors initiates a vectorial signaling cascade that results in the production of an amplified intracellular signal. See text for details. Abbreviations: R, receptor; G, G protein; Eff, effector.

and transduces the presence of an extracellular agonist into intracellular stimulation of G proteins. The G proteins are extrinsic membrane proteins located at the cytoplasmic surface of membranes which, upon activation by receptors, regulate the activity of a wide variety of effector proteins. These effectors range from enzymes that produce soluble second messengers for dispersal into cytosol to ion channels that control ion fluxes across the membrane (Table 19.1).

The G proteins consist of three subunits. The α subunits are the largest (39–46 kDa) and contain the binding site for guanine nucleotides. The β (35–36 kDa) and γ (about 8 kDa) subunits form stable non-covalent complexes and associate more tightly with membranes. The heterotrimeric G proteins can dissociate into their respective α subunits and βγ complexes (see below). The heterotrimeric species are named by their content of α subunit as these appear to be most diverse in both structure and function.

What is the purpose of the G protein? Why do receptors not act directly on effector proteins? There are several advantages for the inclusion of an intervening step. First, the G proteins increase the sensitivity of the system by providing an extra step for signal amplification. Thus, receptors can act catalytically to stimulate multiple G proteins which, in turn, regulate multiple effector proteins. The catalytic action of an effector protein also amplifies the signal and so a few receptors can mobilize production of large amounts of second messenger. Second, as mediators of receptor action, the G proteins can act as integrators of inputs from several stimuli. For example, the presence of several hormones which can stimulate the synthesis of cAMP will all be funneled through the G_s protein. Alternatively, G proteins can act as a point of divergence by regulating more than one effector in response to a stimulus. Divergence can also occur through the action of a receptor on multiple G proteins. This interaction among multiple components allows a dynamic range of responses to stimuli.

What, then, determines the circuitry available in a single cell? Clearly, expression of specific signaling components is one constraint, especially for the spectrum of detection provided by receptors. However, individual cells still express numerous receptors, multiple G proteins and many effector molecules. Specificity for regulation then has to lie in the potential interactions allowed among the various components, either through specificity of their molecular associations or through localization of various components (see below).

The specificity of G proteins for effectors varies from the highly specific action of G_t on cGMP-dependent phosphodiesterase (cGMP-PDE) to the more promiscuous action of the G_i proteins. Some of this action is made more complex by the fact that α subunits and stable βγ complexes can dissociate and act individually on downstream effectors. Table 19.1 presents a summary of some of the properties and functions of G protein α subunits. These subunits are the most variable and can be divided into four subfamilies defined by sequence homology. These subfamilies also differ in function. The known β and γ subunits are less numerous (five and seven genes, respectively), but do have the potential to form multiple unique complexes. The βγ subunits have also been shown to regulate the activities of PLC β isoforms, adenylyl cyclase isozymes, potassium channels, a phosphatidylinositol 3-kinase, G protein receptor kinases and PLA_2 activity. In contrast to the α subunits, little diversity in activity among specific βγ dimers has been observed so far.

Cycles of regulation

The function of heterotrimeric G proteins is characterized by two interdependent cycles involving guanine nucleotide exchange and subunit dissociation (Figure 19.2). Inactive G proteins contain bound GDP and are activated when GTP is exchanged for GDP, a process catalyzed

Table 19.1 G protein α subunits[a]

Class	Members	Toxin[b]	Localization	Effector regulation[c d]	Signal[d]
G_s	α_s	CT	Ubiquitous	↑ AC, ↑ Ca^{2+} channels	↑ cAMP, ↑ Ca^{2+}
	α_{olf}	CT	Olfactory epithelium	↑ AC	↑ cAMP
G_i	α_{i1}	PT	Limited	↓ AC	↓ cAMP
	α_{i2},α_{i3}	PT	Ubiquitous	↑ *K^+ channels*	Δ Voltage
	α_o	PT	Neuronally enriched	↓ *Ca^{2+} channels*	↓ Ca^{2+}
	α_{t1},α_{t2}	PT	Retina	↑ cGMP-PDE	↓ cGMP
	α_{gust}	PT	Taste buds	?	
	α_z	–	Limited	?	
G_q	α_q,α_{11}	–	Ubiquitous	↑ PLC	↑ IP_3, DAG
	α_{14}	–	Limited	↑ PLC	↑ IP_3, DAG
	α_{15},α_{16}	–	Hematopoetic cells	↑ PLC	↑ IP_3, DAG
G_{12}	α_{12}	–	Ubiquitous	*Na^+/H^+ exchange*	
	α_{13}	–	Ubiquitous	↓ *Ca^{2+} currents*	

[a] The α subunits contain the binding site for association with and hydrolysis of GTP. The subunits listed represent unique mammalian genes. Several of the proteins also exist in multiple forms due to splice variation.
[b] Several of the α subunits can be modified by ADP ribosylation; cholera toxin (CT) enhances α_s activity while pertussis toxin (PT) attenuates the action of its substrates.
[c] Activities in italics are examples of regulated activities that may not evolve from direct interaction of the subunit with the effector.
[d] Abbreviations: AC, adenylyl cyclase; cGMP-PDE, cGMP-dependent phosphodiesterase; PLC, phosphatidylinositol-specific phospholipase C; DAG, diacylglycerol.

by activated receptors. G proteins slowly hydrolyze this bound GTP and thus catalyze their own inactivation. The balance between the rates of exchange and hydrolysis determines the ratio of G protein that is activated. The slow hydrolytic activity allows a G protein to act as a timed switch with the potential to remain active for a preset interval (generally several seconds). This timed interval may (1) provide a quantum of activity for each activation, (2) allow the G protein to remain active while a receptor recruits more G proteins or (3) maintain the G protein in an active state until a functional interaction with an effector is achieved. The lifetime of the activated state of some G proteins may be shortened by their interaction with downstream effector molecules (see below).

The heterotrimeric G proteins can dissociate into their respective α subunits and βγ complexes (Figure 19.2b). Dissociation is inhibited by binding of GDP and promoted by association with GTP. Thus, the heterotrimeric form of the protein is preferred in the basal state while dissociation is favored when activated. The roles of the dissociated activated α and βγ subunits in signal transduction have been investigated *in vitro* (see review[10]). However, the potential efficacy of an activated heterotrimer also needs to be considered. Effective attenuation of βγ

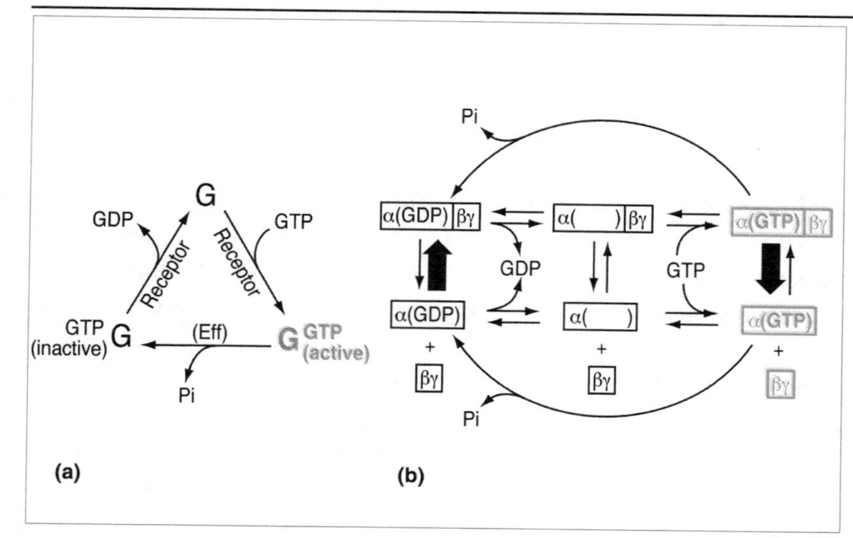

Figure 19.2 Cycles involved in G protein function. Activation of G proteins involves both guanine nucleotide exchange and alteration in subunit affinity. (a) The cycle of guanine nucleotide exchange and GTP hydrolysis. (b) The potential dynamic states of subunit association with respect to nucleotide occupancy. Altered equilibria for subunit association in the presence of GTP (increased dissociation) or GDP (increased association) are indicated. Since GTP is required for activation, the three blue species on the right represent those responsible for downstream action. See text for further discussion.

action by αGDP may indicate that βγ complexes are only effective when dissociated.

The main role of receptors is to facilitate nucleotide exchange; this is achieved by reducing the affinity of G proteins for guanine nucleotides. Both α and βγ subunits are required for efficient exchange, suggesting that the heterotrimeric forms of the protein interact most productively with receptor. In a reciprocal action, nucleotides lower the affinity of receptors for agonists. Since expression of high affinity for agonists by receptors also requires both α and βγ subunits, one deduces that the nucleotide-free form of the heterotrimeric G protein has highest affinity for receptors and represents the form of the G protein that, when complexed with receptors, yields high-affinity binding of agonists. It is presumably the energy of this 'transition state' that accelerates the dissociation of GDP.

Structure of the G proteins

Recent determination of the X-ray crystallographic structures of the G_t[11,12] and G_{i1}[13,14] α subunits provides a more detailed perspective on their mechanism of action. The structure is composed of two domains. One domain is organized around a core structure that defines the guanine nucleotide binding site and which is preserved in several related proteins (like Ras) that use a GTP cycle for function.[15,16] The α subunits also contain a domain composed of interacting α helices which 'masks' the GTP binding site from the surrounding solvent in both the GDP and GTP liganded states. While the helical domain does not interact directly with the nucleotide, it blocks access to the binding site such that significant conformational movement is required for exchange. A dynamic equilibrium between 'closed' and 'open' conformations may be one of the major forces determining affinity and rates of exchange for nucleotide (Figure 19.3). Presumably, receptors facilitate the formation of and stabilization of the 'open' state in addition to weakening direct associations between GDP and the binding site. Perturbation of an equilibrium between 'open' and 'closed' conformations could also be the means by which $\beta\gamma$ increases affinity for GDP. Paradoxically, some of these same interactions with $\beta\gamma$ may play a role in receptor-induced stabilization of the 'open' state.

The three-dimensional structure of α offers an explanation for the slow hydrolytic rates of these proteins. In the activated protein, there are no residues optimally positioned to effect activation of the water that is required for this reaction. Insight into the mechanism of hydrolysis was derived from the structure of α subunits that had been activated by a complex of aluminum and fluoride (AlF_4^-).[14,17] This complex occupies the γ-phosphoryl position of the binding site and, together with GDP, mimics the action of GTP. The crystal structure revealed a planar structure for the coordination of AlF_4^- that represents a transition state for GTP hydrolysis. The interaction of a conserved arginine (178 in α_{i1}) with fluoride atoms and of a conserved glutamine (204 in α_{i1}) with a fluoride atom and the axial water molecule defines their roles in hydrolysis. Slow rates of hydrolysis presumably reflect the difficulty in assuming analogous coordination with GTP.

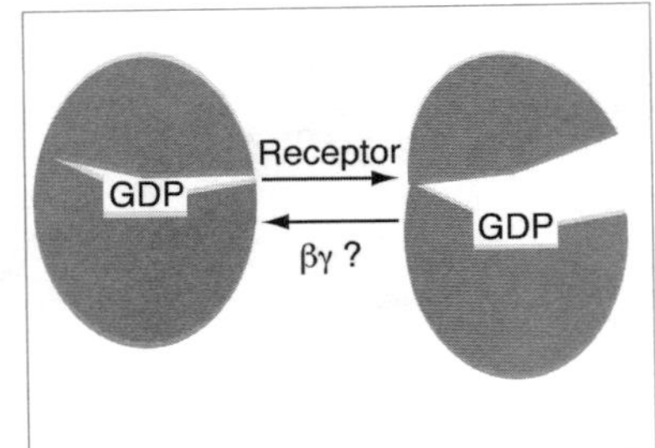

Figure 19.3 The structures of α subunits demonstrate that the nucleotide binding site lies in a deep cleft between the binding domain and a helical domain. The nucleotide in these structures is not exposed to solvent and alteration to an 'open' conformation is required for nucleotide dissociation or binding.

The structures of the conserved β and more variable γ subunits are not yet known. The two polypetides form a stable noncovalent complex that acts as a single functional unit. The existence of five β subunits and at least seven γ subunits predicts the existence of numerous dimeric species. It has been shown that the β_1 and β_2 subunits can both interact with several γ subunits,[18–20] but β_2 and γ_1 do not interact. This selectivity has been mapped to the central portion of the γ subunit.[21,22]

A unique feature of the β subunits is that they are composed largely of seven homologous repeated units that have been referred to as WD repeats. Neer[7] has reviewed this phenomenon and hypothesized that the different repeats form unique but related domains for interaction with numerous other proteins.

Regulation of effectors

Initial studies of the regulation of adenylyl cyclase and the retinal cGMP-dependent phosphodiesterase identified the α subunits of G_s and G_t, respectively, as necessary and sufficient regulators of the enzymes.[5] The $\beta\gamma$ subunits were assigned an inhibitory role as they could facilitate inactivation of the α subunits. It is now clear that $\beta\gamma$ subunits also function as positive regulators of downstream pathways.[10,23,24] Indeed, α and $\beta\gamma$ subunits have complementary activities. In the GTP form, α acts as a stimulator; in the GDP form, α associates with free $\beta\gamma$ subunits to block their activity. Free $\beta\gamma$ can activate downstream effectors or attenuate the stimulatory activity of α subunits by hastening their deactivation or by binding to αGDP to reduce spontaneous nucleotide exchange. It is still not known whether the $\beta\gamma$ subunits exist in only a single state with equal potency for both activating and inhibiting roles or whether the subunits can attain some other more potent but transient form when stimulated by receptors.

The complex nature of the effectors and their regulation by G protein subunits are characterized by three examples. The simplest is regulation of cGMP-PDE by G_t.[25,26] Stimulation of rhodopsin by absorption of light causes the activation of multiple G_t molecules. The activated α_t then stimulates phosphodiesterase by removing the inhibitory constraint of the two PDE-γ subunits on the catalytic PDE-$\alpha\beta$ subunits. While $\beta\gamma_t$ subunits are required for activation of α_t by receptors and facilitate the action of rhodopsin kinase, they do not directly regulate the downstream pathway.

The second example is regulation of phosphatidylinositol-specific phospholipase C (PLC) (see Chapter 12). Isozymes of the PLC-β subclass can be regulated by both α_q and $\beta\gamma$ subunits. However, the regulation of individual isozymes by the respective subunits varies in extent and potency[27] (Figure 19.4). Thus, PLC-β1 and -β3 are highly responsive to stimulation by the α subunit of G_q while the PLC-β2 and -β3 isozymes are readily activated by $\beta\gamma$ subunits. The unique feature of these enzymes is that stimulation can be achieved independently by either subunit. Whereas regulation of PLC-β1 will be largely dependent on the G_q pathway, the β2 and β3 isozymes can respond to the generation of $\beta\gamma$ subunits from other G proteins.

Regulation of adenylyl cyclase can also occur through the action of both α_s and $\beta\gamma$ subunits. This regulation is further complicated in that $\beta\gamma$ subunits can be either stimulatory or inhibitory, depending on the subtype of cyclase.[28] Some subtypes are also the targets of other regulators such as Ca^{2+}, Ca^{2+}/calmodulin and protein kinases. Differential regulation of some of the isozymes of adenylyl cyclase is depicted in Figure 19.5 (see review[29]). In contrast to the independent activation of βγ subunits on PLC, stimulation of type II cyclase by βγ is dependent on co-stimulation by α_s. The potency of α_s is much greater than that of βγ subunits. Therefore, it is presumed that the primary effects of βγ come from G proteins (such as G_i) that are more abundant than G_s and that this bivalent regulation is designed primarily for co-ordinated effects by multiple pathways.

It becomes readily apparent that stimuli can promote different effects on cAMP in cells depending on the type of cyclase expressed. Whereas stimulation of G_i would inhibit cAMP synthesis by cyclase types I, V and VI, it would potentiate the activity of types II or IV. Increased intracellular Ca^{2+} might stimulate the type I cyclase through a calmodulin complex but inhibit types V or VI by direct action. The anticipated effects of stimuli on cAMP accumulation are further complicated by the expression of multiple subtypes of cyclase in a single cell.

A question of balance

The interaction between α and βγ subunits provides a pathway for autoregulation of G proteins. The classic model predicts that βγ subunits deactivate α and α(GDP) turns off βγ. In the simplest scenario, the cell needs to produce stoichiometric amounts of α and βγ subunits so that free βγ is not available and αGDP is fully associated with βγ to suppress autoactivation. However, since each cell produces several different α, β and γ subunits, it seems unlikely that the expression of only the appropriate balance of subunits can be co-ordinated. There must be other mechanisms of control.

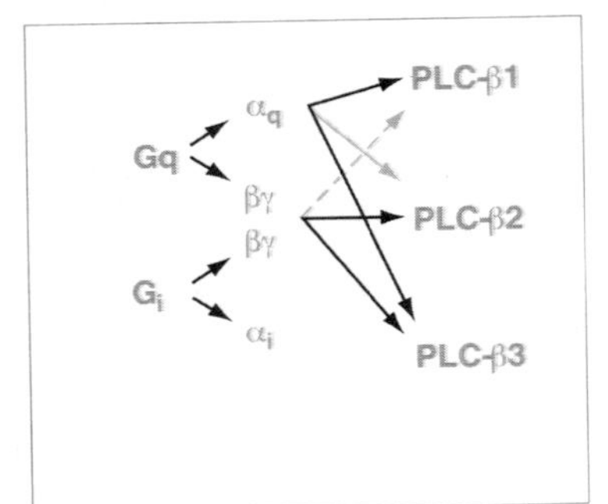

Figure 19.4 Regulation of PLC-β isozymes by G protein subunits. Both α subunits of the G_q subfamily and βγ subunits can independently activate phospholipase C (PLC). The efficacy of the different subunits for PLC-β isozymes is indicated. Regulation of PLC-β4 follows the pattern of PLC-β1.[56]

Subcellular localization may play a role. Signaling units may be localized to specific areas of the membrane and only G proteins within these areas may be available for function. For example, a G_q protein might need to be located near a PLC-β for its activation to be effective. In this scenario, the stoichiometry and types of G proteins in a signaling patch may be tightly controlled. The recent descriptions of G protein enrichment in submembrane structures referred to as caveolae may support this idea.[30,31] The exquisite specificity of G protein subunits for receptors and effectors *in vivo* is not found *in vitro*. For example, all βγ

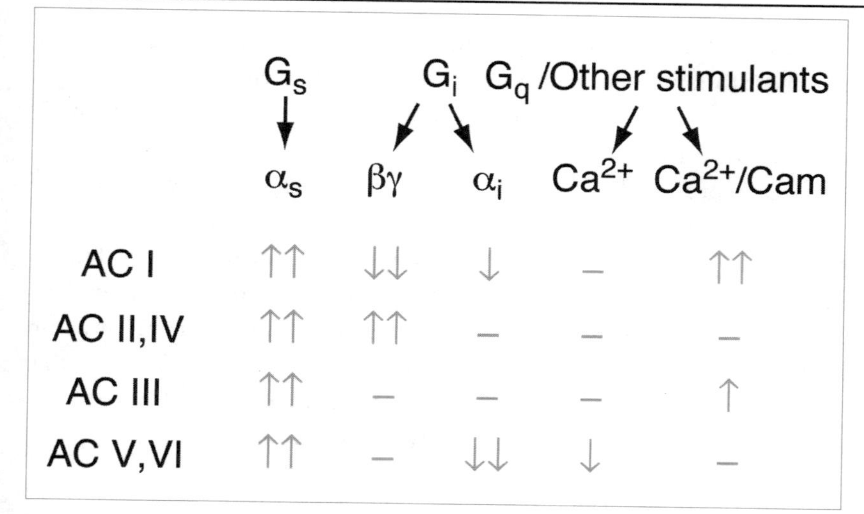

Figure 19.5 Differential regulation of adenylyl cyclase isozymes. The α_s and $\beta\gamma$ subunits of G proteins and other regulators have widely varying effects on different subtypes of the enzyme. This can result in synergy or antagonism by inputs from multiple pathways. The potential modes of regulation available will be determined primarily by expression of the various subtypes. The limited expression of the types I and III enzymes in neural tissue and olfactory epithelium, respectively, will characterize responses in at least some of the cells in these areas. Abbreviations: AC, adenylyl cyclase; Cam, calmodulin.

dimers tested *in vitro* act on various effectors[18,32,33] but very specific βγ subunits are needed for receptor signaling *in vivo*.[34,35] While the specificity of the receptor may account for some of this, localization probably also plays a role. The sorting of G_t, rhodopsin and phosphodiesterase to the outer segments of visual sensory cells is an excellent example of localization. A corollary to such signaling domains is the speculation by Jahangeer and Rodbell[36] that large pools of α subunits exist as oligomers in cells. Such complexes may play a distinct functional role or represent a reservoir of excess α subunits that are excluded from action but can be recruited to functional complexes (i.e. associated with βγ or, perhaps, recruited to a localized 'signaling domain').

GTPase activating proteins might regulate the activity of α subunits. The recent demonstration that the effectors PLC-β and cGMP-PDE could stimulate the GTPase activity of G_q[37] and G_t,[38,39] respectively, demonstrates a mechanism for both a rapid decay of signaling and a means to suppress basal activity by randomly activated α subunits (Figure 19.6). Such self-limiting activities define a system in which effective signaling can only be maintained by rapid activation of the G proteins.

The βγ subunits pose a more difficult problem. They appear to be stable and have no obvious mechanism of inactivation, except for association with αGDP. What prevents their random activation of effectors? One possibility is that free βγ subunits are only effective when freshly generated by receptors. For example, higher concentrations of βγ required for regulation of effectors may only be generated in the vicinity of receptors. Diffusion of free subunits would dilute their efficacy. Alternatively, βγ that is derived during receptor-dependent

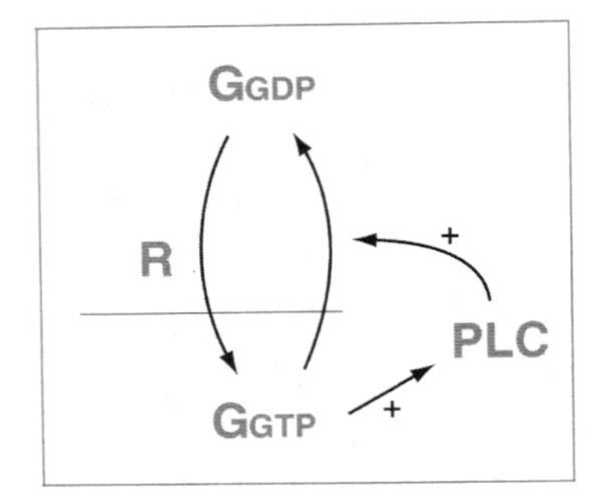

Figure 19.6 Phospholipase C-β (PLC-β) stimulates the intrinsic GTPase activity of G_q, thus acting directly as a feedback inhibitor. Such a system is subject to rapid deactivation but requires sustained activation by receptors for effective signaling.

activation of G proteins may assume a more potent regulatory state that is subject to rapid decay, although there is no evidence for such a state.

A means of controlling free βγ may be through its association with a variety of other cellular proteins (Figure 19.7). In the retinal system, phosducin presumably prevents activation of G_t by sequestering βγ.[40,41] Analogous proteins might regulate free βγ in other systems. Observations that βγ subunits interact with receptor kinases[42,43] has led to evidence that these subunits can interact with proteins that contain PH domains[44] (see Chapter 9). If βγ serves as a more general site for localization of these proteins, such action could either facilitate the function of these moieties or may help to maintain a pool of inactive βγ which can equilibrate with α subunits as required.

Post-translational modification and G protein function

Several modifications affect the lipophilicity and action of G protein subunits (see Table 19.2 and reviews[45,46]). Isoprenylation of the γ subunits via a CAAX motif (where C = cysteine, A = aliphatic and X = any amino acid) at their carboxyl termini anchors the βγ complexes to the membrane. If isoprenylation of γ is prevented, complexes of β and γ subunits are still formed but these exist mostly in cytosol, exhibit reduced affinity for α subunits[47] and have attenuated ability to regulate effectors.[47,48] It is not clear whether diminished activity is simply due to reduced localization on membranes containing effectors or a more direct effect on βγ/effector interaction.

Acylation of α subunits is also important for function (Table 19.2). Several of the α subunits acquire the stable addition of a myristate group to their amino terminus during translation. This modification helps direct these proteins to membranes; at least part of this effect is due to an increased affinity for membrane-associated βγ subunits. This modification is also important for potent regulation of effector proteins.

Palmitoylation via a thioester linkage has been observed on most α subunits. The reversible nature of this modification suggests a potential regulatory role. As with the other lipid modifications, expression of mutated proteins indicates that palmitoylation subserves two significant roles: increased association with membranes and increased ability to interact with effector proteins. Increased palmitoylation of α_s when it is activated *in vivo*[49–51] is indicative of increased turnover of this

Table 19.2 Post-translational modifications of G protein subunits

Modification	Subunit	Proposed function
Carboxy-terminal prenylation[a]:		
Farnesyl (C15)	γ_1	Membrane association (C20 > C15)
Geranylgeranyl (C20)	non-retinal γ subunits	↑ Affinity for α ↑ Stimulation of effectors
Acylation:		
Amino-terminal myristate	α_o, α_i, α_t, α_z	Membrane association ↑ Affinity for $\beta\gamma$ ↑ Potency for effectors
Palmitate	All α (except α_t)	↑ Membrane association ↑ Efficacy for effectors
Arachidonate	α in platelets	Unknown
Phosphorylation	α_i, α_z, $\beta\gamma$	Unknown

[a] Processing of the CAAX motif includes carboxy-terminal proteolysis of the last three amino acids and carboxymethylation of the carboxy-terminal cysteine in addition to the isoprenylation.

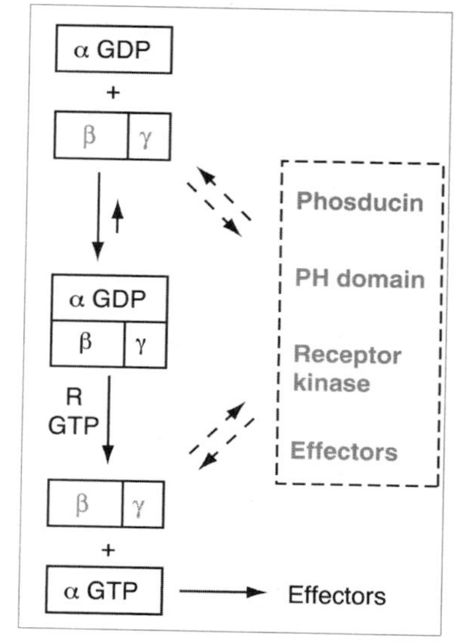

Figure 19.7 The $\beta\gamma$ subunits can interact reversibly with a variety of intracellular proteins. These interactions may serve to regulate the function of these proteins and/or may provide a means of keeping the free pool of $\beta\gamma$ to a minimum.

modification during functional cycling. It is interesting to speculate that regulated palmitoylation will be a means of controlling the activity of specific G proteins (see refs 46 and 52).

Determining a role for phosphorylation of G proteins has been an elusive goal. Such modification is a likely path for feedback inhibition and crosstalk with other signaling pathways. While several α subunits can be phosphorylated by different kinases (see reviews[7,8]), the functional relevance of such modifications has not been established. An interesting modification of $\beta\gamma$ subunits is the phosphorylation of β by the γ phosphate of GTP bound to the α subunit.[53] The potential for regulation by this 'autophosphorylation' is intriguing.[10]

Summary

Signal transduction via the heterotrimeric G proteins occupies a broad role in biology. It provides crucial mediation of cellular responses to the external environment in species from molds to mammals. This chapter has focused on signaling at the plasma membrane. However, heterotrimeric G proteins are also found associated with intracellular structures and several experimental paradigms suggest roles in intracellular regulation. It is anticipated that further exploration will elucidate receptors for detecting intracellular signals and G protein pathways for local regulation of metabolic events within the cell.

The molecular mechanisms by which G proteins effect regulation have been intensely studied. The structures of several α subunits are available and elucidation of the structure of βγ is imminent. These structures will allow more rigorous testing of mechanisms for activation and dissociation of the subunits and the physiological relevance of these mechanisms *in vivo*, in addition to aiding in the appraisal of various G protein deficiencies in the etiology of several diseases.

Our understanding of the circuitry by which G proteins couple receptors to effects in cells is limited. While reconstitutions *in vitro* have begun to provide information on biochemical connections that may be formed, it is clear that the cell imposes more rigorous restrictions. Determination of what is actually allowed in the cellular milieu will be required to understand fully the potential action of a stimulus. One goal is more rigorous evaluation of the connections among receptors, G proteins and effectors and how these connections are controlled *in vivo*.

Future goals also include the elucidation of new pathways that are modulated by this family of proteins. A current emphasis is to determine how several G protein α subunits effect regulation of MAP kinase pathways[54] and how βγ subunits effect regulation of similar pathways in yeast.[55] There are likely to be many proteins other than direct effectors and receptors that also interact with G proteins. These proteins might help to localize the G proteins or more complicated complexes into signaling units or constitute autoregulatory mechanisms (such as GAP proteins or kinases for feedback inhibition). Finally, elucidation of the crosstalk between G proteins and other major signaling pathways will enhance our understanding of the highly varied responses that can frequently be observed from the same stimulus on different cells or on the same cell under different conditions.

References

1 Rall, T.W. and Sutherland, E.W. (1958) Formation of a cyclic adenine ribonucleotide by tissue particles. *J. Biol. Chem.*, **232**, 1065–76.

2 Rall, T.W., Sutherland, E.W. and Berthet, J. (1957) The relationship of epinephrine and glucagon to liver phosphorylase. IV. Effect of epinephrine and glucagon on the reactivation of phosphorylase in liver homogenates. *J. Biol. Chem.*, **224**, 463–75.

3 Rodbell, M., Birnbaumer, L., Pohl, S.L. and Krans, H.M.J. (1971) The glucagon-sensitive adenyl cyclase system in plasma membranes of rat liver. V. An obligatory role of guanyl nucleotides in glucagon action. *J. Biol. Chem.*, **246**, 1877–82.

4 Ross, E.M. and Gilman, A.G. (1977) Resolution of some components of adenylate cyclase necessary for catalytic activity. *J. Biol. Chem.*, **252**, 6966–9.

5 Gilman, A.G. (1987) G proteins: transducers of receptor-generated signals. *Annu. Rev. Biochem.*, **56**, 615–49.
6 Linder, M.E. and Gilman, A.G. (1992) G Proteins. *Sci. Am.*, **267**, 56–65.
7 Neer, E.J. (1995) Heterotrimeric G proteins: organizers of transmembrane signals. *Cell*, **80**, 249–57.
8 Nürnberg, B., Gudermann, T. and Schultz, G. (1995) Receptors and G proteins as primary components of transmembrane signal transduction. Part 2. G proteins: structure and function. *Clin. Invest.*, **73**, 123–32.
9 Birnbaumer, L. and Birnbaumer, M. (1995) Signal transduction by G proteins: 1994 edition. *J. Recept. Res.*, **15**, 213–52.
10 Sternweis, P.C. (1994) The active role of βγ in signal transduction. *Curr. Opinion Cell Biol.*, **6**, 198–203.
11 Noel, J.P., Hamm, H.E. and Sigler, P.B. (1993) The 2.2 Å crystal structure of transducin-α complexed with GTPγS. *Nature*, **366**, 654–63.
12 Lambright, D.G., Noel, J.P., Hamm, H.E. and Sigler, P.B. (1994) Structural determinants for activation of the α-subunit of a heterotrimeric G protein. *Nature*, **369**, 621–8.
13 Coleman, D.E., Lee, E., Mixon, M.B. *et al.* (1994) Crystallization and preliminary crystallographic studies of Giα1 and mutants of Giα1 in the GTP and GDP-bound states. *J. Mol. Biol.*, **238**, 630–4.
14 Coleman, D.E., Berghuis, A.M., Lee, E. *et al.* (1994) Structures of active conformations of $G_{i\alpha 1}$ and the mechanism of GTP hydrolysis. *Science*, **265**, 1405–12.
15 Bourne, H.R., Sanders, D.A. and McCormick, F. (1991) The GTPase superfamily: conserved structure and molecular mechanism. *Nature*, **349**, 117–27.
16 Schweins, T. and Wittinghofer, A. (1994) Structures, interactions and relationships. *Curr. Biol.*, **4**, 547–50.
17 Sondek, J., Lambright, D.G., Noel, J.P. *et al.* (1994) GTPase mechanism of G proteins from the 1.7-Å crystal structure of transducin α-GDP–AlF_4^-. *Nature*, **372**, 276–9.
18 Iñiguez-Lluhi, J.A., Simon, M.I., Robishaw, J.D. and Gilman, A.G. (1992) G protein βγ subunits synthesized in Sf9 cells. *J. Biol. Chem.*, **267**, 23409–17.
19 Schmidt, C.J., Thomas, T.C., Levine, M.A. and Neer, E.J. (1992) Specificity of G protein β and γ subunit interactions. *J. Biol. Chem.*, **267**, 13807–10.
20 Pronin, A.N. and Gautam, N. (1993) Proper processing of a G protein gamma subunit depends on complex formation with a β subunit. *FEBS Lett.*, **328**, 89–93.
21 Spring, D.J. and Neer, E.J. (1994) A 14-amino acid region of the G protein gamma subunit is sufficient to confer selectivity of gamma binding to the β subunit. *J. Biol. Chem.*, **269**, 22882–6.
22 Lee, C., Murakami, T. and Simonds, W.F. (1995) Identification of a discrete region of the G protein gamma subunit conferring selectivity in βγ complex formation. *J. Biol. Chem.*, **270**, 8779–84.
23 Iñiguez-Lluhi, J.A., Kleuss, C. and Gilman, A.G. (1993) The importance of G-protein βγ subunits. *Trends Cell Biol.*, **3**, 230–6.

24 Clapham, D.E. and Neer, E.J. (1993) New roles for G-protein βγ-dimers in transmembrane signaling. *Nature*, **365**, 403–6.

25 Stryer, L. (1991) Visual excitation and recovery. *J. Biol. Chem.*, **266**, 10711–4.

26 Pugh, E.N. and Lamb, T.D. (1993) Amplification and kinetics of the activation steps in phototransduction. *Biochim. Biophys. Acta*, **1141**, 111–49.

27 Smrcka, A.V. and Sternweis, P.C. (1993) Regulation of purified subtypes of phosphatidylinositol specific phospholipase C β by G protein α and βγ subunits. *J. Biol. Chem.*, **268**, 9667–74.

28 Tang, W.J. and Gilman, A.G. (1992) Type-specific regulation of adenylyl cyclase by G protein βγ subunits. *Science*, **254**, 1500–3.

29 Taussig, R. and Gilman, A.G. (1995) Mammalian membrane-bound adenylyl cyclases. *J. Biol. Chem.*, **270**, 1–4.

30 Chang, W.-J., Ying, Y., Rothberg, K.G. *et al.* (1994) Purification and characterization of smooth muscle cell caveolae. *J. Cell Biol.*, **126**, 127–38.

31 Sargiacomo, M., Sudol, M., Tang, Z. and Lisanti, M.P. (1993) Signal transducing molecules and GPI-linked proteins form a caveolin-rich insoluble complex in MDCK cells. *J. Cell Biol.*, **122**, 789–808.

32 Ueda, N., Iñiguez-Lluhi, J.A., Lee, E. *et al.* (1994) G protein βγ subunits: simplified purification and properties of novel isoforms. *J. Biol. Chem.*, **269**, 4388–95.

33 Wickman, K.D., Iñiguez-Lluhi, J.A., Davenport, P.A. *et al.* (1994) Recombinant G-protein βγ-subunits activate the muscarinic-gated atrial potassium channel. *Nature*, **368**, 255–7.

34 Kleuss, C., Scherubl, H., Heschler, J. *et al.* (1992) Different β-subunits determine G-protein interaction with transmembrane receptors. *Nature*, **358**, 424–6.

35 Kleuss, C., Scherübl, H., Hescheler, J. *et al.* (1993) Selectivity in signal transduction determined by γ subunits of heterotrimeric G proteins. *Science*, **259**, 832–4.

36 Jahangeer, S. and Rodbell, M. (1993) The disaggregation theory of signal transduction revisited: further evidence that G proteins are multimeric and disaggregate to monomers when activated. *Proc. Natl. Acad. Sci. USA*, **90**, 8782–6.

37 Berstein, G., Blank, J.L., Jhon, D.-Y. *et al.* (1992) Phospholipase C-β1 is a GTPase-activating protein for $G_{q/11}$, its physiological regulator. *Cell*, **70**, 411–8.

38 Arshavsky, V.Y. and Bownds, M.D. (1992) Regulation of deactivation of photoreceptor G protein by its target enzyme and cGMP. *Nature*, **357**, 416–7.

39 Pagès, F., Deterre, P. and Pfister, C. (1992) Enhanced GTPase activity of transducin when bound to cGMP phosphodiesterase in bovine retinal rods. *J. Biol. Chem.*, **267**, 22018–21.

40 Bauer, P.H., Müller, S., Puzicha, M. *et al.* (1992) Phosducin is a protein kinase A-regulated G-protein regulator. *Nature*, **358**, 73–6.

41 Lee, R.H., Ting, T.D., Lieberman, B.S. *et al.* (1992) Regulation of retinal cGMP cascade by phosducin in bovine rod photoreceptor cells. *J. Biol. Chem.*, **267**, 25104–12.

42 Haga, K. and Haga, T. (1992) Activation by G protein βγ subunits of agonist- or light-dependent phosphorylation of muscarinic acetylcholine receptors and rhodopsin. *J. Biol. Chem.*, **267**, 2222–7.

43 Inglese, J., Koch, W.J., Caron, M.G. and Lefkowitz, R.J. (1992) Isoprenylation in regulation of signal transduction by G-protein-coupled receptor kinases. *Nature*, **359**, 147–50.

44 Inglese, J., Koch, W.J., Touhara, K. and Lefkowitz, R.J. (1995) $G_{\beta\gamma}$ interactions with PH domains and Ras-MAPK signaling pathways. *Trends Biochem. Sci.*, **20**, 151–6.

45 Casey, P.J. (1994) Lipid modifications of G proteins. *Curr. Opinion Cell Biol.*, **6**, 219–25.

46 Wedegaertner, P.B., Wilson, P.T. and Bourne, H.R. (1995) Lipid modifications of trimeric G proteins. *J. Biol. Chem.*, **270**, 503–6.

47 Iñiguez-Lluhi, J.A., Simon, M.I., Robishaw, J.D. and Gilman, A.G. (1992) G Protein β–γ subunits synthesized in Sf9 cells. *J. Biol. Chem.*, **267**, 23409–17.

48 Dietrich, A., Meister, M., Brazil, D. *et al.* (1994) Stimulation of phospholipase C-β2 by recombinant guanine-nucleotide-binding protein βγ dimers produced in a baculovirus/insect cell expression system – requirement of γ-subunit isoprenylation for stimulation of phospholipase C. *Eur. J. Biochem.*, **219**, 171–8.

49 Mumby, S.M., Kleuss, C. and Gilman, A.G. (1994) Receptor regulation of G-protein palmitoylation. *Proc. Natl. Acad. Sci. USA*, **91**, 2800–4.

50 Degtyarev, M.Y., Spiegel, A.M. and Jones, T.L.Z. (1993) Increased palmitoylation of the G_s protein α subunit after activation by the β-adrenergic receptor or cholera toxin. *J. Biol. Chem.*, **268**, 23769–72.

51 Wedegaertner, P.B. and Bourne, H.R. (1994) Activation and depalmitoylation of $G_{s\alpha}$. *Cell*, **77**, 1063–70.

52 Ross, E.M. (1995) Protein modification: palmitoylation in G-protein signaling pathways. *Curr. Biol.*, **5**, 107–9.

53 Wieland, T., Nürnberg, B., Ulibarri, I. *et al.* (1993) Guanine nucleotide-specific phosphate transfer by guanine nucleotide-binding regulatory protein β-subunits. *J. Biol. Chem.*, **268**, 18111–8.

54 Johnson, G.L. and Vaillancourt, R.R. (1994) Sequential protein kinase reactions controlling cell growth and differentiation. *Curr. Opinion Cell Biol.*, **6**, 230–8.

55 Herskowitz, I. (1995) MAP kinase pathways in yeast: for mating and more. *Cell*, **80**, 187–97.

56 Jiang, H., Wu, D. and Simon, M.I. (1994) Activation of phospholipase C β4 by heterotrimeric GTP-binding proteins. *J. Biol. Chem.*, **269**, 7593–6.

Part Three

Nuclear responses

20 Steroid hormone and nuclear receptors

Kelly LaMarco and Maria d.M. Vivanco

Steroid hormone receptors are ligand-activated regulatory proteins that modulate transcription of selected genes under specific developmental and metabolic conditions.[1] In contrast to receptors for peptide hormones, which are located in the cell membrane and evoke a second messenger to deliver the regulatory signal, steroid hormone receptors are present within the cell.

The steroid hormone receptor superfamily consists of receptors for the steroid hormones, which are found both in the cytoplasm and nucleus, and the related nuclear receptors for thyroid hormone, vitamin D and vitamin A, among others.[2–4] These intracellular receptors bind their respective ligands and undergo a conformational change referred to as 'transformation' to yield an active form of the receptor. The ligand-activated receptor can recognize and bind to specific DNA sequences (hormone-response elements or HREs) and bring about activation or repression of transcription initiation by RNA polymerase II at a nearby target gene[5] (Figure 20.1). The effect of ligand binding on steroid hormone and nuclear receptors varies among family members. In some cases it renders the receptor competent to bind in a sequence-specific manner to DNA. Ligand binding can also facilitate release of Hsp90 (steroid hormone receptors, see below), nuclear translocation (glucocorticoid receptor), interaction with transcriptional mediator proteins[6] and, in one case (thyroid hormone receptor), release

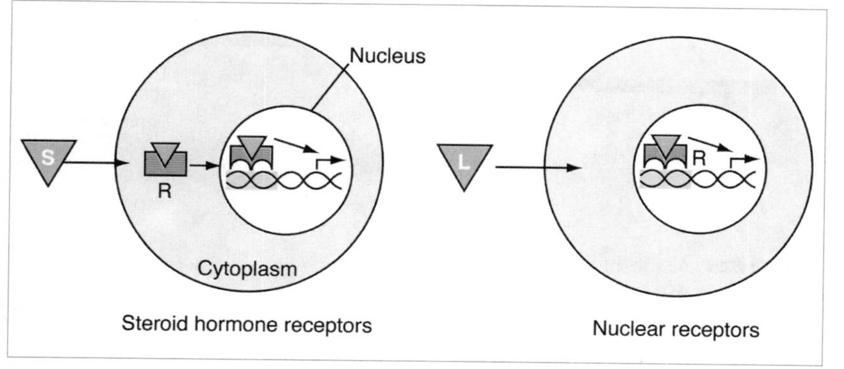

Figure 20.1 Transformation of intracellular receptors. Steroid hormone and nuclear receptors (R) exist in a transcriptionally inactive state. They are activated by a diffusible ligand [either a steroid hormone (S) or another ligand (L)] which binds to and causes a conformational change in the receptor that renders it competent to bind to DNA ('transformation'). The activated receptor can then bind to HREs (tinted region) and modulate transcription of an adjacent target gene (arrow). For GR, translocation from cytosol to nucleus accompanies transformation. Other steroid hormone receptors can exist in both the cytosol and nucleus in the inactive form. Retinoid, thyroid and vitamin D receptors are found in the nucleus regardless of their state of activation.

Signal Transduction. Edited by Carl-Henrik Heldin and Mary Purton. Published in 1996 by Chapman & Hall. ISBN 0 412 70810 8

Box 20.1 Synthesis and functions

Steroid hormones are derived from cholesterol, so that they can be produced at modest metabolic cost through a series of enzymatic reactions. The adrenal steroids are produced in different regions of the adrenal cortex: glucocorticoids in the zona fasciculata, mineralocorticoids in the zona glomerulosa and androgens in the zona reticularis. Glucocorticoids affect sugar and protein metabolism, display anti-inflammatory and immunosuppressive effects and have effects on the central nervous system. Aldosterone is essential for conservation of sodium ions and excretion of potassium ions from the body. The sex steroids (progesterone, estrogens and androgens such as testosterone) are produced in sex organs and control development of the embryonic reproductive system, masculinization or feminization of the brain, development of secondary sex characteristics, reproduction and adult reproductive behavior.

from DNA.[7,8] One result of ligand binding common to all known family members is an observed change in the transcriptional activity of respective target genes.

Molecular studies of steroid hormone receptor function have generally been performed either in purified *in vitro* systems or cultured mammalian cells. It is important to recognize, however, that models of steroid hormone function developed from molecular studies *in vitro* may not yield a completely accurate picture of their mechanisms of action and physiological functions in an intact organism. Thus a major challenge is to test the biological relevance of current models of receptor function in whole-animal experimental systems.

Structure of steroid hormone receptors

Despite the diversity of physiological responses elicited by receptors in this superfamily, all receptors analyzed thus far have a similar domain structure. Each receptor protein can be divided into six regions (A–F) on the basis of amino acid sequence similarity.[9] Each receptor consists of a variable amino-terminal region (A/B), a highly conserved, cysteine-rich central domain (C), followed by the D domain, and a moderately conserved carboxy-terminal region (E/F) (Figure 20.2). The amino-terminal domain varies in length among receptors from ~50 to 500 amino acids and contains a transcriptional regulatory domain that is the target of different kinases. The multifunctional carboxy-terminal region contains the hormone-binding domain, and may have signals for

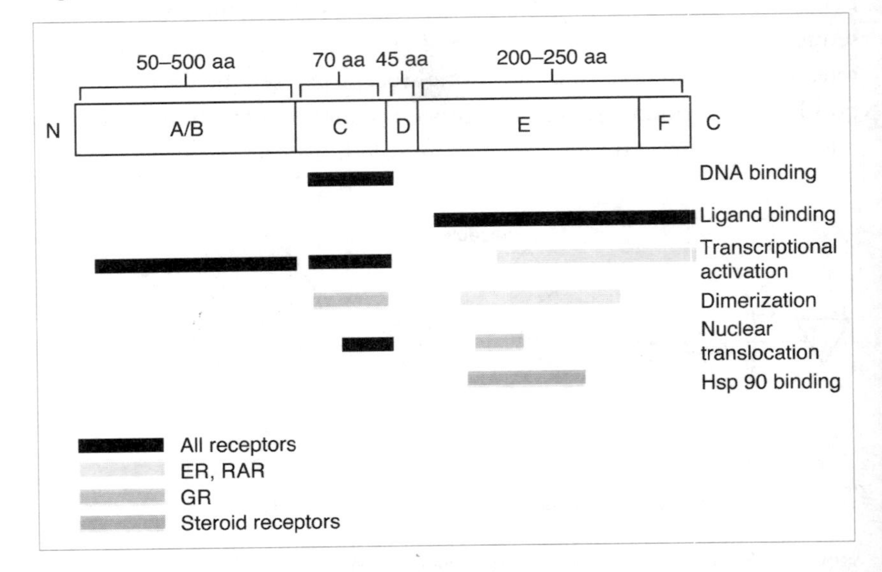

Figure 20.2 Domain structure of steroid and nuclear receptor family members. Functions associated with each domain for various classes of receptors are indicated by bars (see key). The estrogen receptor contains a weak dimerization function in the C domain.[17]

nuclear localization and dimerization, in addition to transcriptional regulatory functions. The C domain contains a zinc finger motif that constitutes the DNA-binding domain (Figure 20.3). This highly conserved domain has been used to isolate related receptors by low-stringency hybridization techniques.[10]

Box 20.2 Related metamorphic hormones

Vitamin D is a precursor of steroid-like hormones produced by specific tissues within the body. Vitamin D_3 (cholecalciferol) is synthesized in the skin by exposure to ultraviolet light, which is finally converted into its hormonal form in the kidney. Vitamin D functions in the regulation of calcium metabolism and development and mineralization of the skeleton. Thyroid hormone is a derivative of thyroxine and a deficiency results in abnormalities in growth, development, reproduction and metabolism. Ecdysteroids have developmental functions in insects and retinoic acid is a vertebrate morphogenic hormone. Retinoids (vitamin A or retinol) can be provided in the diet. Liver stellate cells control storage of retinol and also its mobilization to the bloodstream.

DNA-binding properties

DNA-binding and transcriptional activation experiments have led to a description of palindromic consensus HREs for members of the steroid hormone receptor superfamily. The amino acids in the stem of the first zinc finger of the C domain form an α-helix and make base contacts in the major groove of the DNA.[11] This helix represents the so-called 'P box', which allows distinct receptors to discriminate among different HREs in target gene promoters.[12–14]

The glucocorticoid, progesterone, androgen and mineralocorticoid receptors can all function through HREs found in the mouse mammary tumor virus long terminal repeat (MMTV-LTR) [consensus sequence: GGTACA$(n)_3$TGTTCT, where n is any nucleotide]. The consensus sequence for the retinoic acid and thyroid hormone receptor response elements [TCAGGTCA$(n)_{1-4}$TGACCTGA] is similar to that for the estrogen receptor (ER) [AGGTCA$(n)_3$TGACCT], but the spacing between the half-palindromes can vary and has functional significance (see below). DNA binding and transcriptional modulation are separable processes, as evidenced by the observation that thyroid hormone receptor (TR) can bind to an estrogen response element (ERE), but this binding does not lead to transcriptional activation.

Steroid hormone receptors: the glucocorticoid receptor

The purification and characterization of the glucocorticoid receptor (GR) and identification of glucocorticoid-responsive genes initiated molecular studies to investigate mechanisms of transcriptional regulation utilized by steroid hormone receptors.[5] DNA transfer experiments using cultured mammalian cells stimulated by the synthetic glucocorticoid dexamethasone led to the identification of DNA sequences that confer hormone responsiveness on to heterologous promoters. These studies, coupled with *in vitro* DNA binding experiments using purified receptor protein, resulted in the identification of a 15 bp palindromic consensus

GRE in the MMTV-LTR. This type of GRE was later referred to as 'simple' to distinguish it from more complex 'composite' GREs (see below). Purification to homogeneity of GR and production of receptor-specific antibodies provided the necessary reagents for cloning of receptor coding sequences.[15]

Several functional domains have been described for GR (Figure 20.2). The major transcriptional activation domain resides close to the amino terminus in the A/B domain. Region E/F contains domains for hormone binding and interaction with Hsp90. DNA-binding and dimerization functions have been localized to domain C and a nuclear localization signal resides in domain D.

Structural studies

Consistent with the palindromic structure of the GRE, footprinting and methylation protection studies suggested that GR bound to the GRE as a dimer. The ~70 amino acid DNA binding domain (domain C) can be expressed in bacteria as a functional recombinant protein fragment. This isolated domain, which contains two zinc atoms tetrahedrally co-ordinated by conserved cysteine residues, exhibits sequence-specific binding to GREs and contains amino acid residues necessary for dimerization. The structure of the GR DNA-binding domain obtained by NMR spectroscopy indicates that this domain exists as a monomer in solution.[16]

X-ray crystallographic analyses of DNA-binding domains of steroid receptors bound to HREs[11,17] have aided our understanding of the relationship between structure and function. For GR, two co-crystal structures, one high resolution (2.9 Å) and one lower resolution (4.0 Å), for the DNA-binding domain bound to a GRE have been reported.[11] The high-resolution structure employed a perfectly symmetrical sequence, with a 4 bp spacer between the two half-palindromes; for the lower-resolution structure, the more usual 3 bp spacer was used. A surprising result emerged from these studies: upon binding to DNA the monomeric GR subunit dimerizes, and each subunit interacts specifically with each half-site only when the half-sites are separated by 3 bp. With 4 bp spacing, only one subunit makes specific contacts with one half of the DNA element, while the other subunit is displaced and can make only nonspecific contacts with DNA. These observations suggest that the DNA functions as an 'allosteric effector' such that when the receptor contacts DNA, the two receptor subunits are positioned so as to stabilize dimerization. Figure 20.3 highlights amino acids important for dimerization and DNA contacts. Elucidation of the structures for other domains, such as the hormone-binding domain (with or without bound ligand) and the amino-terminal domain, are now required.

Receptor transformation: association with heat-shock proteins and nuclear localization

Steroid receptors isolated from hormone-free cells exist in a large (9S) heteromeric complex that contains several heat-shock proteins (Hsp90, Hsp70, Hsp56).[18] Heat-shock proteins are involved in protein folding, unfolding and trafficking and act as chaperones, binding to other proteins and preventing their abnormal folding. Hsp90 binds to the hormone-binding domain of the steroid receptors and maintains them in a transcriptionally inactive state. For GR, the 9S form is present in the cytosol. Hormone binding and dissociation of Hsp90 stimulate a conformational change in the receptor that converts it into a transcriptionally active form. The DNA-binding-competent form of GR then proceeds to the nucleus, where it associates with GREs of target genes and modulates transcription.

Both GR and ER contain a nuclear localization domain between the DNA- and hormone-binding domains (D domain).[19] In addition, there is a second, hormone-inducible nuclear localization signal in the hormone-binding domain of GR.[20] Although GR accumulates in the nucleus only in the presence of hormone, hormone binding does not appear to be required for nuclear localization of other steroid hormone receptors.

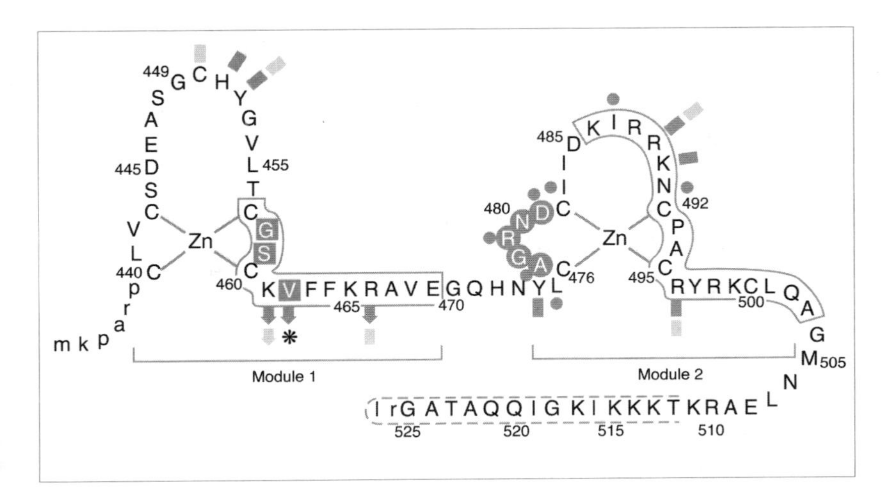

Figure 20.3 Schematic drawing of the GR DNA binding domain. Fragment 440–525, showing the Cys–Zn connectivity and α-helical segments, which are boxed. The numbering convention corresponds to that of the full-length native receptor. The relationship of structure and function is best described in terms of substructures called modules, which are indicated as Modules 1 and 2. Residues that make dimer interface contacts are indicated by the solid dots. Residues making phosphate contacts at the specific and the nonspecific site are indicated by solid and open rectangles, respectively. The solid and open arrows indicate base contacts at the specific and nonspecific site, respectively. The asterisk indicates that the contact between Val462 and the base is not made at the nonspecific site. A disordered section at the carboxyl terminus is indicated by the dashed lines. Three amino acids that direct the discrimination of glucocorticoid and estrogen receptor are indicated by white lettering in the solid squares (458, 459, 462), and those that discriminate between estrogen and thyroid response elements by white lettering in solid circles (478–481). Cloning artefacts from the expression vector construct are indicated by lower-case letters. [Reprinted with permission from Luisi, B.F., Xu, W.X., Otwinowski, Z. *et al.* (1991), *Nature*, 352, 497–505. Copyright 1991 Macmillan Magazines Limited.]

Binding of GR to chromatin

In eukaryotes, transcription occurs while DNA is in a compacted form known as chromatin, where DNA is wound around an octamer of histone proteins to form structures called nucleosomes. How do transcriptional regulatory proteins contend with the inherently repressive effects of chromatin? In general, when promoter sequences are reconstituted into nucleosomes *in vitro*, protein binding to DNA and initiation of transcription are inhibited. However, a number of transcription factors have the ability to recognize their cognate DNA elements in the context of nucleosomal DNA. For example, purified GR can bind specifically to regulatory DNA sequences that have been reconstituted in nucleosomes.[21] Binding of GR causes a disruption of the nucleosomal structure, such that an adjacent regulatory site becomes available for binding to its cognate regulatory protein.[22]

Studies with mammalian GR in the yeast *Saccharomyces cerevisiae* support the notion that GR interacts with a general cellular machine thought to operate by disrupting chromatin.[23] Activation of transcription by mammalian GR in yeast requires the *SWI* (switch) gene products, which have been hypothesized to relieve transcriptional repression by modifying chromatin structure.[24] GR forms a complex *in vitro* with SWI3 protein, and preincubation of an *in vitro* transcription extract with antibodies to SWI3 inhibits activation *in vitro* by mammalian GR. Further studies are required to understand fully the mechanisms by which proteins such as GR enhance transcription in the context of chromatin.

Composite elements: combinatorial regulation of transcription

The concept of composite elements emerged from studies of transcriptional repression by GR.[25] The limited view of a simple response element, where the ligand-activated receptor binds and activates transcription, does not describe accurately the regulatory sequences typically associated with genes transcribed by RNA polymerase II. Response elements that control gene transcription commonly contain clustered recognition sequences for multiple sequence-specific transcriptional regulatory proteins (composite elements). Whether activation or repression of transcription occurs from such an element depends on the specific combinations of factors that recognize the composite elements.

Whereas simple HREs have been characterized in detail, composite

HREs are complex, and no consensus sequences have emerged. In addition, the increasing number of examples of negative regulation by members of the steroid receptor superfamily has revealed that various mechanisms operate to achieve transcriptional repression.[26,27]

An example of transcriptional regulation by GR in the context of a composite element involves the transcription factor AP1. AP1 is typically upregulated in proliferative processes and is activated by phorbol esters via the protein kinase C signal transduction pathway. The transcriptionally active form of AP1 is present as either a homo- or heterodimeric complex composed of proteins of the Jun and Fos families (see Chapter 21). GR antagonizes AP1 activity at the collagenase promoter[28–30] and can either enhance or repress transcription at the proliferin promoter, depending on the protein subunit composition of AP1.[25] Taken together, these observations indicate that two distinct signal transduction pathways can converge at the transcriptional level.

Target genes and knock-out mice

The knowledge that steroid hormone receptors are sequence-specific, ligand-dependent transcription factors does not explain fully how they control cell growth and differentiation. A study of the genes regulated by glucocorticoids provides a better understanding of their physiological roles. For example, inflammatory cytokines, such as interleukin-1 (IL-1), are major regulators of matrix-degrading metalloproteinases such as collagenase. Steroid-activated GR inhibits expression of both the IL-1 and collagenase genes, explaining in part the anti-inflammatory and immune-suppressive effects of glucocorticoids.

The recent development of procedures for homologous recombination in mammalian cells has made it possible to use cloned DNA to alter genomic sequences and derive transgenic mice harboring mutant alleles. This gene knock-out approach allows a systematic analysis of receptor function *in vivo*. Analysis of the disruption of the GR gene has provided insight into the role of glucocorticoid signaling during development. Deletion of the GR gene results in death of the mice within the first few hours after birth owing to impaired lung development and deficient surfactant synthesis. In addition, perinatal induction of gluconeogenic enzymes in liver is impaired and regulation of glucocorticoid synthesis is perturbed, resulting in increased corticosterone and ACTH levels.[57]

Nuclear hormone receptors: the retinoid receptors

The retinoids [vitamin A (retinol) and its natural and synthetic derivatives] are required for many essential biological processes, including vision, reproduction, differentiation, metabolism, hematopoiesis, bone development and pattern formation during embryogenesis.[31,32] Retinoids are also used as therapeutic agents in the treatment of some skin diseases and certain types of cancer. An understanding of the signaling pathways that underlie the large diversity of responses to retinoids underwent a major revolution with the discovery of retinoid receptors that bind to all-*trans*- and 9-*cis*-retinoic acid (RA).

Identification and cloning of the genes encoding RA receptor proteins (RARs)[10,33] and comparative primary sequence analyses revealed that these receptors belong to the steroid/thyroid hormone receptor superfamily.[2–4] The fact that intracellular receptors share a common modular structure (Figure 20.2) facilitated the discovery of new members of this superfamily. One highly exploited approach to characterizing the transcriptional activation and ligand-binding properties of newly discovered receptors is to exchange a DNA-binding domain of known function (e.g. domain C from GR) with the corresponding sequences from the novel receptor to create a derivative with predictable chimeric features that can then be tested in cultured cells.

Discovery of a family of retinoid receptors

Three different subtypes of RARs have been isolated (RARα, -β and -γ), and each can be found in mammals, birds and amphibians. Each gene contains two promoters that generate receptor isoforms with variant amino termini, and several splice variants have been detected.[34] The various RARs are expressed in distinct patterns throughout development and in the adult organism, suggesting that they mediate different functions and explaining, in part, the diverse effects of retinoids observed *in vivo*.

Another nuclear receptor termed the retinoid X receptor (RXR) represents a second class of retinoid-responsive transcription factors.[35] Using different approaches, several investigators identified RXR as the factor that enhances DNA binding of RAR and TR.[36] As with the RARs, three distinct, but related RXR proteins (α, β and γ) have been reported, along with their corresponding patterns of expression. It later became apparent[37,38] that the high-affinity ligand for RXR is 9-*cis*-RA, an isomer of all-*trans*-RA. This important observation led to the definition of two retinoid receptor systems (RARs and RXRs) that are responsive to two distinct retinoid hormones (all-*trans*-RA and 9-*cis*-RA). Furthermore, the ability of RXR to form heterodimers with RARs, TR, VDR, the v-*erbA* oncogene product and an increasing number of orphan receptors[4] expands the number of potential physiological functions for the retinoids and allows their points of regulatory influence to overlap with those of many other important molecules.

Variability among RREs

The nuclear hormone receptors exhibit sequence similarity in their P boxes, implying that they bind to closely related or identical response element half-sites.[14] This raises the question of how specificity of DNA binding is achieved. Retinoid receptors recognize DNA binding motifs that are organized as palindromes, inverted repeats or direct repeats with different spacing between the half-sites.[36] Apparently, distinct combinations of protein surfaces from retinoid receptor homodimers or heterodimers allow the recognition of a specific type of response element. In addition, depending on the presence or absence of the specific ligand for one or both of the receptors, RXRs and RARs can either activate or repress transcription.[39,40] All of these possible permutations increase the combinatorial regulatory possibilities of retinoid receptors.

Thus, when building a regulatory complex that modulates specifically transcription of a retinoid-responsive target gene, potential points of regulation include (1) the subunit composition of the retinoid-responsive receptor complex, (2) the presence or absence of specific ligand(s) necessary to activate the receptor(s) in a given cell type and (3) recognition of the response element by the specific receptor complex. In addition, these receptors may interact with other classes of gene-specific transcription factors or with the basic transcription machinery.[41–43] One can see from this simple example multiple opportunities for regulatory specificity.

Retinoic acid and development

Studies performed nearly a half century ago in which vitamin A was administered maternally during gestation illustrated the importance of retinoids in the process of embryogenesis.[44] In addition, the effects of topical administration of RA on embryonic limb development suggest that RA plays a critical role during organogenesis.[45]

Box 20.3 The search for physiological ligands

Orphan receptors represent a class of putative target proteins for as yet unidentified small molecule ligands. A novel mammalian orphan receptor (FXR) was discovered recently and shown to form a heterodimeric DNA binding complex with RXR.[56] A screen for physiological ligands for this new orphan revealed that farnesol and its metabolites lead to an activation of transcription by the FXR-RXR complex in mammalian tissue culture cells. Farnesyl pyrophosphate is an important precurser in the mevalonate biosynthetic pathway. Farnesol has not been shown to bind directly to FXR. Hence it may activate indirectly transcription by FXR-RXR. Alternatively, another farnesol metabolite may serve as the authentic ligand for FXR.

It has been shown by *in situ* hybridization that RARα transcripts are ubiquitous, whereas RARβ and -γ transcripts are tissue restricted. Several laboratories have now begun to investigate the *in vivo* functions of RA receptors by mutagenesis of the corresponding genes in mice.[34] Surprisingly, the phenotypes of null mutants for individual receptors are relatively modest, implying that there is functional redundancy among members of the RAR family. However, many of the malformations observed in vitamin A-deficient animals are also found in mice carrying various combinations of RAR/RXR double mutations.[34] These genetic studies implicate RARs as essential for the transduction of RA signals during vertebrate development. One recently employed strategy for deciphering tissue-specific functions of RA was to express selectively a dominant-negative form of RAR in the epidermis of transgenic mice. These experiments revealed that RA is indeed required for normal skin development.[46]

Orphan receptors

Orphan receptors constitute an intriguing and ever-growing class of proteins whose defining characteristic is the lack of identification of physiological regulatory ligands. The discovery of physiologically relevant ligands is a priority in this field, as their identity may provide clues as to how aspects of physiology and development are regulated and integrated. It remains possible, however, that certain orphan receptors function as constitutively active transcription factors.

Many orphan receptors have structural domains reminiscent of those found in steroid hormone receptors. Orphans continue to appear as regulatory proteins in a wide variety of biological processes. Here, we limit our discussion to a specific orphan that illustrates seminal concepts in gene regulation.

Adipocyte differentiation: peroxisome proliferator-activated receptor

Adipocyte P2 (aP2) is an intracellular lipid-binding protein expressed exclusively in differentiated fat cells (adipocytes). Differentiation-dependent, tissue-specific transcription is directed by an adipocyte-specific enhancer composed of several regulatory elements. One element, termed ARE6, is necessary and sufficient for adipocyte-specific transcription.

ARE6 exhibits sequence similarity with the direct repeat of HREs spaced by one nucleotide. This motif is a preferred binding site for heterodimers of RXR and the peroxisome proliferator-activated receptors (PPAR). PPARs belong to the steroid hormone receptor superfamily and were discovered as proteins that are activated by agents (hypolipidemic drugs, plasticizers, herbicides) that cause proliferation of peroxisomes in rat liver. The aP2 enhancer is stimulated by peroxisome proliferators, fatty acids and 9-*cis*-RA. Amplification by polymerase chain reaction of PPAR-related sequences from mouse adipocyte cDNA resulted in the identification of an adipocyte-specific PPAR family member (mPPARγ2).[47] In addition, recombinant RXRα and mPPARγ2 form a complex on ARE6 *in vitro*, while transfection of mPPARγ2 and RXRα into cultured fibroblasts results in transcriptional activation of the aP2 enhancer.

If one *in vivo* function of mPPARγ2 is to induce adipocyte differentiation in response to physiological lipid activators, this orphan might provide a molecular means of communication between adipocyte differentiation and lipid metabolism. In a direct test of this possibility,[48] retroviral expression of mPPARγ2 in cultured fibroblasts induced differentiation of these cells into adipocytes, and differentiation was potentiated by addition of known PPAR activators. Coexpression of mPPARγ and C/EBP, a transcription factor that is highly expressed during adipocyte differentiation, results in a synergistic enhancement of this process. Thus, several levels of regulation participate to achieve tissue-specific transcription and adipocyte differentiation: stimulation by lipid and lipid-like compounds (and potentially by an appropriate physiological ligand for mPPARγ), tissue-specific expression and heterodimerization of intracellular receptors and combinatorial regulation by multiple classes of transcription factors.

Hormone receptors and tumorigenesis

The knowledge that intracellular receptors regulate genes involved in normal cell growth, differentiation and homeostasis[49] raises the possibility that steroid hormones and vitamins and their receptors also participate in tumor formation and progression. The cloning of the receptor genes has facilitated an evaluation of their roles in development, physiology and disease. There are clinical syndromes of hormone receptor mutations that result in hormone resistance and/or hormone independence. In addition, several receptors have been associated with various types of cancer.[50,51] Consequently, there is increasing interest in the possibility that the genes encoding these receptors are proto-oncogenes.

Genomic rearrangements can result in dramatic effects on cell growth. The oncogene v-*erbA*, which contributes to an avian erythroleukemia, is a mutated derivative of the gene encoding TR fused in-frame with the viral *gag* gene.[8,52] The protein product of this gene no longer binds thyroid hormones, but retains its ability to bind TREs, such that it functions as a dominant repressor of thyroid hormone responsive genes and appears to block the differentiation activity of endogenous TR. Approximately 90% of acute promyelocytic leukemias (APL) display a reciprocal chromosome 15:17 translocation, which disrupts the RARα gene and a gene on chromosome 15, named PML. Of the two abnormal fusion proteins that are formed, PML–RARα appears to be the active form in the leukemia.[53,54] Patients with APL show a dramatic clinical response to RA, which induces remission presumably because RA stimulates terminal differentiation of the malignant promyeloblasts.

By contrast, there are other tumor types where the intracellular receptor may remain intact. During skin cancer development in the mouse, intracellular levels of GR, in addition to its hormone-dependent nuclear translocation and specific DNA-binding activities, are unaltered throughout cancer progression. However, GR displays only modest transcriptional regulatory activity in preneoplastic cells, whereas it is highly active in the tumor cells.[55] This transition in activation potential of GR represents a molecular parameter that distinguishes early preneoplastic stages from tumor cells, and may suggest a role for GR in the tumorigenic process.

Summary

The progress made in our understanding of steroid and nuclear hormone receptor function is indeed impressive, yet a clear picture of various aspects of hormonal regulation continues to elude us. Genetic studies using gene knock-out techniques should provide a systematic analysis of steroid hormone receptor function *in vivo*. This approach may also be useful for the introduction of mutations that modify, rather than eliminate, receptor function. Finer mutagenesis of receptor sequences should allow assessment of the functions of individual domains and generation of dominant-negative alleles that interfere with wild-type receptor function.

For the retinoid receptors, an obvious challenge is to understand the functional interplay among the various receptor isoforms and ligands. Although retinoids are used with varying success for the treatment of certain cancers, secondary effects commonly result from pleiotropic actions of these ligands. A strategy for new therapies now being investigated is the development of ligand derivatives with selective affinities capable of inducing exclusively particular functions of a given receptor.

Further insights into the molecular properties of intracellular receptors will allow more precise descriptions of the regulatory mechanisms at work in target cells. These findings should, in turn, enable investigators to decipher the functions of intracellular receptors in various diseases and to use this knowledge in the development of sophisticated new therapies.

Acknowledgements

We thank Marie Classon, Robert M. Kypta, Bruce Spiegelman and Keith Yamamoto for insightful comments on the manuscript.

References

1 Milgrom, E. (1990) Steroid hormones, in *Hormones: from Molecules to Disease* (eds Baulieu, E.-E. and Kelly, P.A.) Chapman & Hall, London, pp. 387–442 (review).

2 Evans, R.M. (1988) The steroid and thyroid hormone receptor superfamily. *Science*, **240**, 889–95 (review).

3 O'Malley, B. (1990) The steroid receptor superfamily: more excitement predicted for the future. *Mol. Endocrinol.*, **4**, 363–9 (review).

4 Pfahl, M. (1994) Vertebrate receptors: molecular biology, dimerization and response elements. *Semin. Cell Biol.*, **5**, 95–103 (review).

5 Yamamoto, K.R. (1985) Steroid receptor regulated transcription of specific genes and networks. *Annu. Rev. Genet.*, **19**, 209–52 (review).

6 Halachmi, S., Marden, E., Matin, G. *et al.* (1994) Estrogen receptor-associated proteins: possible mediators of hormone-induced transcription. *Science*, **264**, 1455–8.

7 Damm, K., Thompson, C.C. and Evans, R.M. (1989) Protein encoded by v-*erbA* functions as a thyroid-hormone receptor antagonist. *Nature*, **339**, 593–7.

8 Sap, J., Munoz, A., Damm, K. *et al.* (1986) The c-*erbA* protein is a high-affinity receptor for thyroid hormone. *Nature*, **324**, 635–40.

9 Ham, J. and Parker, M.G. (1989) Regulation of gene expression by nuclear hormone receptors. *Curr. Opinion Cell Biol.*, **1**, 503–11 (review).

10 Petkovitch, M., Brand, N.J., Krust, A. and Chambon, P. (1987) A human retinoic acid receptor which belongs to the family of nuclear receptors. *Nature*, **330**, 444–50.

11 Luisi, B.F., Xu, W.X., Otwinowski, Z. *et al.* (1991) Crystallographic analysis of the interaction of the glucocorticoid receptor with DNA. *Nature*, **352**, 497–505.

12 Umesono, K. and Evans, R.M. (1989) Determinants of target gene specificity for steroid/thyroid hormone receptors. *Cell*, **57**, 1139–46.

13 Mader, S., Kumar, V., de Verneuil, H. and Chambon, P. (1989) Three amino acids of the oestrogen receptor are essential to its ability to distinguish an oestrogen from a glucocorticoid-responsive element. *Nature*, **338**, 271–4.

14 Forman, B.M. and Samuels, H.H. (1990) Dimerization among nuclear hormone receptors. *New Biol.*, **2**, 587–94 (review).

15 Miesfeld, R., Okret, S., Wikstrom, A.C. *et al.* (1984) Characterization of a steroid hormone receptor gene and mRNA in wild-type and mutant cells. *Nature*, **312**, 779–81.

16 Hard, T., Kellenbach, E., Boelens, R. *et al.* (1990) Solution structure of the glucocorticoid receptor DNA-binding domain. *Science* **249**, 157–60.

17 Schwabe, J.W.R., Chapman, L., Finch, J.T. and Rhodes, D. (1993) The crystal structure of the estrogen receptor DNA-binding domain bound to DNA: how receptors discriminate between their response elements. *Cell*, **75**, 567–78.

18 Pratt, W.B. and Welsh, M.J. (1994) Chaperone functions of the heat shock proteins associated with steroid receptors. *Semin. Cell Biol.*, **5**, 83–93 (review).

19 Picard, D., Kumar, V., Chambon, P. and Yamamoto, K.R. (1990) Signal transduction by steroid hormones: nuclear localization is differentially regulated in estrogen and glucocortoid receptors. *Cell Regul.*, **1**, 291–3.

20 Picard, D. and Yamamoto, K.R. (1987) Two signals mediate hormone-dependent nuclear localization of the glucocorticoid receptor. *EMBO J.*, **6**, 3333–40.

21 Perlmann, T. (1992) Glucocorticoid receptor DNA-binding specificity is increased by organization of DNA in nucleosomes. *Proc. Natl. Acad. Sci. USA*, **89**, 3884–8.

22 Cordingly, M.G., Tate, R.A. and Hager, G.L. (1987) Steroid-dependent interaction of transcription factors with the inducible promoter of mouse mammary tumor virus *in vivo*. *Cell*, **48**, 261–70.

23 Yoshinaga, S.K., Peterson, C.L., Herskowitz, I. and Yamamoto, K.R. (1992) Roles of SWI1, SWI2 and SWI3 proteins for transcriptional enhancement by steroid receptors. *Science*, **258**, 1598–604.

24 LaMarco, K. (1994) Dissecting a complex process. *Proc. Natl. Acad. Sci. USA*, **91**, 2886–7 (review).

25 Yamamoto, K.R., Pearce, D., Thomas, J. and Miner, J.N. (1992) Combinatorial regulation at a mammalian composite response element, in *Transcriptional Regulation* (eds McKnight, S. and Yamamoto, K.R.) Cold Spring Harbor Press, Cold Spring Harbor, NY, pp. 1169–92 (review).

26 Burcin, M., Kohne, A.C., Runge, D. *et al.* (1994) Factors influencing nuclear receptors in transcriptional repression. *Semin. Cancer Biol.*, **5**, 337–46 (review).

27 Saatcioglu, F., Claret, F.X. and Karin, M. (1994) Negative regulation by nuclear receptors. *Semin. Cancer Biol.*, **5**, 347–59 (review).

28 Schüle, R., Rangarajan, P., Kliewer, S. *et al.* (1990). Functional antagonism between oncoprotein c-Jun and the glucocorticoid receptor. *Cell*, **62**, 1217–26.

29 Jonat, C., Rahmsdorf, H.J., Park, K.K. *et al.* (1990) Antitumor promotion and anti-inflammation: down-modulation of AP-1 (Fos/Jun) activity by glucocorticoid hormone. *Cell*, **62**, 1189–204.

30 Yang-Yen, H.F., Chambard, J.C., Sun, Y.L. *et al.* (1990) Transcriptional interference between c-Jun and the glucocorticoid receptor: mutual inhibition of DNA binding due to direct protein–protein interaction. *Cell*, **62**, 1205–15.

31 Sporn, M.B., Roberts, A.B. and Goodman, D.S. (1984) *The Retinoids*, Academic Press, Orlando (review).

32 Gudas, L.J., Sporn, M.B. and Roberts, A.B. (1994) Cellular biology and biochemistry of the retinoids, in *The Retinoids* (eds Sporn, M.B., Roberts, A.B. and Goodman DeW.S.) Raven Press, New York, pp. 443–520 (review).

33 Giguere, L.J., Ong, E.S., Segui, P. and Evans, R.M. (1987) Identification of a receptor for the morphogen retinoic acid. *Nature*, **330**, 624–9.

34 Chambon, P. (1994) The retinoid signaling pathway: molecular and genetic analyses. *Semin. Cell Biol.*, **5**, 115–25 (review).

35 Mangelsdorf, D.J., Ong, E.S., Dyck, J.A. and Evans, R.M. (1990) Nuclear receptor that identifies a novel retinoic acid-response pathway. *Nature*, **345**, 224–9.

36 Mangelsdorf, D.J., Umesono, K. and Evans, R.M. (1994) The retinoid receptors in *The Retinoids* (eds Sporn, M.B., Roberts, A.B. and Goodman, DeW.S.) Raven Press, New York, pp. 319–49, and references cited therein.

37 Heyman, R.A., Mangelsdorf, D.J., Dyck, J.A. *et al.* (1992) 9-*cis*-Retinoic acid is a high-affinity ligand for the retinoid X receptor. *Cell*, **68**, 397–406.

38 Levin, A.A., Sturzenbecker, L.J., Kazmer, S. *et al.* (1992) 9-*cis*-Retinoic acid stereoisomer binds and activates the nuclear receptor RXRa. *Nature*, **355**, 359–61.

39 Graupner, G., Willis, K.N., Tzukerman, M. *et al.* (1989) Dual regulatory role for thyroid-hormone receptors allows control of retinoic-acid receptor activity. *Nature*, **340**, 653–6.

40 MacDonald, P.N., Dowd, D.R., Nakajima, S. *et al.* (1993) Retinoid X receptors stimulate and 9-*cis*-retinoic acid inhibits 1,25-dihydroxyvitamin D3-activated expression of the rat osteocalcin gene. *Mol. Cell Biol.*, **13**, 5907–17.

41 Klein-Hitpass, L., Tsai, S.Y., Weigel, N.L. *et al.* (1990) The progesterone receptor stimulates cell-free transcription by enhancing the formation of a stable preinitiation complex. *Cell*, **60**, 247–57.

42 Ing, N.H., Beekman, J.M., Tsai, S.Y. *et al.* (1992) Members of the steroid hormone receptor superfamily interact with TFIIB. *J. Biol. Chem.*, **267**, 17617–23.

43 Berkenstam, A., Vivanco Ruiz, M., Barettino, D. *et al.* (1992) Cooperativity in transactivation between retinoic acid receptor and TFIID. *Cell*, **69**, 401–12.

44 Wilson, J.G., Roth, C.B. and Warkany, J. (1953) An analysis of the syndrome of malformation induced by maternal vitamin A deficiency. Effects of restoration of vitamin A at various times during gestation. *Am. J. Anat.*, **92**, 189–217.

45 Hofmann, C. and Eichele, G. (1994) Retinoids in development, in *The Retinoids* (eds Sporn, M.B., Roberts, A.B. and Goodman, DeW.S.) Raven Press, New York, pp. 389–441 (review).

46 Saitou, M., Sugal, S., Tanaka, T. *et al.* (1995) Inhibition of skin development by targeted expression of a dominant-negative retinoic acid receptor. *Nature*, **374**, 159–62.

47 Tontonoz, P., Hu, E., Graves, R.A. *et al.* (1994) mPPARγ2: tissue-specific regulator of an adipocyte enhancer. *Genes Dev.*, **8**, 1224–34.

48 Tontonoz, P., Hu, E. and Spiegelman, B.M. (1994) Stimulation of adipogenesis in fibroblasts by PPARγ2, a lipid-activated transcription factor. *Cell*, **79**, 1–20.

49 Walsh, D. and Avashia, J. (1992) Glucocorticoids in clinical oncology. *Cleveland Clin. J. Med.*, **59**, 505–15 (review).

50 Miksicek, R.J. (1994) Steroid receptor variants and their potential role in cancer. *Semin. Cancer Biol.*, **5**, 369–79 (review).

51 Scott, G.K., Kushner, P., Vigne, J.L. and Benz, C.C. (1991) Truncated forms of DNA-binding estrogen receptors in human breast cancer. *J. Clin. Invest.*, **88**, 700–6.

52 Weinberger, C., Thompson, C.C., Ong, E.S. *et al.* (1986) The c-*erbA* gene encodes a thyroid hormone receptor. *Nature*, **324**, 641–6.

53 de Thé, H., Lavau, C., Marchio, A. *et al.* (1991) The PML–RARα fusion mRNA generated by the t(15;17) translocation in acute promyelocytic leukemia encodes a functionally altered RAR. *Cell*, **66**, 675–84.

54 Kakizuka, A., Miller, W.H., Umesono, K. *et al.* (1991) Chromosomal translocation t(15;17) in human acute promyelocytic leukemia fuses RARα with a novel putative transcription factor, PML. *Cell*, **66**, 663–74.

55 Vivanco, M.d.M., Jonson, R., Galante, P. *et al.* (1995) A transition in transcriptional activation by the glucocorticoid and retinoic acid receptors at the tumor stage of dermal fibrosarcoma development. *EMBO J.*, **14**, 2217–28.

56 Forman, B.M., Goode, E., Chen, J. *et al.* (1995) Identification of a nuclear receptor that is activated by farnesol metabolites. *Cell*, **81**, 687–93.

57 Cole, T.J., Blendy, J.A., Monaghan, A.P. *et al.* (1995) Targeted disruption of the glucocorticoid receptor gene blocks adrenergic chromaffin cell development and severely retards lung maturation. *Genes Dev.* **1**, 1608–21.

21 Transcription factors

James R. Woodgett

Most, if not all, signal transduction pathways ultimately impinge on gene transcription and alter the expression of genes in response to extracellular and intracellular cues. The response may lead to cell division in the case of mitogenic stimuli (a focus of much transduction research) but more often involves subtle changes in the capacity of a cell to react to its environment. Thus, changes in gene expression play a role in a vast number of cell responses from mast cell secretion to nerve depolarization. In the past decade, the purification and cloning of the regulators of gene transcription (the transcription factors or TFs) have spawned rapid advances in understanding of the mechanisms by which gene expression can be acutely controlled. This chapter reviews some of the recent advances in this fast-moving field and focuses on several pathways that may act as paradigms for the largely uncharted networks that co-ordinate and control nuclear responses to cell signaling. Most of the mechanisms discussed involve control by protein phosphorylation. The reader is referred to Chapter 20 for a discussion of steroid hormone receptor functions which bypass many of the signal amplification events associated with post-translational signaling.

Regulation of AP-1

Tumor viruses provided the first clues to key TFs involved in transcriptional regulation of growth control. Activator protein-1 (AP-1) has been one of the most intensively studied largely owing to two of its components being oncogene products. Thus, AP-1 is a dimer of members of the Fos and Jun protein family.[1,2] These proteins have a 'leucine zipper' dimerization interface carboxy-terminal to a DNA binding domain that recognizes the pseudopalindromic sequence TGA[C/G]TCA found in the regulatory region of numerous genes. Depending somewhat on the cell type, the *fos* and *jun* genes are subject to transcriptional induction and represent part of the cohort of 'immediate–early' genes induced by

Signal Transduction. Edited by Carl-Henrik Heldin and Mary Purton. Published in 1996 by Chapman & Hall. ISBN 0 412 70810 8

stimulation of quiescent cells.[3] However, AP-1 binding and activity can be induced in many cell types in the absence of active protein synthesis, indicating post-translational regulation. In addition, proteins requiring induction, such as c-Fos, also become phosphorylated following synthesis, leading to modulation of activity (see below).

Negative regulation of c-Jun by phosphorylation

Many resting cell types contain a dormant form of c-Jun protein which is highly phosphorylated at sites just amino-terminal to the DNA binding domain (see Figure 21.1).[4] The presence of phosphate groups at these residues interferes with DNA binding, thus preventing AP-1 from interacting with its DNA target. Upon cell stimulation, these sites are dephosphorylated and binding is no longer blocked.[4] Two protein kinases have been implicated in phosphorylating these inhibitory sites *in vitro* and *in vivo*, glycogen synthase kinase-3 (GSK-3)[5,6] and casein kinase-II.[7] GSK-3 has recently been demonstrated to be inactivated following mitogenic stimulation of cells (it is phosphorylated by Rsk 1, a protein kinase activated by the MAP kinase cascade), providing a mechanism whereby the balance of Jun phosphorylation at the inhibitory sites is distorted in favour of dephosphorylation.[8,9] The c-Jun-related proteins JunB and JunD appear to be similarly regulated since reporter gene assays show similar inhibition upon co-transfection with GSK-3.[5] Other proteins such as myogenin, a helix–loop–helix TF, are similarly negatively regulated by phosphorylation.[10]

Positive regulation of c-Jun by phosphorylation

Although interaction with target DNA is clearly a requisite for function,

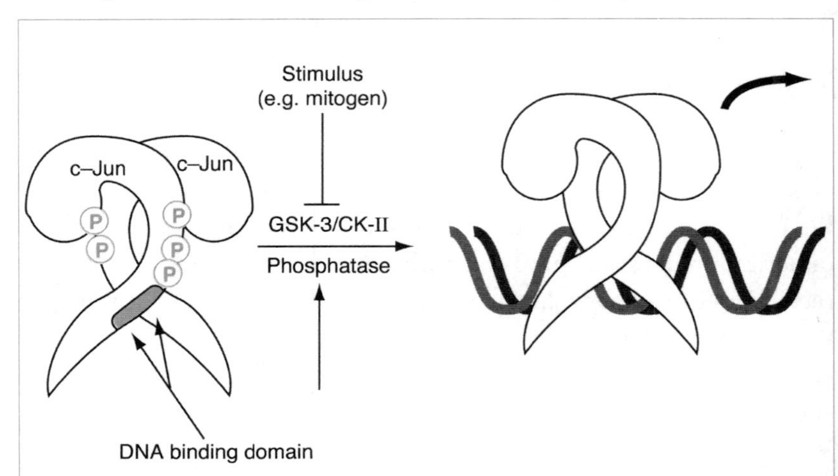

Figure 21.1 Regulation of c-Jun DNA binding by phosphorylation. Three phosphorylation sites proximal to the DNA binding domain of c-Jun are highly phosphorylated in resting cells and prevent association with DNA, rendering the protein inactive. Cellular stimulation causes dephosphorylation of the sites and relief of the block to binding. GSK-3, glycogen synthase kinase-3; CK-II, casein kinase-II.

TFs are regulated at several other levels. Dissection of TF genes has revealed a tremendous degree of modularity such that DNA binding and dimerization domains may be swapped between proteins and still retain function and specificity. This type of analysis revealed a domain required for interacting with the ubiquitous transcriptional machinery, such as the TATA binding proteins and the RNA polymerases. This so-called 'transactivation domain' forms an interaction interface with the other proteins, leading to their stimulation. There are many classes of transactivation domains, some of which are 'constitutive' and active under all circumstances, and others which are regulated. The major transactivation domain of c-Jun falls into the latter category since phosphorylation of two serine residues proximal to the domain is required for efficient transactivation.[11] Phosphorylation of these two residues (Ser63 and Ser73 in human c-Jun) is tightly regulated and stimulated in response to a variety of conditions. In keeping with c-Jun being an oncogene product, several growth-stimulatory molecules increase phosphorylation and activation of the protein (but see below).[12] However, by far the most potent inducers are agents associated with cellular stress such as inflammatory cytokines, thermal shock, UV radiation and metabolic poisons, suggesting a major role for this TF in mediating repair or protection against cellular injury.[13,14]

Elucidation of the major pathway regulating the Jun transactivation domain has been rapid following the cloning of a family of protein serine kinases that specifically target Ser63 and Ser73. This family consists of at least eight proteins derived from three genes by differential splicing which are termed stress-activated protein kinases (SAPKs) since they are primarily activated by cellular stresses but poorly induced by growth stimulants (unlike the MAP kinases).[13–15] Two splice variants of the family have been independently isolated and termed Jnk1 and 2 (Jun amino-terminal kinase).[16,17] These kinases are related to the MAP kinases both structurally and in their dependence on tyrosine and threonine phosphorylation for activity. Recently, the SAPK-activating kinase has been cloned (termed SEK).[18] This enzyme is also regulated by phosphorylation at serine and threonine residues catalyzed by a protein kinase termed MAP kinase kinase kinase or Mek kinase (see Figure 21.2).[19] Interestingly, expression of active Mek kinase in cells, which specifically activates the SAPKs, causes cell growth inhibition in fibroblasts (performed using inducible Mek kinase, since constitutively expressing Mek kinase cell lines are inviable[19]). These data suggest that c-Jun (and other SAPK targets) may be important negative regulators of cell growth, causing cell-cycle arrest in response to damage. The fact that c-Jun is a proto-oncogene does not necessarily mean that its activation is growth stimulatory in all cells, since v-Jun, the oncogenic form of c-Jun, is only a transforming protein in chick cells.[20] Indeed, mutations that make c-Jun a poorer transactivator in chick cells increase its transforming potential, suggesting that the protein may need to repress transcription of certain genes to induce a transformed phenotype.[21]

Several TF substrates other than c-Jun have been identified as SAPK targets, including ATF-2, JunD and Elk-1 (see below for discussion of these factors). While these proteins are all found in the nucleus, it is highly probable that cytoplasmic targets exist since the SAPKs are distributed throughout the cell. Moreover, immunofluorescence and cell fractionation experiments suggest that the SAPKs, unlike the MAP kinases, do not translocate to the nucleus upon stimulation. Perhaps related to this immobility, the SAPKs have the unusual property of binding tightly to their targets in the absence of activation. At least three substrates have 'docking' sites that are distal to the phosphorylation sites. The docking site in c-Jun is deleted in the oncogenic v-Jun which neither binds to the SAPKs nor is phosphorylated by them.[22,23] These observations raise the possibility that the SAPKs play an inhibitory role in Jun function. In resting cells at least a fraction of the SAPKs are tightly complexed with several TFs. Upon cell stimulation, the substrates become phosphorylated, releasing the kinases and relieving the steric inhibition.[23] v-Jun has avoided this level of regulation by preventing any interaction with the kinase.

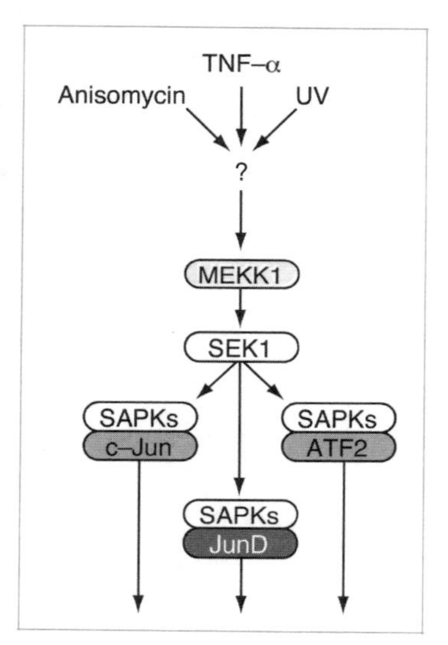

Figure 21.2 Regulation of c-Jun transactivation function by phosphorylation. The major transactivating domain of c-Jun requires phosphorylation at Ser63 and Ser73 for function. This is achieved via a cascade of protein kinases that respond to a variety of cellular stresses. TNF-α, tumor necrosis factor-α; MEKK, Mek kinase 1; SEK1, SAPK-activating kinase-1; SAPKS, stress-activating protein kinases.

Fos phosphorylation

Unlike c-Jun, c-Fos is virtually undetectable in resting cells but is rapidly induced following exposure to mitogens and many other stimuli. The protein is just as rapidly degraded and thus has a very transitory role with cells. Despite this short half life, c-Fos is subject to acute regulation by phosphorylation within cells. Several residues proximal to its carboxyl terminus are targeted by cyclic AMP-dependent protein kinase and phosphorylation of these sites induces transrepression of its own promoter (hence extinguishing its own expression).[24] As with v-Jun, the virally transduced forms of c-Fos contain deletions which remove this auto-suppressive function. Removal of the sites from c-Fos augments its transforming potential.[25]

A further similarity in the regulation of Fos and Jun is manifest by a region of homology between the two proteins that is proximal to the c-Jun transactivation domain and contains Ser73 of c-Jun.[26] In c-Fos, Thr232 has been identified as a target for FRK, a Fos-directed protein kinase that is activated by the Ras pathway but is distinct from the MAP kinases.[27] Phosphorylation of Thr232 stimulates transcriptional activity similar to phosphorylation of c-Jun by the SAPKs. However, FRK does not phosphorylate Jun and the SAPKs do not target Fos.

The CREB and CREM transcription factors

While AP-1 is activated in response to growth stimulants and stress, elevation of cyclic AMP induces a distinct set of transcriptional regulators. Thus agonist-induced synthesis of cyclic AMP causes dissociation of the regulatory subunit of cyclic AMP-dependent protein kinase and the release of its catalytic domain which largely translocates to the nucleus (see Chapter 15). A number of genes are stimulated by elevation of cyclic AMP and share a responsive element related to that of AP-1, TGACGTCA (see above).[28] Several proteins bind to this cyclic AMP response element (CRE) and the first to be identified was termed CREB. Phosphorylation of CREB at Ser133 is necessary for activation.[29] Recently, the basis of this activation was shown to be phosphorylation-dependent binding to a transcriptional accessory factor termed CREB binding protein (CBP) which directly interacts with TFIIB, a central component of the transcriptional machinery (Figure 21.3).[30] The generality of this type of regulation remains to be determined but it is highly likely that other examples of phosphorylation-dependent interactions will emerge, establishing this as a key mechanism via which transcription is tied to signal transduction. CREB contains other potential phosphorylation sites which may modulate its activity, although Ser133 is clearly the dominant regulatory element.

Another cyclic AMP-responsive TF that also binds to CREs (CREM) is subject to more complex regulation. The CREM gene is differentially spliced, generating several distinct proteins some of which are antagonistic.[31] Like CREB, CREM is a phosphoprotein and is phosphorylated at Ser117 by cyclic AMP-dependent protein kinase. This site is also targeted by other protein kinases including p70 S6 kinase and Ca^{2+}/calmodulin-dependent protein kinases.[32] CREM may therefore integrate multiple signals impinging on CRE-containing genes.

CREs are also responsive to expression of adenoviral proteins, specifically E1A protein, which interacts with and activates a TF termed ATF2. Like c-Jun, ATF2 requires phosphorylation by SAPKs for

Figure 21.3 Phosphorylation of cyclic-AMP response element binding protein (CREB) recruits an intermediary protein to the transcriptional machinery. Cyclic AMP-dependent protein kinase phosphorylates Ser133 of CREB which facilitates an interaction with CREB binding protein (CBP), a 300 kDa protein that also interacts with proteins required for initiation of transcription.

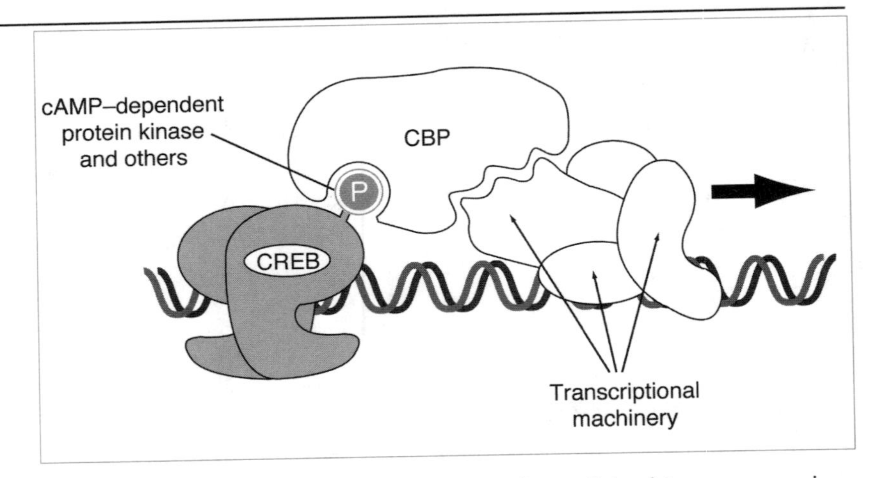

activation (at Thr69 and Thr71) and binding of the kinase occurs in a region distal to the phosphorylation site.[33]

Regulation of the *fos* promoter

Owing to its exquisite sensitivity to induction by a variety of signals, the c-*fos* promoter has been a focus of attention in determining mechanisms of transcriptional control. As might be expected, this promoter is a complex amalgam of control elements. At the core is the serum response element (SRE) which mediates many of the mitogenic signals acting on the promoter (although there are additional growth factor elements). The major SRE binding protein is serum response factor (SRF) which recruits accessory factors to the *fos* promoter such as Elk-1 (TCF) and SAP1 and SAP2.[34] These TFs are structurally related and contain transcriptional activation and DNA-binding domains. However, these proteins require DNA-bound SRF for productive interaction with DNA and, although SRF is itself phosphorylated,[35] it is these accessory factors which transduce the activatory signals. Upon mitogenic stimulation, Elk-1 becomes phosphorylated at a series of serines proximal to its carboxyl terminus which stimulates its trans-activating function[36] and also stabilizes its interaction with SRF and DNA.[37,38]

The major Elk-1 kinases appear to be the MAP kinases[36] which are downstream elements of the Ras pathway and are thus potently induced by growth factors (see Chapter 11 and Figure 21.4). Blockade of Ras, or other downstream components of this pathway such as Raf, attenuates induction of the c-*fos* gene, providing evidence for the physiological relevance of the pathway. However, anisomycin, which potently activates the SAPK pathway, but has no effect on the MAP kinases, induces *fos* gene expression. Thus, Elk-1 and/or the other ternary complex factors such as SAP1 and SAP2 may be targets for additional pathways (P. Shaw, personal communication).

NF-κB and regulated subcellular distribution

The TFs described above are predominantly nuclear and thus the signaling pathways must directly extend into the nucleus. In contrast, the proteins that comprise the NF-κB TF are found in the cytoplasm in resting cells, complexed with inhibitory proteins (IκB).[39] Upon receipt of an appropriate stimulus, IκB dissociates from the complex, probably in response to a phosphorylation event, and is rapidly degraded.[40] Loss of IκB unmasks a nuclear localization signal on the NF-κ B subunits allowing translocation to the nucleus and activation of specific genes (Figure 21.5).

Although the molecular details of the activation process and its components in mammals are unknown, a related pathway has been delineated genetically in *Drosophila*. During early development an NF-κB homolog, Dorsal, is maintained in the cytoplasm by its interaction with Cactus, an IκB cognate (see review[41]). Three genes have been identified

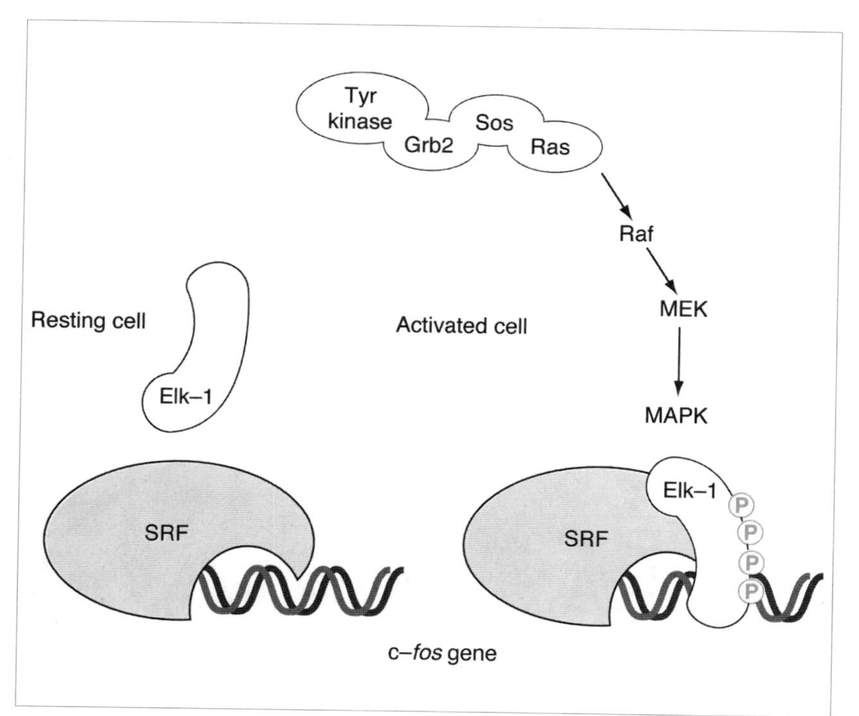

Figure 21.4 Regulation of c-*fos* promoter transcription by serum response factor (SRF) and accessory proteins. SRF interacts with the promoter via the serum response element (SRE) in resting cells. Phosphorylation of ternary accessory proteins such as Elk-1 by MAP kinases allows the formation of a transcriptionally productive complex via stabilization of interactions with DNA and the transcriptional machinery.

that act upstream of *dorsal*. One of these (*toll*) encodes a receptor and is upstream of *tube* and *pelle*, the latter encoding a protein serine kinase.[42] While Pelle may not directly interact with Cactus, it is, at present, the strongest candidate for an IκB kinase. NF-κB is activated in response to a wide variety of agents including mitogens, inflammatory cytokines and oxidative stress. Elucidation of the signaling components upstream of mammalian IκB is a major focus of research.

Cutting out the middlemen: the cytokine pathway

The TFs described in the preceding sections all involve a series of cytoplasmic mediators for their regulation. This presumably allows both signal amplification and efficient crosstalk between the pathways. By contrast, nuclear signaling elicited by a variety of cytokines is a model of elegant simplicity in which TFs are activated at the plasma membrane and directly translocate to the nucleus. Of course, nothing is that simple:

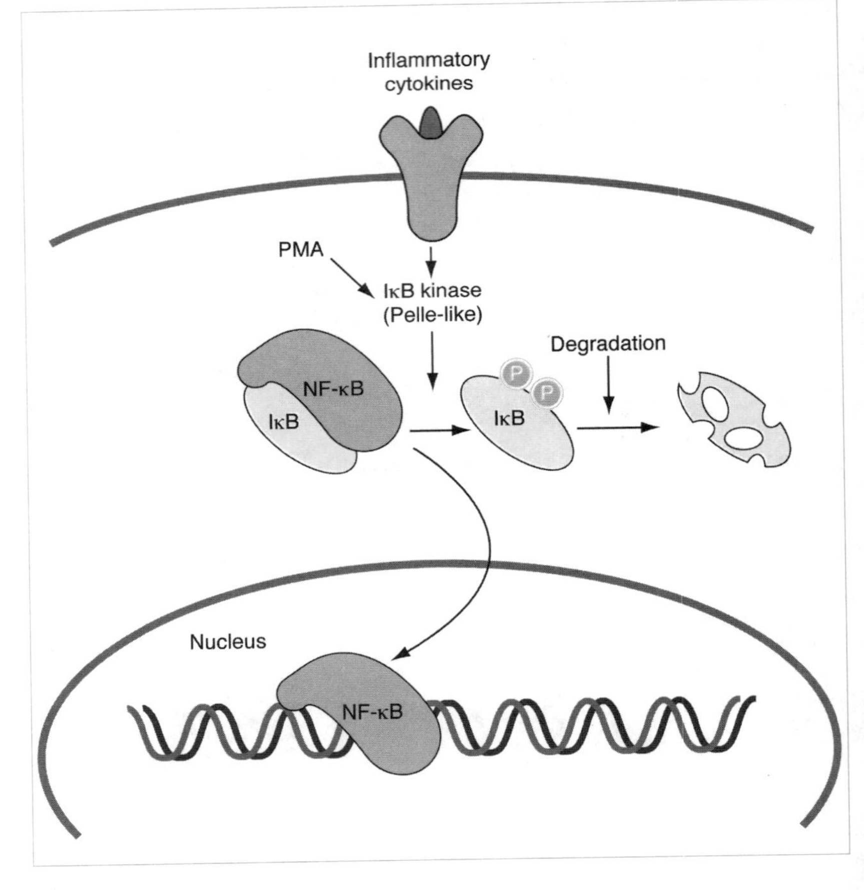

Figure 21.5 Regulation of nuclear accessibility. The inactive NF-κB/IκB complex is located in the cytoplasm of resting cells. Upon stimulation, a signal, probably phosphorylation of IκB by a Pelle-like kinase, causes dissociation of this inhibitor and subsequent degradation. A nuclear localization signal on the NF-κB subunit is unmasked, allowing translocation to the nucleus and interaction with specific genes. PMA, phorbol myristate acetate.

there are multiple cytokines with distinct cellular effects and there are multiple components of the signaling pathway itself.

A major breakthrough in this field emerged from studies of mutagenized cell lines that were selected for deficiencies in signaling via interferons α/β or γ.[43,44] Complementation of these lines yielded a series of genes that play key roles in the signaling process. Thus, binding of interferon to its receptor on the plasma membrane activates certain members of the Jak tyrosine kinase family, of which there are four (see Chapter 3 and Figure 21.6).[45,46] Specific Jaks associate with the intracellular tail of the interferon receptors in addition to receptors for erythropoietin, interleukins-2, -3, -4, -5, -6, -7, -11 and -12, G-CSF, GM-CSF, CNTF, LIF, growth hormone and prolactin (see Chapter 2), underlining the breadth of cytokines that utilize this pathway.[45,46]

Following ligand-induced dimerization of the receptor, the associated Jaks transphosphorylate on tyrosine residues causing the recruitment, via their SH2 domains (see Chapter 9), of members of a family of TFs called STATs (signal transducer and activator of transcription;

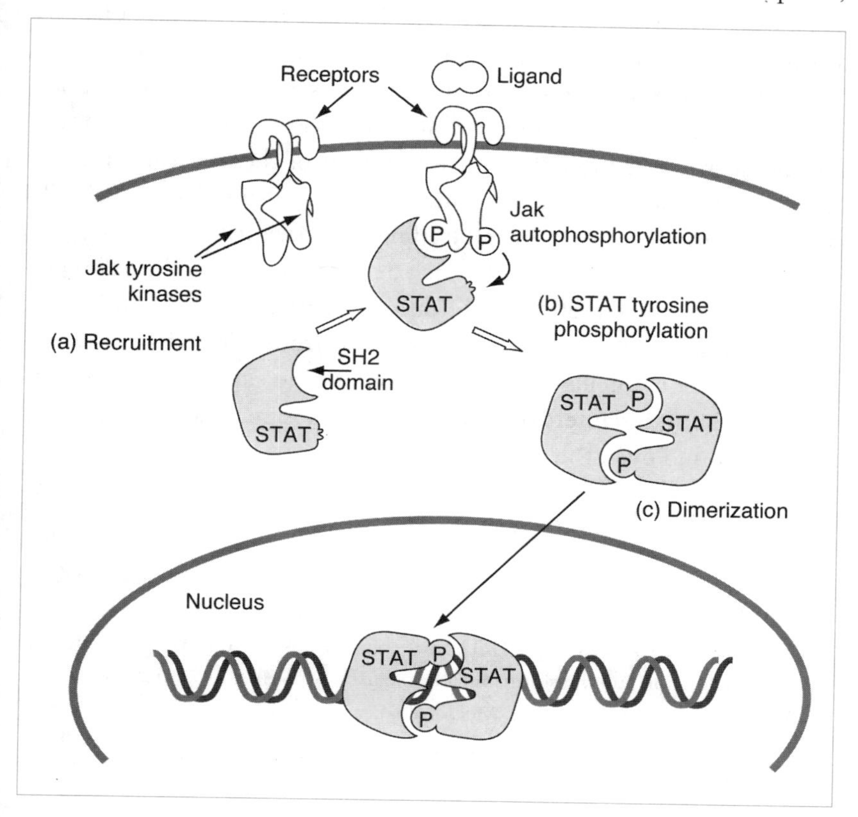

Figure 21.6
Dimerization-dependent nuclear translocation. In response to a variety of cytokines, the Jak tyrosine kinases autophosphorylate and thence recruit the STAT transcription factors via their SH2 domains. Direct Jak tyrosine phosphorylation of the STATs allows intermolecular dimerization of the STATs via their SH2 domains and subsequent translocation to the nucleus.

at least six different members).[47,48] In resting cells, the STATs are monomers, but when phosphorylated by the Jaks they dimerize via their SH2 domains and translocate to the nucleus, where they interact with specific DNA sequences and transactivate the associated genes.[47]

The basis of specificity in this pathway is not fully understood but relies in part on the SH2 domains within the STATs which direct them to particular receptor types.[49,50] It is also noteworthy that the cytokines can activate additional pathways such as MAP kinases via interaction of the Jaks with other SH2-domain containing proteins such as vav, PTP1D and Shc. STAT3 has an additional requirement for serine phosphorylation providing a further level of co-ordinate control.[51] Thus, the response to any given ligand is probably modulated by the presence of other receptors and the efficacy of interactions mediated by the Jak kinases.

Summary

This chapter has described several modes of regulation of TFs. The specific examples described act as paradigms for a variety of other factors too numerous to mention. Together, the mechanisms include regulation of DNA binding, transactivation, interaction with other components, nuclear accessibility and dimerization. What has not been detailed, for reasons of space, is the degree of interaction of these factors and the convergence and crosstalk of pathways. Thus, stimulation of cells induces a complex wave of events leading to contextual changes in gene expression, the consequences of which depend entirely on the cell type, be it proliferation, secretion, cell death, differentiation or a myriad of other responses. The molecular dissection of the regulatory pathways promises to allow intelligent manipulation of gene expression, although the intertwining and interdependence of nuclear signal transduction will likely confound even the wisest gene therapist!

References

1 Woodgett, J.R. (1990) *Fos* and *jun*: two into one will go. *Semin. Cancer Biol.* **4**, 389–97.

2 Angel, P. and Karin, M. (1991) The role of Jun, Fos and the AP-1 complex in cell proliferation and transformation. *Biochim. Biophys. Acta*, **1072**, 129–57.

3 Almendral, J.M., Sommer, D., McDonald-Bravo, H. *et al.* (1988) Complexity of the early genetic response to growth factors in mouse fibroblasts. *Mol. Cell. Biol.* **8**, 2140–8.

4 Boyle, W.B., Smeal, T., Defize, L.H.K. *et al.* (1991) Activation of protein kinase C decreases phosphorylation of cJun at sites that negatively regulate its DNA binding activity. *Cell*, **64**, 573–84.

5 Nikolakaki, E., Coffer, P., Hemelsoet, R. *et al.* (1993) Glycogen synthase kinase-3 phosphorylates Jun-family members *in vitro* and negatively regulates their trans-activating potential in intact cells. *Oncogene*, **8**, 833–40.

6 de Groot, R.P., Auwerx, J., Bourouis, M. and Sassone-Corsi, P. (1993) Negative regulation of Jun/AP-1: conserved function of glycogen synthase kinase 3 and the *Drosophila* kinase shaggy. *Oncogene*, **8**, 841–7.

7 Lin, A., Frost, J., Deng, T. *et al.* (1992) Casein kinase II is a negative regulator of c-Jun DNA binding and AP-1 activity. *Cell*, **70**, 777–89.

8 Sutherland, C., Leighton, I.A. and Cohen, P. (1993) Inactivation of glycogen synthase kinase-3β by phosphorylation: new kinase connections in insulin and growth-factor signaling. *Biochem. J.*, **296**, 15–9.

9 Stambolic, V. and Woodgett, J.R. (1994) Negative regulation of glycogen synthase kinase 3 in intact cells. *Biochem. J.*, **303**, 701–4.

10 Li, L., Zhiu, J., James, G. *et al.* (1992) FGF inactivates myogenic helix–loop–helix proteins through phosphorylation of a conserved protein kinase C site in their DNA-binding domains. *Cell*, **71**, 1181–94.

11 Franklin, C.C., Sanchez, V., Wagner, F. *et al.* (1992) Phorbol ester-induced amino terminal phosphorylation of c-Jun but not JunB regulates transcriptional activation. *Proc. Nat. Acad. Sci. USA*, **89**, 7247–51.

12 Binetruy, B., Smeal, T. and Karin, M. (1991) Ha-Ras augments c-Jun activity and stimulates phosphorylation of its activation domain. *Nature*, **351**, 122–7.

13 Pulverer, B., Kyriakis, J., Avruch, J. *et al.* (1991) Phosphorylation of c-*jun* by MAP kinases. *Nature*, **353**, 670–4.

14 Kyriakis, J.M., Banerjee, P., Nikolakaki, E. *et al.* (1994) The stress-activated protein kinase subfamily of c-Jun kinases. *Nature*, **369**, 156–60.

15 Pombo, C.M., Bonventre, J.V., Woodgett, J.R. *et al.* (1994) The stress-activated protein kinases (SAPKs) are major c-Jun amino terminal kinases activated by ischemia and reperfusion. *J. Biol. Chem.*, **269**, 26546–50.

16 Derijard, B., Hibi, M., Wu, I.H. *et al.* (1994) JNK1: a protein kinase stimulated by UV light and Ha-Ras that binds and phosphorylates the c-Jun activation domain. *Cell*, **76**, 1025–37.

17 Sluss, H.K., Barrett, T., Derijard, B. and Davis, R.J. (1994) Signal transduction by tumor necrosis factor mediated by JNK protein kinases. *Mol. Cell. Biol.*, **14**, 8376–84.

18 Sanchez, I., Hughes, R., Mayer, B. *et al.* (1994) SAP/ERK kinase-1 (SEK1) defines the SAPK pathway regulating c-Jun N-terminal phosphorylation. *Nature*, **372**, 794–8.

19 Yan, M., Dai, T., Deak, J. *et al.* (1994) MEKK1 activates the stress activated protein kinase (SAPK) *in vivo*, not MAP kinase, via direct phosphorylation of the SAPK activator SEK1. *Nature*, **372**, 798–800.

20 Bos, T.J., Monteclaro, F.S., Mitsunobu, F. *et al.* (1992) Efficient transformation of chicken embryo fibroblasts by c-Jun requires structural modification in coding and noncoding sequences. *Genes Dev.* **4**, 1677–87.

21 Morgan, I.M., Havarstein, L.S., Wong, W.Y. *et al.* (1994) Efficient induction of fibrosarcomas by v-*jun* requires mutations in the DNA binding region and the transactivation domain. *Oncogene*, **9**, 2793–97.

22 Black, E.J., Catling, A.D., Woodgett, J.R. *et al.* (1994) Transcriptional activation by the v-Jun oncoprotein is independent of positive regulatory phosphorylation. *Oncogene*, **9**, 2363–8.

23 Dai, T., Rubie, E.A., Franklin, C.C. *et al.* (1995) SAP kinases bind directly to the δ domain of c-Jun in resting cells: implications for repression of c-Jun function. *Oncogene*, **10**, 849–55.

24 Ofir, R., Dwarki, V.J., Rashid, D. and Verma, I.M. (1990) Phosphorylation of the C-terminus of Fos protein is required for transcriptional transrepression of the *c-fos* promoter. *Nature*, **348**, 80–2.

25 Tratner, I., Ofir, R. and Verma, I.M. (1992) Alteration of a cyclic AMP-dependent protein kinase phosphorylation site in the c-Fos protein augments its transforming potential. *Mol. Cell. Biol.*, **12**, 998–1006.

26 Bannister, A.J., Brown, H.J. and Kouzarides, T. (1994) Phosphorylation of the c-Fos and c-Jun HOB1 motif stimulates its activation capacity. *Nucleic Acids Res.*, **22**, 5173–6.

27 Deng, T. and Karin, M. (1994) c-Fos transcriptional activity stimulated by H-Ras-activated protein kinase distinct from JNK and ERK. *Nature*, **371**, 171–5.

28 Brindle, P.K. and Montminy, M.R. (1992) The CREB family of transcription factors. *Curr. Opinion Genet. Dev.*, **2**, 199–204.

29 Gonzales, G.A. and Montminy, M.R. (1989) Cyclic AMP stimulates somatostatin gene transcription by phosphorylation of CREB at serine 133. *Cell*, **59**, 675–80.

30 Chrivia, J.C., Kwok, R.P.S., Lamb, N. *et al.* (1993) Phosphorylated CREB binds specifically to the nuclear protein CBP. *Nature*, **365**, 855–9.

31 Foulkes, N.S., Borrelli, E. and Sassone-Corsi, P. (1992) CREM gene: use of alternative DNA-binding domains generates multiple antagonists of cAMP-dependent transcription. *Cell*, **64**, 739–49.

32 de Groot, R.P., Ballou, L.M. and Sassone-Corsi, P. (1994) Positive regulation of the cAMP-responsive activator CREM by the p70 S6 kinase: an alternative route to mitogen-induced gene expression. *Cell*, **79**, 81–91.

33 Livingstone, C., Patel, G. and Jones, N. (1995) ATF2 contains a phosphorylation dependent transcriptional activation domain. *EMBO J.*, **14**, 1785–97.

34 Dalton, S. and Treisman, R. (1992) Characterisation of SAP-1, a protein recruited by serum response factor to the *c-fos* serum response element. *Cell*, **68**, 597–612.

35 Marais, R.M., Hsuan, J.J., McGuigan, C. *et al.* (1992) Casein kinase II phosphorylation increases the rate of serum response factor-binding site exchange. *EMBO J.*, **11**, 97–105.

36 Marais, R., Wynne, J. and Treisman, R. (1993) The SRF accessory protein Elk-1 contains a growth factor-regulated transcriptional activation domain. *Cell*, **73**, 381–93.
37 Gille, H., Sharrocks, A.D. and Shaw, P. (1992) Phosphorylation of p62TCF by MAP kinase stimulates ternary complex formation at cFos promoter. *Nature*, **358**, 414–7.
38 Gille, H., Kortenjann, M., Thomae, O. *et al.* (1995) Erk phosphorylation potentiates Elk-1-mediated ternary complex formation and transactivation. *EMBO J.*, **14**, 951–62.
39 Baeuerle, P.A. and Baltimore, D. (1988) IκB: a specific inhibitor of the NF-κB transcription factor. *Science*, **242**, 540–6.
40 Brown, K., Gerstberger, S., Carlson, L. *et al.* (1995) Control of IκB-α proteolysis by site-specific, signal-induced phosphorylation. *Science*, **267**, 1485–8.
41 St Johnston, D. and Nuslein-Volhard, C. (1992) The origin of pattern and polarity in the *Drosophila* embryo. *Cell*, **68**, 201–19.
42 Shelton, C.A. and Wasserman, S.A. (1993) *pelle* encodes a protein kinase required to establish dorsoventral polarity in the *Drosophila* embryo. *Cell*, **72**, 515–25.
43 Velazquez, L., Fellous, M., Stark, G.R. and Pellegrini, S. (1992) A protein tyrosine kinase in the interferon α/β signaling pathway. *Cell*, **70**, 313–22.
44 Muller, M., Briscoe, J., Laxton, C. *et al.* (1993) The protein tyrosine kinase Jak1 complements defects in interferon-α/β and -γ signal transduction. *Nature*, **366**, 129–35.
45 Darnell, J.E., Jr, Kerr, I.M. and Stark, G.R. (1994) Jak-STAT pathways and transcriptional activation in response to IFNs and other extracellular signaling proteins. *Science*, **264**, 1415–21.
46 Ihle, J.N. and Kerr, I.M. (1995) Jaks and Stats in signaling by the cytokine receptor superfamily. *Trends Genet.*, **11**, 69–74.
47 Shuai, K., Horvath, C.M., Tsa Huang, L.H. *et al.* (1994) Interferon activation of transcription factor Stat91 involves dimerization through SH2–phosphotyrosyl peptide interactions. *Cell*, **76**, 821–8.
48 Zhong, Z., Wen, Z. and Darnell, J.E., Jr (1994) Stat3 and Stat4: members of the family of signal transducers and activators of transcription. *Proc. Natl. Acad. Sci. USA*, **91**, 4806–10.
49 Heim, M.H., Kerr, I.M., Stark, G.R. and Darnell, J.E., Jr (1995) Contribution of STAT SH2 groups to specific interferon signaling by the Jak-STAT pathway. *Science*, **267**, 1347–9.
50 Stahl, N., Farruggella, T.J., Boulton, T.G. *et al.* (1995) Choice of STATs and other substrates specified by modular tyrosine-based motifs in cytokine receptors. *Science*, **267**, 1349–53.
51 Zhang, X., Blenis, J., Li, H.-C. *et al.* (1995) Requirement of serine phosphorylation for formation of STAT-promoter complexes. *Science*, **267**, 1990–4.

22 The p53 tumour suppressor

Jo Milner

The protein p53 is found in all tissues of the body and has been highly conserved through evolution. It is not essential for life but plays a crucial role in maintaining genetic stability at the cellular level. This is of utmost importance because genetic damage leads progressively to tumor development and cancer. The fundamental importance of p53 as a tumor suppressor is underscored by the statistical prediction that at least one in six of the population will develop cancer due to defective p53 function.

The p53 protein is activated in response to DNA damage and then operates as a transcription factor, able to bind specific DNA sequences (p53-response elements) and up-regulate the expression of target genes.[1] Under these conditions, p53 induces either cell growth arrest or programmed cell death (apoptosis) (Figure 22.1). Both responses guard against replication and amplification of genetic damage within a population of cells.

DNA damage and the p53 response

Remarkably, a single double-strand break in DNA is sufficient to induce a p53 response.[2] This was shown by microinjection of known numbers of DNA molecules into the nuclei of normal cells growing in culture. Using DNA with precise types of damage, it was also demonstrated that

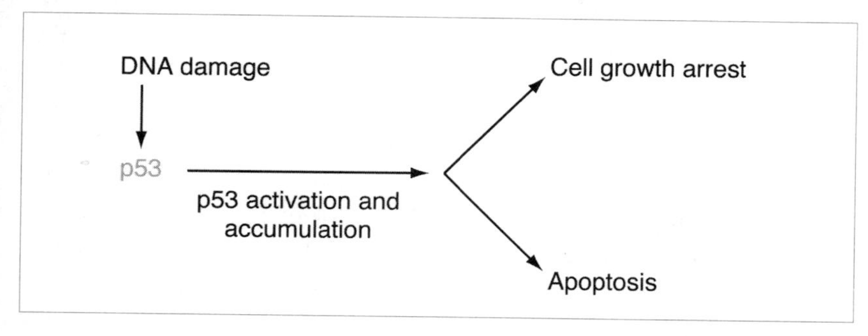

Figure 22.1 A single double-strand break in DNA is sufficient to initiate a p53 response. The concentration of p53 protein is raised by post-transcriptional control mechanisms. The activated p53 protein induces growth arrest by transactivating the expression of cell cycle regulators, such as $p21^{WAF-1/CIP1}$, or inducing apoptosis, depending on cell type.

Signal Transduction. Edited by Carl-Henrik Heldin and Mary Purton. Published in 1996 by Chapman & Hall. ISBN 0 412 70810 8

blunt-ended breaks or breaks with overhangs at the 3′ or 5′ termini work equally well. In contrast, a single nick in one strand is not sufficient to activate and initiate a p53 response.

The cellular concentration of p53 protein increases in response to DNA damage. The effect is independent of gene transcription and involves post-transcriptional control mechanisms,[1] with increased translation of pre-existing p53 mRNA. Post-transcriptional mechanisms are important in the overall regulation of gene expression and, in particular, may permit a more rapid response than transcriptional control. The process involves cytoplasmic mRNA–protein interactions[3] but the precise mechanism(s) by which the concentration of p53 is increased are unknown.

Protein stabilization may also contribute to the rise in p53 protein concentration in response to DNA damage. p53 is normally very short-lived, with a half-life of around 5–20 min, depending on cell type. Degradation is energy-dependent and may involve the ubiquitin proteolytic system.[4] This would be consistent with its rapid turnover and explain p53 accumulation in cells defective for components of the ubiquitin pathway.[5]

Induction of cell growth arrest

Raising the concentration of p53 protein orchestrates a number of cellular responses to DNA damage (see review[6]). The expression of specific target genes can be down-regulated or up-regulated, depending on promoter type.

Down-regulation of gene expression

Genes under control of transcription factors other than p53 can be down-regulated. In some cases this is due to p53 complexing with, and inactivating, specific transcription factors. Genes down-regulated by p53 include those under control of minimal 'TATA box' promoters; suppression may be mediated by the binding of p53 to the TATA-binding protein (TBP), a subunit of the basal transcription factor TFIID. Studies *in vitro* show that the p53–TBP interaction represses transcription from minimal promoters. Domains at each end of p53 can interact with a common domain on TBP[7] (see Figure 22.2). The E1A protein of adenovirus can disrupt the p53-TBP interaction through the carboxyl terminus of p53 and relieve p53-mediated repression of transcription.

P53 also interacts with other transcription factors, including the CCAAT binding factor (CBF) and SP-1. This suggests a scenario in which p53 down-regulates the expression of certain groups of genes by effectively sequestering their respective transcription factors (Box 22.1).

Up-regulation of gene expression

Specific interaction of p53 with DNA, on the other hand, up-regulates the expression of p53 target genes. The first to be identified was GADD45 (growth arrest and DNA damage) whose induction following exposure of cells to ionizing radiation is dependent upon p53. The cell cycle regulator $p21^{WAF-1/CIP1}$ is also induced by p53 and causes arrest in G1 of the cell cycle by inhibiting cyclin-dependent protein kinases (CDKs) specific for this phase of the cell cycle. Both GADD45 and $p21^{WAF-1/CIP1}$ act co-ordinately to arrest the growth of cells following DNA damage.

It has now been demonstrated that $p21^{WAF-1/CIP1}$ also plays a role in the exit of cells from the division cycle during terminal differentiation.[8,9] Here the expression of $p21^{WAF-1/CIP1}$ is independent of p53, and this is consistent with normal tissue development in p53-null mice (first shown by Donehower and co-workers, reviewed in ref. 1).

Mdm2 is also up-regulated by p53. The product of this gene may play a role in the recovery of cells from p53-induced growth arrest since it forms protein–protein complexes with the amino terminus of p53 and inhibits p53 transactivating functions.[10] Thus the Mdm-2 protein may serve as part of a negative feed-back loop, regulating p53 suppressor function to determine the length of cell growth arrest. The *Mdm2* gene was originally isolated as an oncogene amplified on a mouse double minute chromosome. Its human counterpart, *HDM2*, is over-expressed in several types of human sarcoma expressing wild-type p53,[11,12] and may play a role in tumor development by overriding normal wild-type p53 function.

Box 22.1 Genes with p53-responsive elements and genes suppressed by p53

Genes with p53-response elements	Genes suppressed by p53
GADD45	Proliferating cell nuclear antigen (PCNA)
WAF-1	BCL-2
MDM2	DNA polymerase a
Ribosomal gene cluster (RGC)	Myb
Muscle creatine kinase	fos/jun
Mouse endogenous retrovirus-like element (GLN LTR)	Interleukin 6
	Retinoblastoma (RB)

See ref. 6 for more extensive lists and for references.

Apoptosis versus cell growth arrest

In certain cell types, wild-type p53 initiates apoptosis instead of G1 arrest. Many of us whose skin has peeled following sunburn have unknowingly witnessed an end-point of p53-induced apoptosis.[13,14] Wild-type p53 is important in protecting against UV-induced damage and the onset of skin cancer. If the p53 response in a single skin cell is reduced by p53 mutation, sunburn can select for clonal expansion of the defective cell and promote tumor progression. The apoptotic response to p53 is not restricted to the skin and is observed in different tissues.

Apoptosis is a form of programmed cell death by which single cells are destroyed and removed from living tissues without causing tissue damage or inflammation.[15] It plays an essential role in the balance between cell proliferation and cell death, and is a regulable process. However, despite the importance of apoptosis in tissue development and homeostasis, its molecular mechanisms are poorly understood. The system is complex and multiple pathways may be recruited. At least one pathway is p53-dependent.

The nature of the p53 response to DNA damage is determined by the cell type involved.[1] Thus, exposure of normal murine fibroblasts to ionizing radiation induces p53-dependent G1 arrest but no apoptosis. Under the same conditions, murine thymocytes rapidly undergo p53-dependent apoptosis. Thymocytes from mice in which both p53 alleles have been knocked out by genetic recombination completely lack the apoptotic pathway induced by ionizing radiation. Nonetheless, the same cells retain the apoptotic response to corticosteroids and other agents known to induce apoptosis in lymphoid cells.[16–18] This clearly demonstrates the presence of more than one apoptotic pathway in a single cell, only one of which is dependent on p53.

The very fact that p53 knock-out mice develop normally indicates that apoptotic events required for tissue differentiation are independent of p53. In these mice, the absence of p53 is manifested by the early development of malignancies, most commonly of lymphoid cells (see ref.1).

Susceptibility to apoptosis is influenced by many proteins and families of proteins, including Bcl-2, Bcl-x, Bax, Bcr-abl and c-Myc (see ref. 19, for recent reviews). For example, Bcl-2 is a survival factor and can prevent apoptosis. The natural antagonist of Bcl-2 is Bax. Studies *in vitro* and *in vivo* have revealed that p53 down-regulates Bcl-2 and up-regulates the expression of Bax, and this may be one mechanism by which p53 initiates apoptosis.

Sequence-specific DNA binding

Sequence-specific DNA binding and transcriptional activation are essential for wild-type p53 suppressor function.[20] The p53 protein is modular, and specific DNA binding maps to the central core domain (Figure 22.2). Transcriptional activation by p53 may be mediated by interaction with co-activators $TAF_{II}40$ and $TAF_{II}60$ (TBP-associated factors).[24]

The best studied p53 target sequence contains two repeats of the DNA consensus sequence 5′-PuPuPuC[A/T][T/A]GPyPyPy-3′. Each repeat represents a half-site and, *in vivo*, each half-site may be separated by long stretches of base pairs without apparent loss of function.[25] Mass measurements by scanning transmission electron microscopy (STEM) indicate that p53 binds a complete consensus sequence as a tetramer, with subsequent tetramer stacking and DNA loop formation.[26] Binding of tetrameric p53 to DNA was first demonstrated *in vitro* by size fractionation studies and gel shift analysis. However, the situation in cells may be more complex, since monomeric p53 has transactivating functions even though it is unable to bind DNA *in vitro*.[27]

Sequence-specific DNA binding by p53 is dependent upon protein conformation. The 1620^+ form (reactive with the conformation-dependent monoclonal antibody PAb1620) correlates with wild-type p53 suppressor function and is a prerequisite for binding to the DNA consensus p53-CON *in vitro*. Competition experiments show that murine p53 replaces human p53 (hp53) in p53–DNA complexes. This correlates with the greater lability observed for hp53–DNA complexes at a given temperature.[28] Mixed human–murine p53 oligomers are also competent for DNA binding *in vitro*, with an estimated affinity around 5×10^{-10} M, similar to that observed for either human or murine p53 alone.

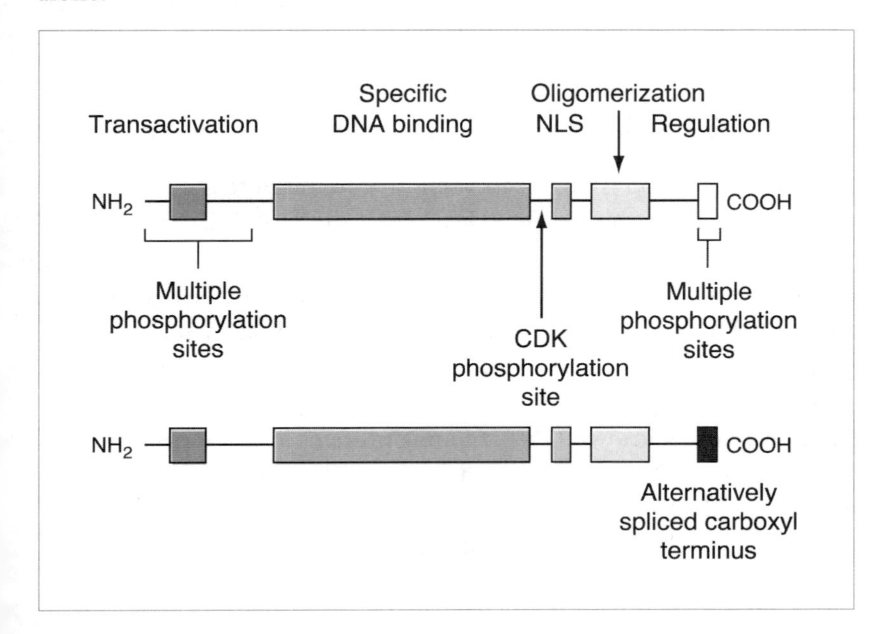

Figure 22.2 Like most transcription factors, the p53 protein is divisible into functional domains or modules. The molecular structures of the central core and oligomerization domains have been resolved.[21,22] The central core domain binds to specific DNA sequences. It has intrinsic conformational flexibility,[23] which may be important for the normal regulation of p53 function. Non-specific binding to damaged DNA maps to the carboxyl terminus.[51,52] Structural mutants of p53 are metabolically stable and react with the monoclonal antibody PAb240; the PAb240 epitope is partially buried within a hydrophobic core domain on wild-type p53 (see Figure 22.3). NLS, nuclear localizastion sequence.

Loss of p53 function

A single missense point mutation is sufficient to inactivate p53 tumor suppressor function. The p53 gene seems disconcertingly susceptible to point mutations which, in human cancer, predominate within the central 'core' domain. Indeed, p53 is famed for being the most frequently affected gene in human cancer, independent of tumor type.

In cancer, loss of one p53 allele is commonly associated with missense mutations within the second allele. Although mutations may affect some 25% of the coding sequence, there are many 'hot spots.' Certain hot spots are also linked with specific tumor types. For example, codons 175, 248 and 273 are major hot spots for missense point mutation in general, whereas hepatocellular carcinoma caused by exposure to aflatoxin B is associated with mutation at codon 249. Germ-line mutation of p53 is associated with cancer predisposition, as observed in the Li–Fraumeni syndrome (see ref. 6).

In addition to mutation, wild-type p53 can also be inactivated by specific protein–protein interactions, as seen with Mdm-2 (see above) and with transforming proteins of DNA tumor viruses[29] (Box 22.2). Indeed, this phenomenon led to the original description of p53 as a cellular protein in complex with the large T antigen of simian virus 40 (SV40).[31–34] The E1B protein of adenovirus type 5 and E6 of human papilloma virus types 16 and 18 (HPV-16 and -18) also selectively interact with wild-type p53. Large T and E1B appear to sequester p53 in relatively stable protein–protein complexes. HPV E6, on the other hand, targets p53 for rapid proteolysis by the ubiquitin system.[35,36]

Box 22.2 Examples of p53 protein–protein interactions

TATA binding protein (TBP)	Large T of SV40
CCAAT binding factor (CBF)	E1B of adenovirus type 5
Transcription factor ERCC3	E6 of human papilloma virus types 16 and 18
Transcription factor SP1	
MDM2[a]	
TBP-associated factors TAF$_{II}$40[a] and TAF$_{II}$60[a]	
Replication protein A (RPA)	
Cellular E6-associated protein (E6-AP)	

See ref. 6 for more extensive lists and for references.

[a]Mutation of p53 residues 20 and 23, within the transactivating domain, inhibits binding by MDM2[30] and by TAFs,[24] suggesting that these proteins compete for a similar binding domain on p53. Transcriptional activation by p53 is mediated by TAF coactivators, and this may be blocked by MDM2 during recovery from G1 arrest.

Mutant p53 lacks specific DNA-binding capacity

The function of p53 is remarkably sensitive to missense point mutation and the reason for this became clear when the molecular structure of the central core domain in complex with a DNA target was resolved by X-ray crystallography[21] (Figure 22.3). It was already well established that mutation results in loss of specific DNA binding by p53 and that many mutations also cause a structural change in the protein.

The molecular structure shows that the p53 DNA-binding domain differs from other known DNA-binding proteins in that it is a relatively open structure, requiring a large β-sandwich to position and orientate the structural elements that interact with DNA.[21] This β-sandwich provides scaffolding for a loop–sheet–helix motif and for two large

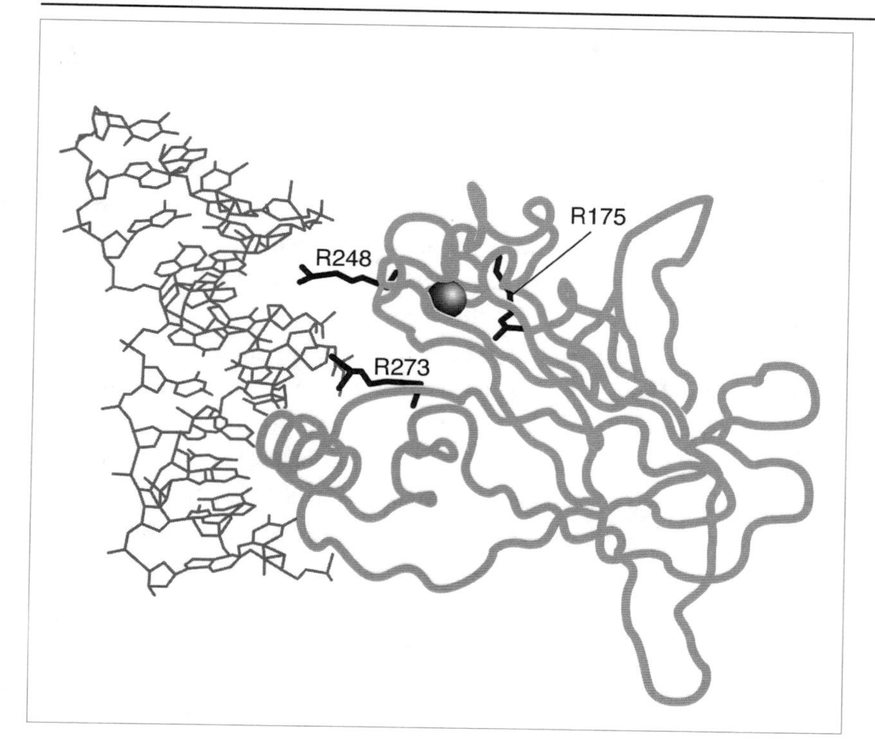

Figure 22.3 Molecular structure of the p53 core domain in complex with DNA. Ribbon model of the p53 core (colored blue) with some of the most frequently mutated residues indicated in black, including Arg248 and Arg273 (which contact the DNA) and Arg175 (important for stabilizing the tertiary structure of the p53 core domain). The DNA is colored grey and the zinc atom is represented by a sphere, grey. (Crystal structure co-ordinates[21] were kindly made available by Nikola Pavletich, Memorial Sloan-Kettering Cancer Center, New York, USA).

loops (L2 and L3), orientated and stabilized by side-chain interactions and by tetrahedral coordination of a zinc atom (via three cysteines and one histidine). This orientation is crucial because the loop–sheet–helix and loops L2 and L3 form the DNA-binding surface of p53. Specific DNA recognition involves both major and minor grooves of the target DNA.

Missense mutations are largely scattered over the central DNA-binding domain, with hot spots at codons 175, 248 and 273, each encoding arginine. The crystal structure shows that side-chain interactions of Arg175 are important for maintaining the precise conformation (tertiary structure) required for specific DNA binding. Mutation at this site destabilizes the tertiary structure of p53 with loss of DNA binding and tumor suppressor function. Substitution of Arg248 and Arg273 also causes loss of DNA binding, but here the effect is independent of tertiary structure, and is due to the fact that these residues make direct contact with the target DNA. Thus, *structural mutations* abolish DNA binding through their effects on p53 tertiary structure, whereas *DNA contact mutations* affect crucial residues involved in contacting the DNA target sequence, without necessarily have a discernible effect on tertiary structure.[21,23]

p53 is a multifunctional protein

The multiple functions of p53 protein partly explain its chequered history, first described as an oncoprotein, subsequently as a tumor suppressor (Figure 22.4). In fact, wild-type p53 adopts at least three immunologically distinct forms and it has been argued that each may correlate with a defined function in cell growth control.[38] Distinct functional forms of p53 may be generated by (1) alternative conformations, (2) post-translational modification(s) and (3) alternative splicing.

Conformational flexibility, which can provide the molecular mechanism for switching a protein between different functional forms, has been proposed for the regulation of p53 during the normal cell growth response[39] and appears to be an intrinsic property of wild-type p53. It has been suggested that many structural mutations may stabilize p53 in the conformation normally induced by cell growth stimulation. Studies *in vitro* indicate that p53 structure is subject to modulation by redox[40] and by copper, a redox metal.[41] It is possible that these systems operate in the cellular regulation of p53 conformation, both in response to cell growth control signals and also in response to DNA damage.

The predominant post-translational modification of p53 is phosphorylation, and there are multiple sites for specific kinases in the amino- and carboxy-terminal regions[42] (Figure 22.2). Phosphorylation may have a knock-on effect on protein conformation. Alternatively, it may affect p53 function in some other way, such as activation of specific DNA-binding activity.[43] Given the multiple phosphorylation sites on

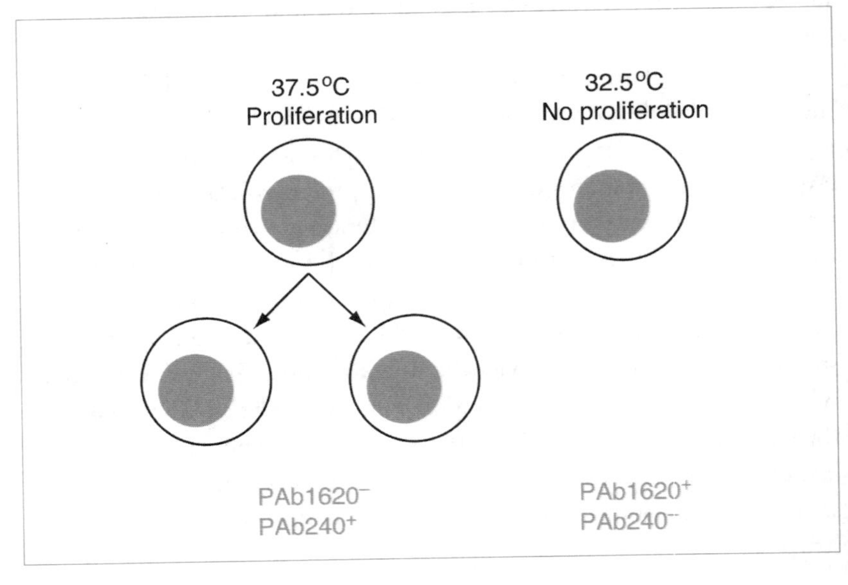

Figure 22.4
Temperature-sensitive alleles of p53 (first discovered by Oren and co-workers[22]) exhibit either mutant or wild-type functions, at 37.5 and 32.5 °C, respectively. The different effects of p53 on cell proliferation correlate with alternative conformations of the protein, determined by reactivity with monoclonal antibodies, as indicated. Many structural mutants of human p53 retain the capacity to revert to apparently wild-type form, and this represents one possible approach to anti-cancer therapy.[23]

p53 and the dynamic nature of phosphorylation/dephosphorylation reactions, unraveling their functional implications for p53 is likely to prove a formidable task. There is indirect evidence that phosphorylation is involved in the response to DNA damage, and p53 is a known substrate for DNA-dependent protein kinase (DNA-PK)[44] and the stress-induced c-Jun kinase (JNK).[45] There is good evidence that DNA-PK activity is involved in the repair of double-strand DNA breaks.[46] As yet, the role of phosphorylation of p53 is unknown.

Alternative splicing[47] of murine p53 has been reported (Figure 22.2), with substitution of 17 new amino acid residues in place of 26 carboxy-terminal residues of the non-spliced protein.[48] The two different p53 proteins are detectable in mouse epidermal cells and may be differentially regulated at different points of the cell cycle. Functional differences between the two splice variants of p53 have also been detected, in that alternative splicing renders p53 unable to promote re-annealing of single-stranded RNA or DNA.[49]

Summary

Wild-type p53 in some way senses and responds to DNA damage. The response involves accumulation of p53 protein by post-translational control mechanisms. p53 affects the transcription of many genes, either by interacting with proteins that are involved in gene regulation or by direct binding to DNA and transactivation of genes. p53 induces either cell growth arrest or apoptosis, depending on cell type. By preventing the proliferation of DNA-damaged cells, the p53 protein plays a crucial role in the maintenance of genomic integrity and tumor suppression.

Loss of p53 function is linked with the development of over half of all human cancers, and p53 represents a prime agent/target for novel anticancer therapies. One approach is to use gene therapy to restore wild-type p53 function.[50] Future areas important for the development of p53-based anticancer therapies include determination of the mechanisms by which p53 responds to and is activated by DNA damage, and characterization of the mechanisms determining the choice between p53-induced apoptosis or cell-growth arrest. Much research is understandably driven by the ability of p53 to suppress cancer. However, there is evidence that p53 is multifunctional and may even have cell-growth-promoting functions. A comprehensive understanding of the functional potential of p53 may have additional implications for non-cancer related disease, tissue repair and regeneration.

References

Many of these references are to reviews, from which details of original published works can be obtained.

1 Canman, C.E. and Kastan, M.B. (1995) Induction of apoptosis by tumor suppressor genes and oncogenes. *Semin. Cancer Biol.*, **6**, 17–25.
2 Wahl, G.M., Huang, L.-C., Linke, S. and Alasan, A. (1995) Cell cycle control of genetic stability. Presented at the Workshop on p53 and the Cell Cycle, York, April 1995.
3 McCarthy, J.E.G. and Kollmus, H. (1995) Cytoplasmic mRNA–protein interactions in eukaryotic gene expression. *Trends Biochem. Sci.*, **20**, 191–7.
4 Ciechanover, A. (1994) The ubiquitin–proteasome proteolytic pathway. *Cell*, **79**, 13–21.
5 Chowdary, D.R., Dermody, J.J., Jha, K.K. and Ozer, H.L. (1994) Accumulation of p53 in a mutant cell line defective in ubiquitin pathway. *Mol. Cell. Biol.*, **14**, 1997–2003.
6 Hesketh, R. (1995) *The Oncogene Facts Book*. Academic Press, New York.
7 Horikoshi, N., Usheva, A., Chen, J., Levine, A.J. *et al.* (1995) Two domains of p53 interact with the TATA-binding protein, and adenovirus 13S E1A protein disrupts the association, relieving p53-mediated transcriptional regulation. *Mol. Cell. Biol.*, **15**, 227–34.
8 Halevy, O., Novitch, B.G., Spicer, D.B. *et al.* (1995) Correlation of terminal cell cycle arrest of skeletal muscle with induction of p21 by MyoD. *Science*, **267**, 1018–21.
9 Parker, S., Eichele, G., Zhang, P. *et al.* (1995) p53-independent expression of p21Cip1 in muscle and other terminally differentiating cells. *Science*, **267**, 1024–7.
10 Chen, J., Marechal, V. and Levine, A.J. (1993) Mapping of the p53 and mdm-2 interaction domains. *Mol. Cell. Biol.*, **13**, 4107–14.
11 Ladanyl, M., Cha, C., Lewis, R. *et al.* (1993) MDM2 gene amplification in metastatic osteosarcoma. *Cancer Res.*, **53**, 16–8.
12 Oliner, J.D., Kinzler, K.W., Meltzer, P.S. *et al.* (1992) Amplification of a gene encoding a p53-associated protein in human sarcomas. *Nature*, **358**, 80–3.
13 Ziegler, A., Jonason, A.S., Leffell, D.J. *et al.* (1994) Sunburn and p53 in the onset of skin cancer. *Nature*, **372**, 773–6.
14 Kamb, A. (1994) Sun protection factor p53. *Nature*, **372**, 730–1.
15 Bellamy, C.O.C., Malcolmson, R.D.G., Harrison, D.J. *et al.* (1995) Cell death in health and disease: the biology and regulation of apoptosis. *Semin. Cancer Biol.*, **6**, 3–16.
16 Lowe, S.W., Schmitt, E.M., Smith, S.W.*et al.* (1993) p53 is required for radiation-induced apoptosis in mouse thymocytes. *Nature*, **362**, 847–9.
17 Clarke, A.R., Purdie, C.A., Harrison, D.J. *et al.* (1993) Thymocyte apoptosis induced by p53-dependent and independent pathways. *Nature*, **362**, 849–52.

18 Lotem, J. and Sachs, L. (1993) Hematopoietic cells from mice deficient in wild-type p53 are more resistant to induction of apoptosis by some agents. *Blood*, **82**, 1092–6.

19 Eastman, A. (Guest Ed.) (1995) *Apoptosis in Oncogenesis and Chemotherapy* (*Seminars in Cancer Biology* Vol. 6), Academic Press, New York.

20 Pietenpol, J.A. *et al.* (1994) Sequence-specific transcriptional activation is essential for growth suppression by p53. *Proc. Natl. Acad. Sci. USA*, **91**, 1998–2002.

21 Cho, Y., Gorina, S. Jeffrey, P.D. and Pavletich, N.P. (1994) Crystal structure of a p53 tumour suppressor–DNA complex: understanding tumorigenic mutations. *Science*, **265**, 346–55.

22 Michalovitz, D., Halevy, O. and Oren, M. (1990) Conditional inhibition of transformation and of cell proliferation by a temperature-sensitive mutant of p53. *Cell*, **62**, 671–80.

23 Milner, J. (1995) Flexibility: the key to p53 function? *Trends Biochem. Sci.*, **20**, 49–51.

24 Thut, C.J., Chen, J.L., Klem, R. and Tjian, R. (1995) p53 transcriptional activation mediated by coactivators $TAF_{II}40$ and $TAF_{II}60$. *Science*, **267**, 100–4.

25 Deb, S.P., Munoz, R.M., Brown, D.R. *et al.* (1994) Wild type human p53 activates the human epidermal growth factor promoter. *Oncogene*, **9**, 1341–9.

26 Stenger, J.E., Tegtmeyer, P., Mayr, G.A. *et al.* (1995) p53 oligomerisation and DNA looping are linked with transcriptional activation. *EMBO J.*, **13**, 6011–20.

27 Shaulian, E., Zauberman, A., Milner, J. *et al.* (1993) Tight DNA binding and oligomerisation are dispensible for the ability of p53 to transactivate target genes and suppress transformation. *EMBO J.*, **12**, 2789–97.

28 Hall, A.R. and Milner, J. (1995) Structural and kinetic analysis of p53–DNA complexes and comparison of human and murine p53. *Oncogene*, **10**, 561–7.

29 Vousden, K.H. (1995) Regulation of the cell cycle by viral oncogenes. *Semin. Cancer Biol.*, **6**, 109–16.

30 Lin, J., Chen, B. Elenbaas, A. and Levine, A.J. (1994) Several hydrophobic amino acids in the p53 N-terminal domain are required for transcriptional activation, binding to mdm-2 and the adenovirus E1B. *Genes Dev.*, **8**, 1235–46.

31 Lane, D.P. and Crawford, L.V. (1979) T antigen is bound to a host protein in SV40-transformed cells. *Nature*, **278**, 261–3.

32 Linzer, D.I.H. and Levine, A.J. (1979) Characterisation of a 54K dalton cellular SV40 tumour antigen present in SV40-transformed cells and uninfected embryonal carcinoma cells. *Cell*, **17**, 43–52.

33 Kress, M., May, E., Cassingena, R. and May, P. (1979) Simian virus 40-transformed cells express new species of proteins precipitable by anti-simian virus 40 tumour serum. *J. Virol.*, **31**, 472–83.

34 Chang, C., Simmons, D.T., Martin, M.A. and Mora, P.T. (1979) Identification and partial characterisation of new antigens from simian virus 40-transformed mouse cells. *J. Virol.*, 463–71.

35 Scheffner, M., Werness, B.A., Huibregtse, J.M. *et al.* (1990) The E6 oncoprotein encoded by human papilloma virus types 16 and 18 promotes the degradation of p53. *Cell*, **63**, 1129–36.

36 Kessis, T.D., Slebos, R.J., Nelson, W.G. *et al.* (1993) Human papilloma virus 16 E6 disrupts the p53-mediated cellular response to DNA damage. *Proc. Natl. Acad. Sci. USA*, **90**, 3988–92.

37 Clore, G.M., Omichinski, J.G., Sahaguchi, K. *et al.* (1994) High-resolution structure of the oligomerization domain of p53 by multi-dimensional NMR. *Science*, **265**, 386–91.

38 Milner, J. (1994) Forms and functions of p53. *Semin. Cancer Biol.*, **5**, 211–9.

39 Milner, J. (1991) A conformation hypothesis for the opposing functions of p53 in cell growth control and in cancer. *Proc. R. Soc. London, Ser. B* **245**, 139–45.

40 Hainaut, P. and Milner, J. (1993) Redox modulation of p53 conformation and sequence-specific DNA binding *in vitro*. *Cancer Res.*, **53**, 4469–73.

41 Hainaut, P., Rolley, N., Davies, M. and Milner, J. (1995) Modulation by copper of p53 conformation and sequence-specific DNA binding: role for Cu(11)/Cu(1) redox mechanism. *Oncogene*, **10**, 27–32.

42 Meek, D. (1994) Post-translational modification of p53. *Semin. Cancer Biol.*, **5**, 203–10.

43 Hupp, E., Meek, D.W., Midgley, C.A. and Lane, D.P. (1993) Regulation of the specific DNA binding function of p53. *Cell*, **71**, 875–86.

44 Fiscella, M., Ullrich, S.J., Zambrano, N. *et al.* (1993) Mutation of the serine 15 phosphorylation site of human p53 reduces the ability of p53 to inhibit cell cycle progression. *Oncogene*, **8**, 1519–28.

45 Milne, D.M., Cambell, L.E., Cambell, D.G. and Meek, D.W. (1995) p53 is phosphorylated *in vitro* and *in vivo* by an ultraviolet radiation induced protein kinase, characteristic of the c-jun kinase, JNK1. *J. Biol. Chem.*, **270**, 5511–8.

46 Lees-Miller, S.P., Godbout, R., Chan, D.W. *et al.* (1995) Absence of p350 subunit of DNA-activated protein kinase from a radiosensitive human cell line. *Science*, **267**, 1183–5.

47 Sharp, P.A. (1994) Split genes and RNA splicing: Nobel Lecture. *Cell*, **77**, 805–15.

48 Kulesz-Martin, M.F., Lisafeld, B., Huang, H. *et al.* (1994) Endogenous p53 protein generated from wild-type alternatively spliced p53 RNA in mouse epidermal cells. *Mol. Cell. Biol.*, **14**, 1698–1708.

49 Wu, L., Bayle, J.H., Elenbaas, B. *et al.* (1995) Alternatively spliced forms in the carboxy-terminal domain of p53 protein regulate its ability to promote annealing of complementary single strands of nucleic acids. *Mol. Cell. Biol.*, **15**, 497–504.

50 Zhang, W.-W., Fang, X., Mazur, W. *et al.* (1994) High-efficiency gene transfer and high level expression of wild-type p53 in human lung cancer cells mediated by recombinant adenovirus. *Cancer Gene Ther.*, **1**, 5–13.

51 Lee, S., Elenbaas, B., Levine, A.J. *et al.* (1995) p53 and its 14kDa C-terminal domain recognize primary DNA damage in the form of insertion– deletion mismatches. *Cell*, **81**, 1013–20.

52 Jayarman, L. and Prives, C. (1995) Activation of p53 sequence-specific DNA binding by short single strands of DNA requires the p53 C-terminus. *Cell*, **81**, 1021–9.

23 Cell cycle regulation

Steve Coats and
Jim Roberts

A defining property of cells is the capacity for self-duplication. Ordinarily this involves cell growth, chromosome replication and cell division, where cell growth generally refers to the duplication of everything in the cell other than chromosomal DNA. The mitotic cell cycle is the output of the molecular events and specific regulatory networks that coordinate to produce two daughter cells which not only closely resemble their parent and each other, but who can also go on and repeat the process.

The cell cycle can conveniently be divided into four phases. Chromosome duplication occurs during S phase and chromosome segregation plus cytokinesis during M phase. A G1 phase intervenes between M and S, while a G2 phase separates S from M (Figure 23.1). A critical feature of the cell cycle is that it precisely duplicates the cell, and this requires that cell cycle events be executed in proper sequence. The strict ordering of cell cycle events is determined by a group of regulatory pathways that prevent the execution of certain cell cycle events until preceding ones have been successfully completed. For example, chromosome replication cannot occur until a cell has completed mitosis (see review[1]) and mitosis cannot occur until chromosome replication has been completed (see review[2]). The pathways that render one cell cycle event dependent upon completion of another are called checkpoints.[3] Checkpoints can be thought of as the set of intracellular conditions that must be satisfied for cell cycle progression to continue. There are many checkpoints during the cell cycle, each monitoring the execution of a particular cell cycle event (DNA synthesis, spindle assembly, DNA repair, etc.; see Figure 23.1). Checkpoints safeguard against mistakes in cell duplication by stopping the cell cycle should errors not be corrected.

The cell cycle is also responsive to extracellular conditions. In single-cell organisms this generally serves to co-ordinate cell proliferation with the supply of extracellular nutrients. In multicellular organisms the extracellular influences on proliferation are more complex; cell prolifer-

Signal Transduction. Edited by Carl-Henrik Heldin and Mary Purton. Published in 1996 by Chapman & Hall. ISBN 0 412 70810 8

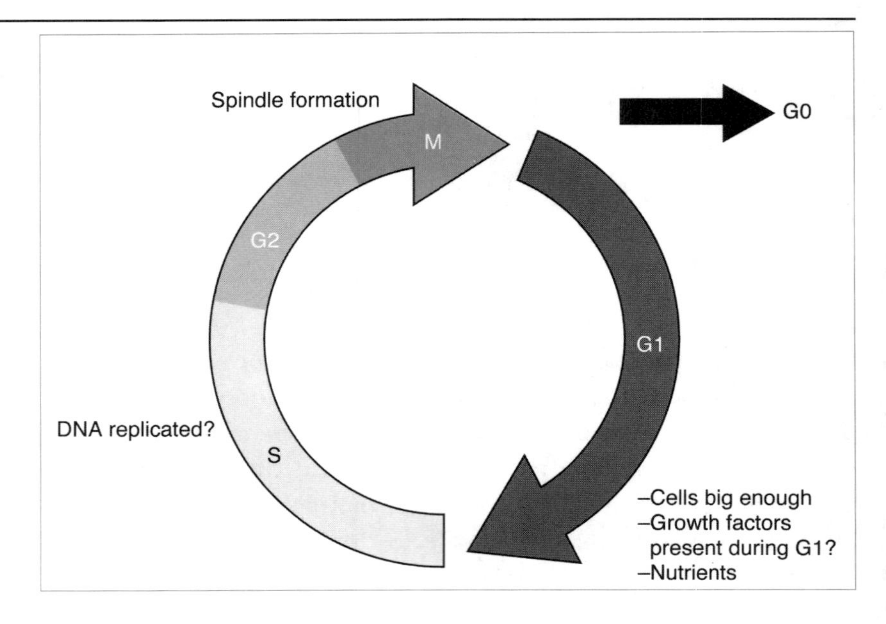

Figure 23.1 The eukaryotic cell cycle. Diagram depicting the four phases of the cell cycle and the checkpoints that control transit through the cell cycle.

ation must be co-ordinated among many cell types to allow for proper embryonic development, and restricted proliferation of specific lineages is also necessary in adult organisms. When these controls break down there can be a variety of pathological consequences, including tumorigenesis. Extracellular signals, such as hormones and growth factors, usually control cell proliferation exclusively during the G1 phase of the cell cycle. Mammalian cells, for example, require growth factors until 2 h prior to entry into S phase. Once cells have passed this point (the restriction point), they no longer require growth factors and are committed to complete one cell cycle.[4] In single-cell organisms such as the yeast *Saccharomyces cerevisiae*, the nutrient environment of the cell similarly controls cell proliferation during G1. This point of commitment (Start) is physiologically analogous to the mammalian cell restriction point.[5] Moreover, the molecules that control the restriction point are homologous to the molecules that control Start. In fact, the basic elements of cell cycle control have been highly conserved during evolution, and this extends not only to G1 but also to the mechanisms of cell cycle control during S, G2 and M. This chapter describes the molecules that control the cell cycle of all eukaryotes, and also discusses how their actions are regulated by signals that determine whether the cell cycle should stop or start.

Protein kinases that regulate the cell cycle

The first insights into mechanisms of cell cycle regulation came from genetic experiments in yeast. Using the yeast *S. cerevisiae*, Hartwell *et al.*[6] showed that the Start transition was controlled by the product of the *CDC28* gene and, using the yeast *Schizosaccharomyces pombe*, Nurse.[7] showed that the transition from G2 into M phase was controlled by the product of the *cdc2* gene. At the same time, control of the G2/M transition was also being studied in marine invertebrates and frogs (specifically *Xenopus laevis*). Investigators noticed the periodic nature of a factor that stimulated progression through the meiotic cell cycle in oocytes. This protein (or protein complex) was termed maturation-promoting factor (MPF) or mitosis-promoting factor, the latter aptly describing its role in the cell cycle. The periodic activation of MPF is required for oocytes in transit from G2 into M phase of meiosis. Inactivation of MPF is required for anaphase and cytokinesis. As described below, these diverse areas of research were unified when it was discovered that the yeast *CDC28* and *cdc2* genes were homologous to each other and, furthermore, that an essential component of MPF was the *Xenopus* homolog of cdc2/CDC28.[6,7]

The *cdc2/CDC28* gene encodes a 32–34 kDa protein kinase. In yeast, a single cdc kinase is required for completing both Start and mitosis, although there are additional *CDC28*-related genes that are not directly involved in cell cycle control. There are more than a dozen mammalian genes in the *cdc2/CDC28* family, and at least four of them have been shown to have important roles at different stages of the cell cycle. The products of this family of genes have been named the cyclin-dependent kinases (CDKs), because they must assemble into a holoenzyme with a cyclin subunit to become catalytically active (see below). In mammalian cells CDK2, −3, −4 and −6 all have roles in G1 and/or S phase while the role of cdc2 appears to be restricted to mitosis.[8]

The periodic activation of MPF was first explained by the discovery of a component of MPF (cyclin) that was purified from sea-urchin eggs and clams and which oscillated in abundance and activity during the cell cycle. While the MPF cyclin (cyclin B) regulates entry of cells into mitosis, there are additional cyclins required for each phase of the cell cycle. In general, all organisms express three types of cyclins. In mammals, for instance, cyclin B is essential for mitosis, cyclin A regulates S phase and cyclins D and E regulate the progression of mammalian cells through G1 (letters denote the order in which cyclins were discovered). Yeasts express comparable cyclins: Cln1–3 function at Start, Clb5 and −6 in S phase, and Clb1–4 during G2 and M.[9] All cyclins contain a highly conserved region of 100 amino acid residues called the cyclin box which is required for their interaction with CDKs.[10] Figure 23.2 demonstrates the basic structural domains of the cyclins.

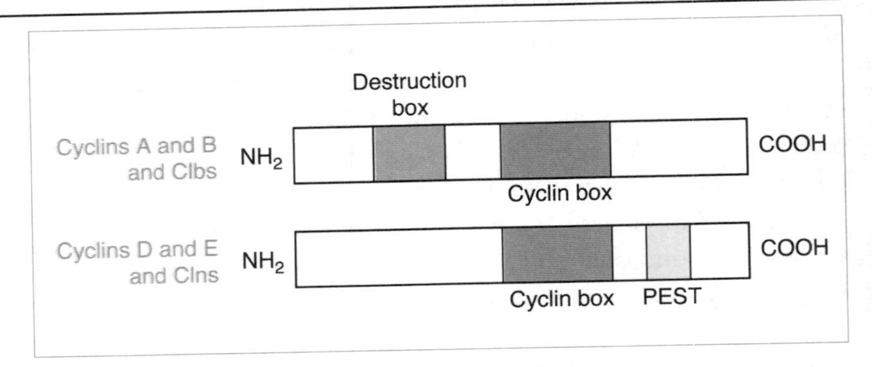

Figure 23.2 Structural domains of the cyclins. The A- and B-type cyclins contain a destruction box that is required for their ubiquitination and degradation. Cyclins D and E and the Clns possess a PEST domain, rich in proline, glutamic acid, serine and threonine residues, that may target them for proteolysis. All cyclins have a structural motif, termed the cyclin box, that is required for their interaction with CDKs.

Regulation of cyclin–CDK complexes by phosphorylation

CDKs are regulated post-translationally by phosphorylation and cyclins are regulated by proteolysis. The key amino acids phosphorylated on CDK2 are Thr160, Tyr15 and Thr14 (we shall use the amino acid numbering for human CDK2). The phosphorylation on Thr160 is required for CDK2 activity, whereas the Tyr15/Thr14 phosphorylations are dominant and inactivate cyclin–CDK complexes (Figure 23.3).

In fission yeast, wee1, originally identified as a temperature-sensitive mutation that altered cell size,[11] was found to be the tyrosine kinase that catalyzes the phosphorylation at Tyr15. Wee1 normally restrains cells during G2 until a critical size is obtained for cell division. Wee1 mutants were unable to impose this size checkpoint and cells continued through the cell cycle at a reduced mass. After its original identification in *S. pombe*, homologs of wee1 were then identified in many other eukaryotes, and most data suggest that it is part of a universal pathway for restraining cdc2 activity until conditions are appropriate for entry into mitosis.[12,13] Wee1 itself is also subject to control by other kinases, including cdc2 and nim1/cdr, a dual-specificity kinase that phosphorylates and inactivates wee1 during late G2.[14] Cdc25 is a tyrosine phosphatase that counteracts the action of wee1 on cdc2 by

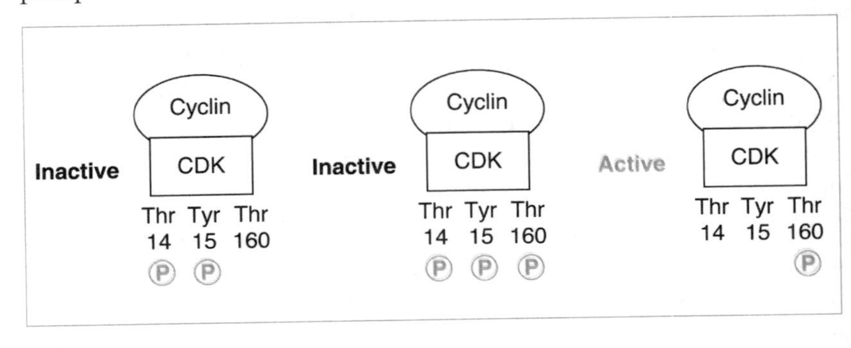

Figure 23.3 Regulation of cyclin–CDK complexes by phosphorylation. Cyclin–CDK2 complexes that are phosphorylated on Thr14 and Tyr15 either in the presence or absence of phosphorylation at Thr160 are inhibited and unable to phosphorylate substrates. Dephosphorylation at Thr14 and Tyr15 by cdc25 and phosphorylation at Thr160 by CAK activates cyclin–CDK2 complexes.

dephosphorylating Tyr15.[15,16] Both cdc25 and wee1 are regulated during the cell cycle; the activity of cdc25 increases as cells enter mitosis whereas that of wee1 decreases. The concerted activation of cdc25 and inhibition of wee1 combine to promote activation of cdc2 and irreversibly drive cells into mitosis.

An additional cyclin–CDK complex [CDK7–cyclin H or CDK-activating kinase (CAK)] is responsible for the activating phosphorylation at Thr160.[17] Three-dimensional structural analysis has shown that the region of CDK2 adjacent to Thr160 blocks an active site required for protein substrates to bind.[18] Presumably, the binding of cyclin and the subsequent phosphorylation of Thr160 by CAK induces a conformational change that allows protein substrates to interact with cyclin–CDK complexes. Cyclins can also be phosphorylated, but the specific role that this may play in regulating cyclin–CDK kinase activity is unknown. The degradation of cyclins will be addressed later.

Yeast cell cycle

G1 to S phase regulation

In unicellular organisms such as yeast, the available supply of nutrients normally regulates the cell cycle. Budding yeasts that have an abundant supply of nutrients reach a critical size during late G1, at which point they simultaneously initiate parallel processes leading to DNA replication, budding and duplication of the spindle pole body. Initiation of these processes reflects completion of Start, and commitment to complete the cell cycle.

The G1 cyclins, CLN1, -2 and -3, are necessary and rate-limiting for traverse of G1. These G1 cyclins are transcriptionally regulated during G1 by a positively acting feedback loop in which cyclin–cdc28 kinase activates the Swi4/Swi6 heterodimeric transcription factor necessary for transcription of the Cln genes. In addition to Cln1, -2 and -3, two of the B-type cyclins (Clb5 and -6) are expressed during late G1 and are important for determining the onset of S phase.[19,20]

Activation of B-type cyclin–cdc28 kinase complexes is also negatively regulated during G1 by an inhibitory protein (Sic1), which binds directly to these complexes and blocks their kinase activity. Sic1 is induced as cells complete mitosis and is destroyed just after cells complete Start, probably by ubiquitin-dependent proteolysis. Destruction of Sic1 is necessary for activation of Clb5, 6–cdc28 and entry into S phase.[20] Cln–cdc28 complexes can be similarly inactivated by an inhibitory binding protein, Far1, which is induced by mating pheromones and causes cell cycle arrest at Start, allowing haploid cells to conjugate.[21] Rum1 is a cdc2 inhibitor discovered in fission yeast, and may play a role similar to that of Sic1.[22]

The discovery of replication origins [also referred to as autonomously replicating sequence (ARS)] in budding yeast led to the identification of proteins involved in the initiation of DNA synthesis. A multiprotein complex [termed origin recognition complex (ORC)] binds ARS sequences throughout the cell cycle. It is thought that signals occurring after start are required for activation of the ARS–ORC complex. While direct activation of ORC by cdc28 has not been demonstrated, it is possible that this kinase phosphorylates a component of the ORC–ARS complex, thereby converting the complex into structures that initiate DNA replication (see ref. 1 for a more extensive review of yeast DNA replication).

Regulation of G2/M

In yeast, a single CDK regulates passage through Start, entry into S phase and entry into mitosis. However, its cyclin partners are different at each cell cycle transition. In budding yeast, the cyclins that regulate mitosis are Clb1, -2, -3 and -4. Deletion or inactivation of all four genes arrests cells with duplicated spindle pole bodies and large buds in G2.[23] Transcription of Clb3 and -4 peaks during G2 while Clb1 and -2 transcription peaks as cells enter mitosis. All four proteins are degraded as cells exit mitosis and enter G1. The Clbs contain conserved domains (called destruction boxes) at their amino termini which regulate their proteolysis during late anaphase (Figure 23.2). The destruction box is recognized by a mitosis-specific enzyme that ubiquitinates the cyclins, leading to their proteolytic destruction in the proteosome.

Mammalian cell cycle regulation

G0/G1 transition

The entry of mammalian cells into the cell cycle is typically mediated by growth factors or mitogens present in serum. Quiescent cells stimulated to enter the cell cycle by addition of growth factors maximally express D-type cyclins (D1–D3) in mid G1, cyclin E during late G1, cyclin A during S phase and B-type cyclins during late G2 and mitosis (see Figure 23.4). The CDK partners for the cyclins are either constitutively expressed or only slightly upregulated following serum stimulation.

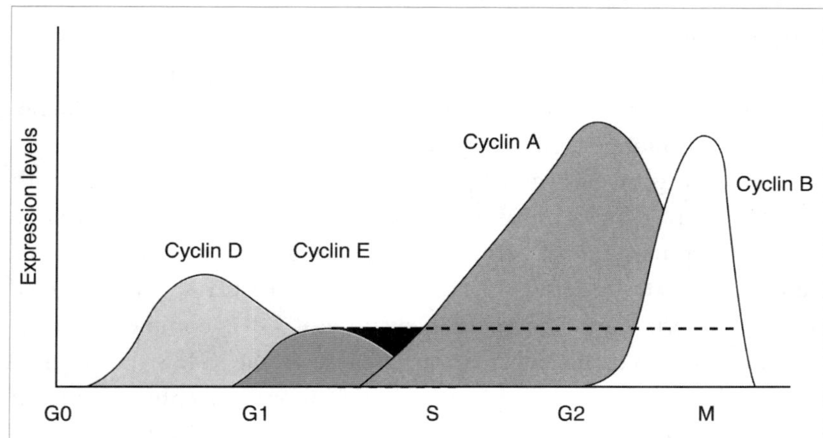

Figure 23.4 Expression of cyclins during the mammalian cell cycle. This graph illustrates the temporal order and relative levels of cyclins A, B, D and E during the cell cycle. While expression of cyclin E and B occurs during discrete periods of the cell cycle, the concentration of cyclin D peaks during mid G1 and stays at a constant low level during the remainder of the cell cycle (dashed line). The concentration of cyclin A increases during late G1 and remains elevated through G2 and mitosis when ubiquitin targeted destruction occurs.

Cyclin D1 is a nuclear protein whose expression is both necessary and rate-limiting for transit through G1. Thus, overexpression of cyclin D1 shortens G1 and microinjection of cyclin D1 antibodies during early G1 blocks cells from entering S phase.[24] These results suggest that cyclin D1 regulates the traverse of mid G1, enabling cells to reach late G1 where additional signals stimulate S phase entry. CDK4 and CDK6 are the partners for cyclin D1 and its activity is regulated by CAK.

The major, and perhaps only, substrate for cyclin D1–CDK4 complexes is the retinoblastoma gene product (Rb).[25] At least in part, Rb is thought to control G1 progression by regulating the activity of a transcription factor, E2F.[26] E2F activates transcription of many genes necessary for DNA replication and is sequestered into an inactive complex by association with Rb. Cyclin D1 binds directly to Rb through a motif of five amino acid residues located near its amino terminus.[27] Cyclin D1–CDK4 kinase complexes phosphorylate Rb during mid to late G1. Hyperphosphorylated Rb is unable to bind E2F, which enables free E2F to promote expression of genes necessary for S phase. Cells expressing nonfunctional Rb do not require cyclin D1 to proliferate. Therefore, the primary function of cyclin D1–CDK4 complexes may be to inactivate Rb.

G1/S phase transition

The expression of cyclin E in mammalian cells occurs later than cyclin D1, and cyclin E–CDK2 kinase activity occurs shortly before cells enter S phase. Mammalian fibroblasts overexpressing cyclin E are smaller than normal cells, have a shorter G1 and microinjection of cyclin E antibodies during G1 blocks entry of these cells into S phase.[28,29] Thus, cyclin E appears to have some properties in common with cyclin D1, but cyclins E and D1 appear to control different G1 events. In support of this idea, cells lacking functional Rb still require cyclin E for entry into S phase, while they no longer require cyclin D1.[29]

The expression of cyclin E, like cyclin D1, is regulated by growth factors. Cells that constitutively express cyclins E or D1 have a reduced requirement for serum factors, indicating that growth factors function in part by regulating the expression of G1-specific cyclins. Recent reports have shown that cyclin E–CDK2 complexes phosphorylate and activate cdc25A phosphatase at the G1/S transition.[30] Since cdc25A dephosphorylates the inhibitory phosphotyrosine at residue 15 on CDK2, its activation by cyclin E–CDK2 potentially creates an autoregulatory loop. Alternatively, cdc25A might promote activation of the S phase cyclin A–CDK2 complex (see below), thereby creating a link between the sequentially acting cyclin E and cyclin A kinases.

S phase

The expression and activity of cyclin A increases during late G1 and early S phase and then abruptly declines during mitosis. Cyclin A is required for DNA replication both *in vivo* and in cell-free model systems. The specific substrates of the cyclin A–CDK2 complex are not known, but one candidate is replication protein A (RP-A), a multisubunit protein that binds single-stranded DNA and unwinds DNA at replication origins. In addition, specific complexes containing cyclin A–CDK2 and proliferating cell nuclear antigen (PCNA) have been localized to origins of DNA replication.[31] While these data suggest a role for cyclin A in DNA replication, there is no direct evidence that phosphorylation of replication factors by cyclin A–CDK2 enhances their activity *in vivo*.

Recently, it has been shown that cyclin A–CDK2 complexes phosphorylate and inactivate the E2F transcription factor *in vitro*.[32,33] This suggests that cyclin A–CDK2 complexes negatively regulate at least one protein involved in G1/S phase transition, and this may be important for turning off G1/S phase-specific gene expression once DNA replication begins.

G2/M

B-type cyclins regulate the transition from G2 to M phase. While CDK2 is required for the G1/S phase transition, cdc2 mediates the traverse of

G2 and mitosis. Cyclin B expression increases during early G2 and peaks during late G2 and early M phase. Cyclin B–cdc2 complexes have been shown to interact with numerous proteins that regulate structural changes during mitosis. The events regulated by cyclin B–cdc2 indicate that active complex is required through metaphase and inactivation of cyclin B–cdc2 may be required for cells to complete anaphase. Expression of cyclin B mutants lacking the destruction box allows cells to enter but not exit from anaphase. The proteolytic machinery that degrades cyclin B may also degrade other proteins during mitosis, including those involved in sister chromatid association.

The degradation of cyclin B during anaphase is mediated by the ubiquitin degradation pathway. Both cyclin A and B contain a structural motif (destruction box) that is required for degradation. They are targeted for degradation by the conjugation of multiple ubiquitin molecules to a lysine residue located carboxy-terminal to the destruction box. Ubiquinated lysines are recognized by proteolytic enzymes that specifically degrade ubiquinated proteins.[34] Cyclin B–cdc2 complexes have been shown to stimulate the ubiquitination pathway, thereby regulating their own degradation.[35]

The phosphorylation of myosin light chain during anaphase by cyclin B–cdc2 complexes inhibits its interaction with actin and prevents cytokinesis.[36] These results demonstrate that inactivation of cyclin B–cdc2 is required for myosin light chain to bind actin and initiate the contraction of a myosin/actin ring that physically separates the two daughter cells. Therefore, the degradation of cyclin B may be a rate-limiting event for cytokinesis. By making chromosome segregation dependent upon the initiation of cyclin proteolysis and cytokinesis dependent upon cyclin B–cdc2 inactivation, cells ensure that separation of chromosomes occurs prior to division.

CDK inhibitors: all good things must come to an end

Mammalian cells express two families of low-molecular-mass proteins that bind to and inhibit cyclin–CDK complexes: the p21/p27 family and the p16 family. The p16 family of inhibitors is thought to be specific for cyclin D–CDK complexes, while the p21/p27 family inactivates all known cyclin–CDK complexes. In general, expression of these cyclin-dependent kinase inhibitors (CKIs) is induced by factors that negatively regulate cell proliferation and during normal states of growth arrest (such as quiescence or terminal differentiation). In yeast, proteins such as Far1 perform analogous roles to inhibit cell cycle progression in response to extracellular signals. It is thought that cdc2 activity is primarily regulated by phosphorylation, but it is possible that CKIs also contribute to cdc2 regulation during G2/M.

CKIs appear to link the cell cycle machinery to signals that either promote or inhibit cell cycle progression. Thus, p27 is repressed by mitogenic growth factors and induced by specific anti-mitogenic signals such as TGF-β.[37] p21 responds to other signals. For example, p21 concentrations increase in cells exposed to DNA-damaging agents. This response is dependent upon p53, and is thought to be an essential component of the pathway by which p53 coordinates cell cycle progression with DNA repair[38,39] (see Chapter 22). The concentration of p21 is also elevated in cells that have permanently withdrawn from the cell cycle, such as senescent cells and terminally differentiated cells; this effect is independent of p53. Thus, high concentrations of p27 are found in cells reversibly arrested in a quiescent state, whereas high concentrations of p21 are more characteristic of cells that have either irreversibly exited the cell cycle or contain damaged DNA.

In normal proliferating cells, the cyclin–CDK holoenzyme is assembled into a quaternary complex that also contains p21 and PCNA. These quaternary complexes are catalytically active, apparently because inhibition of CDKs by p21 (or p27) requires binding of more than one p21 molecule to the cyclin–CDK complex.[40] It remains to be determined whether the single p21 molecule present in active cyclin–CDK complexes is affecting their enzymatic or biological activities in other ways. p21 not only inhibits CDKs, but can also prevent the replicative activity of PCNA, which is a polymerase δ processivity factor involved in replication of the leading strand of DNA. However, p21 does not inhibit the gap filling repair/synthesis function of PCNA.[41] Thus, p21 may co-ordinate the replicative and repair activities of PCNA by blocking DNA synthesis, but allow DNA repair in response to DNA damage.

The p16 family of CDK inhibitors interact exclusively with CDK4 and CDK6. Since the major substrate of these kinases is the Rb protein, the ultimate effect of p16 and related inhibitors is to modulate the phosphorylation state of Rb. Members of the p16 family are induced by TGF-β and this contributes to cell cycle arrest in G1 by preventing Rb phosphorylation. In contrast, many primary tumors have suffered deletions in p16 and this leads to deregulated CDK4 activity, constitutive Rb phosphorylation and aberrant regulation of cell proliferation.[42]

Summary

This review has shown that the co-ordinated expression of specific cyclins and cyclin-dependent kinases (CDKs) regulates the progression of cells through the cell cycle (Figure 23.5). While additional cyclins and CDKs may be isolated in the future, the fundamental mechanisms that regulate their activity will likely remain the same. The interplay between cyclin synthesis and degradation coupled with post-translational regulation of the CDKs enables cells to control exquisitely their transit through the cell cycle.

Although progress has been made in identifying cyclins and CDKs, very little is known about the *in vivo* targets of these kinase complexes. The various cyclin–CDK complexes presumably activate specific substrates during the cell cycle. Each activated substrate or set of substrates would stimulate distinct events that are required for cells to progress to the next phase of the cell cycle. While Rb has been identified as a substrate for cyclin–CDK complexes during G1, additional cyclin–CDK substrates must exist for cells to traverse S, G2 and M phases. Therefore, one of the major challenges over the next few years will be to identify and characterize substrates that interact with the various cyclin–CDK complexes.

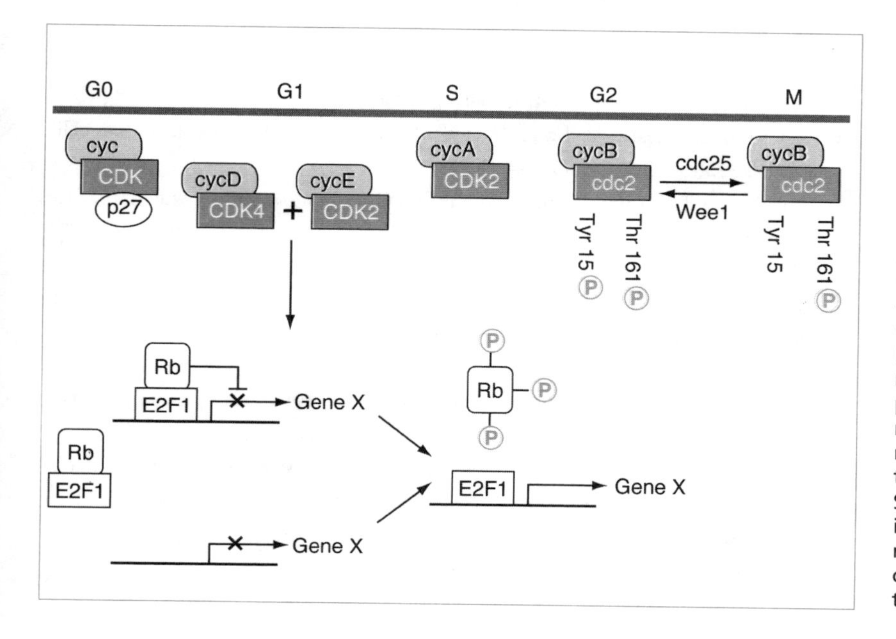

Figure 23.5 Some important regulatory events that occur during the mammalian cell cycle. The interaction between distinct cyclins (cyc) and CDKs regulates the progression of cells through the cell cycle. The types of cyclin–CDK complexes and their relative positions during the cell cycle are illustrated. Increased levels of p27 during G0 may create a threshold that cyclin–CDK complexes have to overcome in order to exit G0 and progress through G1. cdc2 is activated during late G2 by phosphorylation at Thr161 and dephosphorylation at Tyr15. Although not depicted in this diagram, quaternary complexes containing cyclins, CDKs, PCNA and p21 have been found in non-transformed cells. It is possible that p21 may be required to regulate DNA repair and serve as a factor that tethers cyclins to CDKs. See text for additional information regarding the regulation and function of cyclin–CDK complexes during the cell cycle.

References

For an excellent overall cell cycle review, see: Murray, A. and Hunt, T. (1993) *The Cell Cycle: an Introduction*, Freeman, San Francisco.

1 Heichman, K.A. and Roberts, J.M. (1994) Rules to replicate by. *Cell*, **79**, 557–62.

2 Nurse, P. (1994) Ordering of S phase and M phase in the cell cycle. *Cell*, **79**, 547–50.

3 Hartwell, L. and Weinert, T. (1989) Checkpoints: controls that ensure the order of cell cycle events. *Science*, **246**, 629–34.

4 Rossow, P., Riddle, V. and Pardee, A. (1979) Synthesis of labile, serum-dependent protein in early G1 controls animal cell growth. *Proc. Natl. Acad. Sci. USA*, **76**, 4446–50.

5 Johnston, G., Pringle, J. and Hartwell, L. (1977) Coordination of growth with cell division in the yeast *S. cerevisiae*. *Exp. Cell Res.*, **105**, 79–98.

6 Hartwell, L., Culotti, J., Pringle, J. and Reid, B. (1974) Genetic control of the cell division in yeast. *Science*, **183**, 46–51.

7 Nurse, P. (1990) Universal control mechanism regulating onset of M-phase. *Nature*, **344**, 503–7.

8 Sherr, C. (1993) Mammalian G1 cyclins. *Cell*, **73**, 1059–65.

9 Nasmyth, K. (1993) Control of the yeast cell cycle by the Cdc28 protein kinase. *Curr. Opinion Cell Biol.*, **5**, 166–79.

10 Poon, R. and Hunt, T. (1992) Identification of the domains in cyclin A required for binding to, and activation of, p34cdc2 and p32cdk2 protein kinase subunits. *Mol. Biol. Cell*, **3**, 1279–94.

11 Nurse, P. (1975) Genetic control of cell size at cell division in yeast. *Nature*, **256**, 457–61.

12 Mueller, P.R., Coleman, T.R. and Dunphy, W.G. (1995) Cell cycle regulation of a Xenopus Wee1-like kinase. *Mol. Biol. Cell*, **6**, 119–34.

13 Igarashi, M., Nagata, A., Jinno, S. *et al.* (1991) Wee1^{+}-like gene in human cells. *Nature*, **353**, 80–3.

14 Coleman, T.R., Tang, Z. and Dunphy, W.G. (1993) Negative regulation of the Wee1 protein kinase by direct action of the Nim1/Cdr1 mitotic inducer. *Cell*, **72**, 919–29.

15 Kumagai, A. and Dunphy, W.G. (1991) The cdc25 protein controls tyrosine dephosphorylation of the cdc2 protein in a cell-free system. *Cell*, **64**, 903–14.

16 Strausfield, V., Labbe, J.C., Fesquat, O. *et al.* (1991) Dephosphorylation and activation of a p34^{cdc2}/cyclin B complex *in vitro* by human cdc25 protein. *Nature*, **35**, 242–5.

17 Fisher, R.P. and Morgan, D.O. (1994) A novel cyclin associates with MO15/cdk7 to form the cdk-activating kinase. *Cell*, **78**, 713–24.

18 DeBondt, H., Rosenblatt, J., Jarncarik, J. *et al.* (1993) Crystal structure of the cyclin-dependent kinase 2. *Nature*, **363**, 595–602.

19 Epstein, C. and Cross, F. (1992) CLB5: A novel B cyclin from budding yeast with a role in S phase. *Genes Dev.*, **6**, 1695–1706.

20 Schwob, E., Bohm, T., Mendenhall, M. and Nasmyth, K. (1994) The B-type cyclins kinase inhibitor p40^{sic1} controls the G1 to S phase transition in *S. cerevisiae. Cell*, **79**, 233–44.

21 Peter, M. and Herskowitz, I. (1994) Joining the complex: cyclin-dependent kinase inhibitory proteins and the cell cycle. *Cell*, **79**, 181–4.

22 Moreno, S., and Nurse, P. (1994) Regulation of progression through the G1 phase of the cell cycle by the rum1 + gene. *Nature*, **367**, 236–42.

23 Amon, A., Tyers, M., Futcher, B. and Nasmyth, K. (1993) Mechanisms that help the yeast cell cycle clock tick: G2 cyclins transcriptionally activate their own synthesis and repress G1 cyclins. *Cell*, **74**, 993–1007.

24 Quelle, D., Ashmun, R., Shurtleff, S. *et al.* (1993) Overexpression of mouse D-type cyclins accelerates G1 phase of rodent fibroblasts. *Genes Dev.*, **7**, 1559–71.

25 Kato, J., Matsushime, H., Hiebert, S.W. *et al.* (1993) Direct binding of cyclin D to the retinoblastoma product (pRb) and pRb phosphorylation by the cyclin-dependent kinase cdk4. *Genes Dev.*, **7**, 331–42.

26 Nevins, J. (1992) E2F: a link between the Rb tumor suppressor protein and viral oncoproteins. *Science*, **258**, 424–9.

27 Dowdy, S., Hinds, P., Lovic, K. *et al.* (1993) Physical interaction of the retinoblastoma protein with human D cyclins. *Cell*, **73**, 499–511.

28 Ohtsubo, M. and Roberts, J.M. (1993) Cyclin-dependent regulation of the G1 in mammalian fibroblasts. *Science*, **259**, 1908–12.

29 Ohtsubo, M., Theodores, A., Schumacher, J. *et al.* (1995) Human cyclin E, a nuclear protein essential for the G1 to S phase transition. *Mol. Cell. Biol.*, **15**, 2612–24.

30 Hoffman, I., Draetta, G. and Karsent, E. (1994) Activation of human cdc25A by a cdk2–cyclin E dependent phosphorylation at the G1/S transition. *EMBO J.*, **13**, 4302–10.

31 Cardosa, M., Leonhardt, H. and Nadel-Ginard, B. (1993) Reversal of terminal differentiation and control of DNA replication: cyclin A and cdk2 specifically localize at subnuclear sites of DNA replication. *Cell*, **74**, 979–92.

32 Dynlacht, B., Flores, O., Lees, J. and Harlow, E. (1994) Differential regulation of E2F transactivation by cyclin/cdk2 complexes. *Genes Dev.*, **8**, 1772–86.

33 Krek, W., Ewen, M., Shirodkar, S. *et al.* (1994) Negative regulation of the growth-promoting transcription factor E2F-1 by a stably bound cyclin A-dependent protein kinase. *Cell*, **78**, 161–72.

34 Glotzer, M., Murray, A. and Kirschner, M. (1991) Cyclin is degraded by the ubiquitin pathway. *Nature*, **349**, 132–8.

35 Felix, M., Labbe, J., Doree, M. *et al.* (1990) Triggering of cyclin degradation in interphase extracts of amphibian eggs by cdc2 kinase. *Nature*, **346**, 379–84.

36 Slatterwhite, L., Lohka, M., Wilson, K. *et al.* (1992) Phosphorylation of myosin-II regulatory light chain by cyclin–p34^{cdc2}: a mechanism for the timing of cytokinesis. *J. Cell Biol.*, **118**, 595–605.

37 Polyak, K., Kato, J., Solomon, M. *et al.* (1994) $p27^{kip1}$, a cyclin–cdk inhibitor, links transforming growth factor beta and contact inhibition to cell cycle arrest. *Genes Dev.*, **8**, 9–22.

38 El-Deiry, W., Tukino, T., Velculesseu, V. *et al.* (1993) WAF1, a potential mediator of p53 tumor suppression. *Cell*, **75**, 817–25.

39 Xiong, Y., Zhang, H. and Beach, D. (1993). Subunit rearrangement of the cyclin-dependent kinases is associated with cellular transformation. *Genes Dev.*, **7**, 1572–83.

40 Zhang, H., Hannon, G. and Beach, D. (1994) p21-containing cyclin kinases exists in both active and inactive states. *Genes Dev.*, **8**, 1750–8.

41 Li, R., Waga, S., Hannon, G. *et al.* (1994) Differential effects by the p21 cdk inhibitor on PCNA dependent DNA replication and DNA repair. *Nature*, **371**, 534–7.

42 Sheaff, R. and Roberts, J. (1995) Lessons in p16 from phylum Falconium. *Curr. Biol.*, **5**, 28–31.

Index

Page numbers appearing in **bold** refer to figures and page numbers in *italic* refer to tables.